OILWELL DRILLING ENGINEERING

DON W. DAREING
Professor Emeritus, Mechanical Engineering
The University of Tennessee

ASME PRESS

Library of Congress Cataloging-in-Publication Data

Names: Dareing, Don W., author.
Title: Oilwell drilling engineering / by Don W. Dareing, Professor Emeritus, Mechanical Engineering, The University of Tennessee.
Description: New York : ASME Press, [2019] | Includes bibliographical references.
Identifiers: LCCN 2018053258 | ISBN 9780791861875
Subjects: LCSH: Oilwell drilling.
Classification: LCC TN871.2 .D3323 2019 | DDC 622/.3381--dc23 LC record available at https://lccn.loc.gov/2018053258

To: Tron and Hannah

TABLE OF CONTENTS

PREFACE

The early 1960s marked the beginning of major research efforts to improve drilling technology. Also, major efforts were underway to develop technology to explore outer space. The slide rule was still an engineering tool, and there was some use of hand calculators. Illinois's computer (ILLIAC), using vacuum tubes, was one of several automatic computers emerging as an engineering and research tool.

The evolution of drilling and space technologies ironically are similar. Both require engines of thrust, navigation, and the ability to make course changes. The Rocketdyne F-1 engine became, and still is, the most powerful engine ever built, producing thrust up to 2 M lbs. Similarly, drill bits require thrust plus rotational power, navigation, and the ability to drill along a prescribed well path to intersect a reservoir at a specified point. Even though there are similarities, navigating through geo-space offers the challenge of unpredictable forces while space travel navigates through variable gravity fields and environmental drag. Both technologies have matured so that trajectories and targets are achievable with high degrees of accuracy and reliability.

The key to drill bit navigation is the ability to monitor drill bit parameters in real time, while drilling several thousand feet into the earth, allowing well path monitoring and control. This was first accomplished in 1960 by a group of research engineers at Jersey Production Research Company (Tulsa, Oklahoma), who designed and tested the first measurement-while-drilling (MWD) system.

A goal of this book is to present the state of the art of oil well drilling technology, much of which has been developed since the early 1960s. This book explains drilling practices in terms of engineering fundamentals. It is designed for undergraduate students, graduate students and entry level engineering professionals who are interested in learning the language of the industry and the technology of oil well drilling.

An early discussion of optimum drilling practices sets the stage for more detailed coverage of key aspects of drilling. Drill bit types and features are discussed in terms of application, performance, footage cost and wear characteristics. Individual cutter load and heat generation is related to cutter density and type in explaining wear in rigid body drill bits.

Hydraulics of drilling starts with fluid power delivered by mud pumps at the surface. Pump power is partially consumed by parasitic fluid losses leaving the remaining portion for bit cleaning, drilling motors and other equipment located within the bottom-hole assembly. Optimizing available hydraulic power within the BHA is an important part of best drilling practices. It is a necessity for proper application of drilling motors and turbines.

Drillstring dynamics and vibration control are special features of the book. Each mode of vibration (axial, torsion and lateral) is discussed in terms of natural frequencies and sources of excitation. Methods for alleviating severe vibrations, especially, in bottom-hole assemblies are explained. The negative impact of premature failures on footage costs is discussed along with factors that affect reliability of drilling tools and drilling cost.

Casing design, following standard practice is covered as well as well control, including kick detection and methods of kick removal from the well bore.

The mathematics of directional navigation, tools, and methods of well path control, including the use of rotary steerable tools is explained. This discussion is made against a backdrop of historical advances in drill bit directional control, navigation and downhole tools.

Every author approaches this subject differently, relying heavily on personal experiences. I have learned much by working for major oil companies and contractors, in addition to academia. Each provided opportunity to develop professionally and to learn from worldwide operations. Of special notice are: Exxon Production Research Company (Houston), Conoco North Sea (London) and Norton-Christensen (Salt Lake City). I am most fortunate to have the opportunity to develop a career in this wonderful professional environment.

Personal friends, who have enriched my career, are Dr. William C. Maurer and Dr Kirk E. Boatright. Dr. Maurer advanced the science of rock mechanics as relates to chip formation and chip hold-down and developed his theory of perfect cleaning. His testing included rock failure caused by an impact chisel in a high-pressure chamber in Exxon's laboratory, field testing, and raised boring in mine cavities. His fundamental work is now a point of reference for measuring efficiencies of drill bit performance.

Dr. Kirk E. Boatright was one of a team that developed and tested the very first downhole tool for measuring real time drilling parameters directly above drill bits while drilling at several thousand feet. Others on this pioneering team included George H. Paff, Paul M. Ferguson, Jack M. Kellner and D. Joe Meador. This initial effort paved the way for a more advanced tool, which I had the pleasure of using. Bottom hole drilling data was subsequently obtained while drilling in Texas, Oklahoma, Louisiana and Canada. This field data was vital in developing mathematical models of drillstring dynamics as discussed in my book.

I am grateful to ASME for the opportunity to publish this technology.

Don W. Dareing
Professor Emeritus, Mechanical Engineering
University of Tennessee
Life Fellow Member, ASME

UNITS OF MEASURE USED IN OILWELL DRILLING

1 bbl = 42 gal
1 ft^3 = 7.48 gal
1 gal = 231 in.^{3}
1 cm = 0.3937 in.
1 cc = 1 mL
1 gal = 3.785 lit

$\gamma_{water} = 62.4\ \text{lb/ft}^3$ (fresh water)

$$\gamma_{water} = 62.4\ \text{lb/ft}^3 \left| \frac{1\ \text{ft}^3}{7.48\ \text{gal}} \right| = 8.342\ \text{ppg}$$

$\gamma_{salt} = 1.025 \times 62.4 = 63.96\ \text{lb/ft}^3$ (sea water; nominal)

$$\gamma_{salt} = 63.96\ \text{lb/ft}^3 \left| \frac{1\ \text{ft}^3}{7.48\ \text{gal}} \right| = 8.551\ \text{ppg}$$

$$\gamma_{steel} = \frac{490\ \text{lb}}{\text{ft}^3} \left. \frac{1\ \text{ft}^3}{7.48\ \text{gal}} \right| = 65.5\ \text{ppg}$$

1 cp = 1.45×10^{-7} reyns

Units of viscosity in reyns are: $\dfrac{\text{lb sec}}{\text{in}^2}$

Units of viscosity in poise are: $\dfrac{\text{dynes sec}}{\text{cm}^2}$

1 lb = 448,200 dynes
1 lb = 4.4482 N

Viscosities of Different Fluids

	cp	reyns
Water	1	$1.45\ 10^{-7}$
Ethylene Glycol	20	$29\ 10^{-7}$
Olive Oil	100	$145\ 10^{-7}$
SAE 30	300	$435\ 10^{-7}$
Honey	1500	$2175\ 10^{-7}$

Chapter 1

HISTORICAL PERSPECTIVE OF OIL WELL DRILLING

1.1 PETROLEUM GEOLOGY

The earth is estimated to be 4.5 billion years old. It spun off of the sun and settled into its current orbit accounting for the four seasons. Initially, the earth was a molten mass having a rotation of its own, accounting for day and night. With time, a giant land mass called Pangaea formed, split-up, and gradually drifted apart forming the continents.

Movement of the continents is explained by the theory of plate tectonics. These plates have been identified around the globe. It is their movement that causes earthquakes. Plate tectonics also explains the existent of nearly all mountain ranges, such as the Himalayas, Alps, Rocky Mountains and Appalachian Mountains in North American.

The geologic calendar is divided into four eras [1]:

- Cenozoic (recent life—65 Million years ago)
- Mesozoic (65 M to 225 M years ago)
  - About 60 to 70 M years ago, the dinosaur became extinct
- Paleozoic (225 M to 570 M years ago)
  - Giant land mass, Pangaea, began to drift apart about (225 M years ago) to form current continents
  - Sufficient plant and animal life to form organic material (570 M years ago)
- Precambrian (570 M to 4500 M years ago)
  - Insufficient plant and animal life to result in the formation of hydrocarbons

The end of one era and the beginning of another indicates a major change in plant and animal life; types of species and population. Some became extinct while others became more prolific.

The aspects of geology that are of interest in our discussion of oil recovery are: rock types and origin of hydrocarbons.

Three types of rocks:

1. Igneous — This rock was once molten and solidified as the earth cooled. The most common igneous rock is <u>granite</u>. It is commonly

referred to as the basement or base rock. We do not expect to find hydrocarbons in igneous rock.

2. Sedimentary — They are formed by sediments collecting in sea beds. With time unconsolidated layers of sediments are pushed down deep into the earth and through pressure, temperature, and chemical action solidify to form sedimentary rock. This process is called lithification (conversion into rock). There are three different types of sedimentary rock.
   A. Sandstone — Sand is formed from the erosion of igneous rock by wind and water. Over time sand flowed into lake beds or basins (large bowls of granite). Larger grains settled out first followed by smaller grains and silt (and mud). The lithification process followed after layers of sand were pushed deep into the earth. Multiple layers of sedimentary rock are displayed in the Grand Canyon and southern Utah (Moab).
   B. Limestone — Limestone comes from two sources: biogenic (living matter) and chemistry. Coral reefs are good examples of biogenic sources. Limestone can also be formed by chemical action. Over time limestone grains bonded to form limestone rock through the process of lithification.
   C. Shale — Shale originated with silt and mud. It usually carried with it the remains of small animal and plant life. After lithification, it became tight grained and impermeable. It is fissile (readily splits into thin pieces) and laminated.

   Beneath sedimentary rock there will always be a basement rock (granite). A basin is a large depression where sedimentary material is captured over geologic time.
3. Metamorphic — This type of rock is formed from recycling preformed rock through heat and pressure, usually caused by tectonic plate movement. An example is the reheating of shale forming slate or gneiss. We do not expect to find hydrocarbons in this type of rock.

1.2 PROPERTIES OF HYDROCARBONS

Standards used throughout the petroleum industry are set by the American Petroleum Institute, which is supported by the petroleum industry and various professional societies; the Society of Petroleum Institute (SPE) and the American Society of Mechanical Engineers (ASME).

Some time ago, the industry established a formula to indicate the quality of crude oil. The formula defines API gravity by

$$^{0}API = \frac{141.5}{SG} - 131.5 \tag{1-1}$$

where SG is specific gravity of the crude oil. The purpose of using API gravity instead of standard gravity is to produce whole numbers instead of decimals. The two numbers are in reverse, i.e., the lower the SG the higher the API gravity. For example, if $\gamma_0 = 52.7$ lb/ft^3, SG = 0.845 and ^{0}API = 36, while a SG of 0.8 gives ^{0}API = 45.38.

As a general rule, the higher the API gravity, the better the quality of the crude oil. API gravity usually increases with depth where older hydrocarbons are found.

1.2.1 Chemical Bonding

Hydrocarbon molecules are made up of C and H atoms. Hydrocarbons from two different reservoirs are never the same. There are hundreds of possible combinations with methane being the simplest.

Three basically different oil chemistries are:

A. Paraffin Base Oil — This is a light (and often clear) oil typically having an API > 30. Their hydrocarbon chain is defined by C_nH_{2n+2}. Propane, Butane, and Pentane are examples.
B. Naphthene Base Oil — This is a heavy oil with API < 25 and is usually black. Its hydrocarbon chain is cyclopental, C_nH_{2n}. A 5-carbon atom ring is smallest naphthene ring. Intermediate Base Oil — mixed characteristic between Paraffin and Naphthene. API gravity between 25 and 30. Examples are Ethylene (C_2H_4) and benzene (C_6H_6).

API gravity is now telling us about chemistry. The traditional use of API gravity is an indicator of the quality of crude oil. World oil prices (according to the Stock Market) are based on:

- Brent light (world reference)
- Saudi Arabian light
- West Texas Intermediate (WTI)

Older hydrocarbons are cleaner, more refined. In some locations, such as the North Slope of Alaska, recovered oil out of the well is good enough to put directly into vehicles. Younger hydrocarbons, such as tar sands, are found in shallow formations and are chemically more complex and unrefined by nature and time. As a general rule, older hydrocarbons are found in deep formations and are more refined.

1.3 TYPES OF HYDROCARBON TRAPS

Hydrocarbons are believed to originate from the decay of small micro-organisms and plant life. These elements were deposited along with silt and mud at the bottom of lakes and covered by multiple layers of sediments. During the formation of shale, they were squeezed out of by high levels of overburden pressure on top of the shale and deposited into adjacent sandstone. Shale is referred to as the *source rock* and sandstone the *reservoir rock*.

The oil then migrated as the earth's formations shifted and tilted. During migration, hydrocarbons displaced connate water as it moved upward along tilted porous and permeable sedimentary rocks. The migration of oil stopped when it became trapped by an impermeable formation barrier. This migration took place over millions of years.

Three events must happen for there to be an oil deposit. There must be a source rock (shale), a reservoir rock (primarily sandstone; can be limestone) and an impermeable cap rock (shale or limestone or chalk).

Four (4) types of oil traps are:

1. Anticline trap
2. Fault trap
3. Salt Dome
4. Stratigraphic trap

Each of these traps are illustrated in Figures 1-3 to 1-6, courtesy of Dr. Kirk E. Boatright, Training Consultants International [2].

In some cases, these subsurface oil traps may be manifested by surface geography, such as bulges and faults. However, in general, seismic surveys are required (Figure 1-1). Geologists look for these types of structures when examining surface geography and seismic maps of subterranean formations.

Seismic surveys alone cannot confirm the presence of oil. They give a graphic picture of geologic formations and show rock structures capable of containing oil deposits. Exploratory drilling into promising structure confirms the presence of oil (Figure 1-2).

Three or four exploratory wells are usually required to delineate the total reservoir. Needed information includes, reservoir rock properties (porosity and permeability), amount of crude oil in-place, the gas-oil ratio (GOR) and total volume of oil under surface conditions. This information is collected during exploratory drilling. Prior to geophysical technology, geologist selected drilling sites based on surface geography. The tops of salt domes along the gulf coast are a good example.

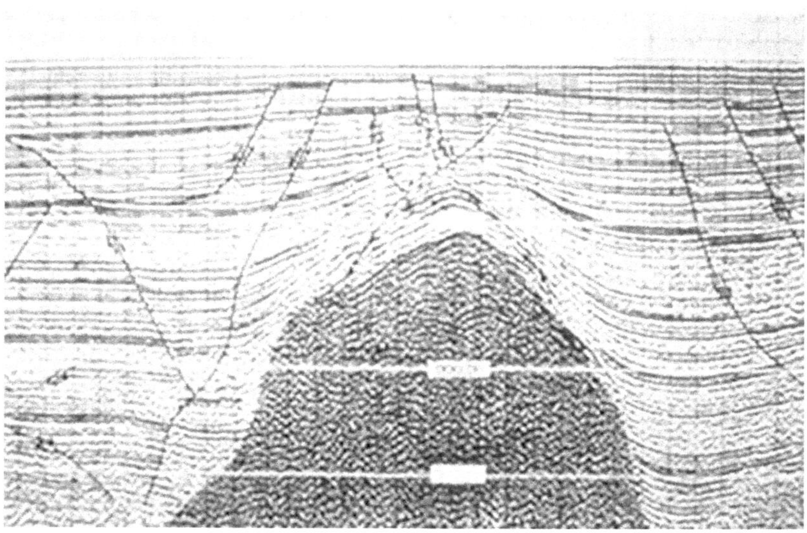

Figure 1-1. Seimic survey of sub teranian formations.

Figure 1-2. Exploratory drilling near mouth of the Mackenzie River (Canada).

Subsurface fluids over time migrate and separate according to density. Since hydrocarbon fluids originated in sedimentary formations located within sea floors, hydrocarbon reservoirs naturally contain salt water. As sedimentary rocks reoriented from horizontal layers, liquid and gas hydrocarbons migrated upward. Over geologic time, water, oil, and gas travel and separated according to densities until they were trapped by an impermeable boundary or cap rock. Shale, Limestone or Chalk are common cap rocks.

A certain sequence of events must occur in geologic time for an oil and gas reservoir to exit. There must be a reservoir rock (usually sandstone) and an impermeable boundary or trap.

1.3.1 Anticline Traps

Sedimentary rock is initially formed in horizontal layers. Limestone (or shale) may be alternated with sandstone (reservoir rock). Over time these sedimentary layers are pushed down deep into the earth and go through a lithification process caused by pressure, temperature, and chemical action. Over time, side forces on these horizontal layers cause them to fold.

Initially, sandstone was saturated with salt water or connate water. Over geologic time, the lighter hydrocarbons migrate upward. As it migrates upward, oil pressure is decreases. If oil pressure drops below the bubble point pressure of the original hydrocarbon, gas is released from the hydrocarbon. Oil, gas, and connate water separate according to density and become trapped by the shale forming an anticline trap as shown. This type of trap can be a simple fold or a dome shape (Figure 1-3). This migration and separation took place over millions of years until equilibrium was established within the trap.

If the final reservoir pressure is greater than the bubble point pressure of the original hydrocarbon, a gas cap will not form.

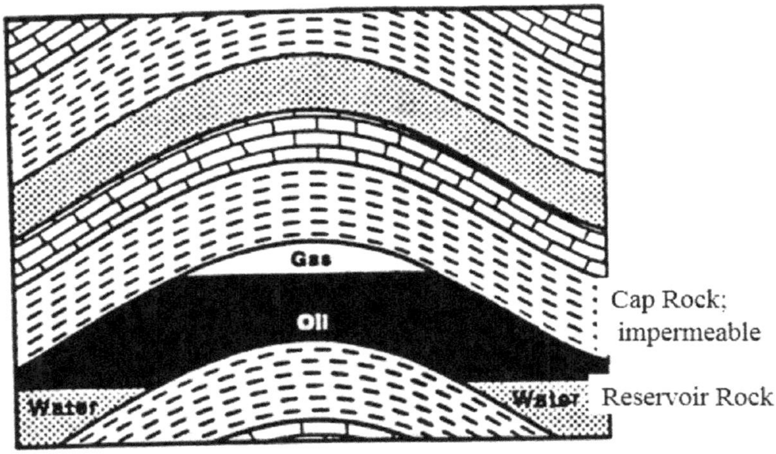

Figure 1-3. Anticline trap.

1.3.2 Fault Traps

Movement of the earth's crust causes complex stress within sedimentary rock. This stress may cause the formations to shear as well as fold (Figure 1-4). In geologic time oil migrates and is trapped within this fault line. An important aspect of this trap is the impermeable rock above the sandstone, which existed prior to the fault or shearing, must also provide an impermeable boundary along the shear plane. Without this side blockage, the oil would keep going upward and possibly be lost to the surface.

It is estimated that, of the hydrocarbons generated in the source rock of the North Sea, less than 10% was trapped. This implies that 90% of the original crude oil made its way to the surface and was lost to the environment.

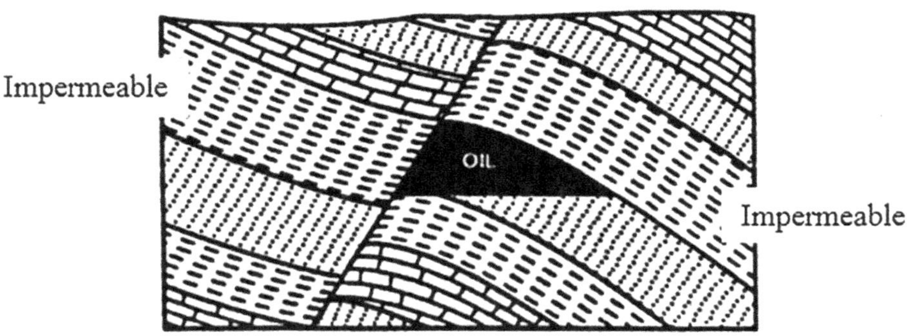

Figure 1-4. Fault trap.

1.3.3 Salt Dome

Several events occurred over geologic time to cause the existents of domes of salt as depicted in Figure 1-5. These salt domes may be several thousand feet tall and strong enough to cause the surface to bulge forming islands in swamp areas as found along the Gulf coast. They are significant in our discussion of hydrocarbon traps because salt is impermeable.

The origin of these domes is most interesting. At some point in geologic time, inland seas of salt water were trapped. With changes in climate, these inland seas evaporated leaving behind layers of salt as residue. Later, layers of sand blew over the salt layers and protected them. Geologic events that followed resulted in these regions dropping below sea level with millions of years of sedimentation deposited on top of these salt layers. With time, lithification occurred, and salt became rock salt. After lithification, the salt beds became impermeable.

Salt rock has two properties different from shale, sandstone, or limestone:

- It is much less dense
- It is a plastic solid (deformable)

These properties promote a buoyant effect, which caused salt layers to move upward through over laying rock structures, lifting, deforming, and fracturing them as plastic flow continued upward.

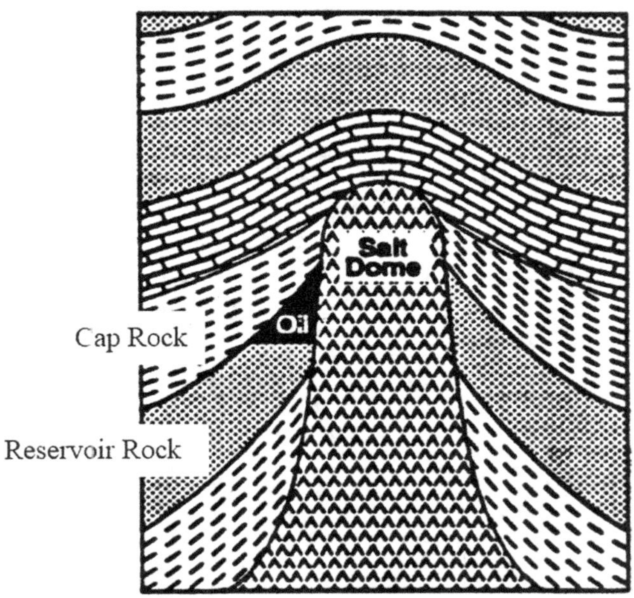

Figure 1-5. Salt dome trap.

It is estimated that this extruding process took place at the rate of 1 in. per 100 years, a very slow rate of travel. However, in geologic time this rate of upward travel is significant. For example, over 1 million years a salt dome would move upward 833 ft. Over 10 million upward travel would be 8330 ft. The North Dayton Salt Dome located northeast of Houston, Texas has reached a few thousand feet from the surface, making the dome over 45,000 ft high.

It is apparent that regions around a salt dome provide an ideal trap for migrating hydrocarbons. Regions around these domes also provide opportunity for the existence of anticline and fault traps.

1.3.4 Stratigraphic Traps

Hydrocarbons are formed in horizontal sedimentary rock as in the previous cases. With geologic time reservoir rock is tilted upward as shown. The upward ends of these formations are eroded away with time and a new sedimentary layer is deposited across the exposed ends (Figure 1-6). After lithification, the new layer may form an impermeable (shale, chalk, or limestone) cap, providing an opportunity to trap oil.

Fluid pressures inside oil and gas reservoirs are much higher than fluids outside of reservoirs. Fluid pressure within the pores of formation layers above oil and gas reservoirs is called formation pressure. Formation pressures that increase hydrostatically with depth are considered to be normal pressures. Formation pressures that do not increase hydrostatically with depth are abnormal pressures.

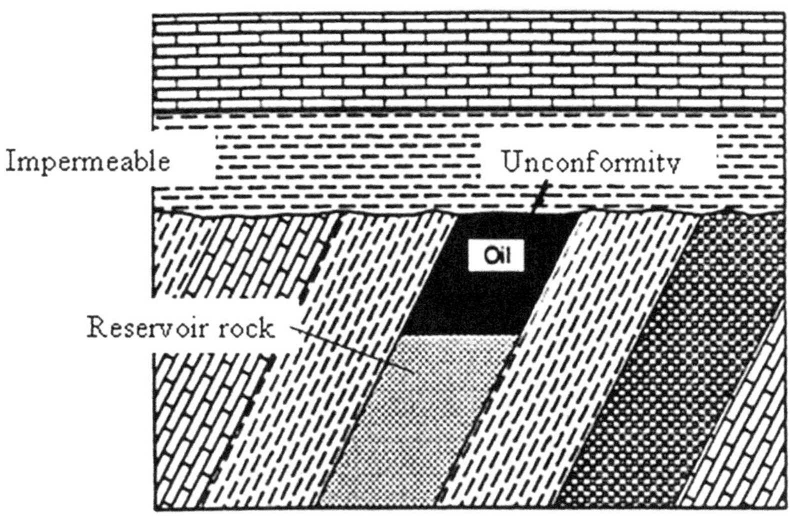

Figure 1-6. Stratigraphic trap.

1.4 RESERVOIR EVALUATION

Oil exploration starts with mapping geologic formations. This is done by seismic surveys. Current geo-physical techniques map and identify possible oil traps by giving three-dimensional pictures of underground terrain. Seismic surveys alone cannot confirm the presence of oil. They give a graphic picture of geologic formations and identify zones of possible oil deposits. Exploratory drilling into the promising structure must be conducted to establish the presence of oil.

Three or four exploratory wells are usually required to delineate the total reservoir. Needed information includes, reservoir rock properties (porosity and permeability), amount of in-place crude oil, the gas-oil ratio (GOR) and total volume of oil under surface conditions. This information is collected during exploratory drilling operations. Prior to this geophysical technology, geologist selected drilling sites based on surface geography. The tops of salt domes along the gulf coast are good examples.

Samples of the crude are obtained within the reservoir by a bottom hole sampler; sometimes called a *pressure bomb* (Figure 1-7). The dimensions of the bottom hole sampler are 4 in. to 6 in. in diameter (OD) and 18 in. to 24 in. long. Total sample volume is about 600 cc (cm^3) and initially in a vacuum. The sampler is lower into the reservoir at a datum level.

Once in place, a timer opens an access valve permitting reservoir oil to fill the sampler. In this manner, reservoir oil is captured under reservoir temperature and pressure conditions. The oil sample is then retrieved and transferred to a high-pressure transport cylinder and taken to a laboratory where P-V-T analysis is made [3]. Laboratory tests give information necessary to conduct an economic analysis of the reservoir and its oil. Information gathered from the PVT test includes:

- Oil Formation volume factor (B_0)
- Bubble point pressure (BPP)
- Gas-Oil ratio (GOR)
- API gravity

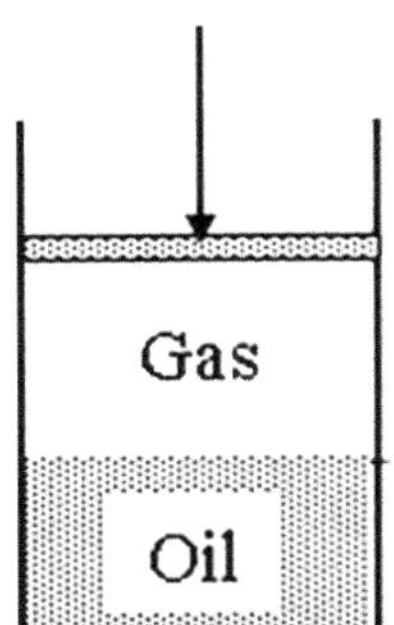

Figure 1-7. Bottom hole sampler.

Table 1-1. PVT test proceedure.

	Step #1	Step #2	Step #3	Step #4	Step #5
Oil Volume (bbl)	1.31	1.333	1.21	1.04	1.00
Gas Volume (ft³)	0.0	0.0	2.99	676	567
Pressure (psi)	3,500	2,500	1,200	14.7	14.7
Temperature (°F)	160	160	160	160	60
	(in situ)	(BPP)			(STB)

1.4.1 Formation Volume Factor

The oil formation volume factor is defined as

$$B_0 = \frac{RB(t_r, p_r)}{STB(t_s, p_s)} \tag{1-2}$$

where

RB = reservoir barrels
STB = stock tank barrels (60°F and 14.7 psia)

This is determined by bringing the pressure and temperature of the sample down to standard pressure and temperature (14.7 psia and 60°F) and measuring the liquid volume of the sample. Consider a reservoir oil having in situ temperature of t_r = 160°F and pressure of p_r = 3500 psi. A typical PVT test is outlined in Table 1-1.

The far-left diagram (#1) represents reservoir oil conditions with sample volume (600 cc) scaled to 1.31 bbl. Pressure is gradually reduced (maintaining reservoir temperature) until gas begins to form as depicted in (#2), identifying a bubble point of 2500 psia. Pressure is further reduced as as shown by diagrams (#3) and (#4) to standard pressure of 14.7 psia still maintaining reservoir temperature. Finally, the temperature is lowered to 60°F creating stock tank temperature and pressure conditions (#5). According to this example the "oil formation volume factor" is B_0 = 1.31.

1.4.2 Bubble Point Pressure

Bubble point pressure is determined by keeping the sample at reservoir temperature while dropping pressure slowly. At some pressure gas begins to escape (bubble) from the liquid oil. This pressure defines the bubble point, which in the above example is 2,500 psi (Step #2). During production operations, bottom hole pressure is maintained above the bubble point by a production choke at the surface to prevent gas from forming in the reservoir.

1.4.3 Gas-Oil Ratio

The gas-oil ratio is determined under stock tank conditions.

$$GOR = \frac{SCF}{STB} \tag{1-3}$$

where

SCF — Standard Cubic Feet (MSCF = 1000 SCF)
STB — Stock Tank Barrels (p = 14.7 psia, T = 60°F)

The GOR for this example is GOR = 567 SCF/STB.

Gas is difficult and costly to transport, especially from offshore production systems. As a general rule the energy equivalence of oil and gas is approximately

$$16{,}000 \text{ SCF (gas)} \sim 1 \text{ STB (oil)}$$

Oil is the more valuable of the two.

Gas that comes out of solution is not refined; it has the same chemistry as the oil. Natural gas, on the other hand, is highly refined and is typically found in deep reservoirs. Methane gas is a good example of a natural gas.

1.4.4 Reservoir Rock Properties

Core samples of reservoir rock are taken also to determine porosity and permeability. Porosity is used to determine volume of "in-place" oil. Permeability is used to predict production rates and developing a reservoir production plan.

1.4.5 Reservoir Economics

Consider a reservoir that is 5 miles by 3 miles and 200 ft thick. Total volume of the reservoir is

$$Vol_T = 5 \text{ mi}\left|\frac{5280 \text{ ft}}{1 \text{ mi}}\right| 3 \text{ mi}\left|\frac{5280 \text{ ft}}{1 \text{ mi}}\right| 200 \text{ ft} = 8.36 \times 10^{10} \text{ ft}^3$$

Converting this volume to barrels

$$Vol_T = 8.36 \times 10^{10} \text{ ft}^3\left|\frac{7.48 \text{ gal}}{1 \text{ ft}^3}\right|\left|\frac{1 \text{ bbl}}{42 \text{ gal}}\right| = 1.489 \times 10^{10} \text{ bbl}$$

Assuming a reservoir rock porosity of $\phi = 0.13$, the amount of crude oil within the reservoir (RB, reservoir barrels) is

$$RB = 1.489 \times 10^{10}(0.13) = 1.936 \times 10^9 \text{ bbl}$$

We assume here that the total pore space is occupied by oil. This is not exactly true as connate water is always present in reservoir rock. In reality, the pore space is saturated partly by the oil and partly by the connate water.

Now assume the oil formation volume factor is $B_0 = 1.31$. The amount of oil when brought to the surface under stock tank conditions is

$$STB = \frac{1.936 \times 10^9}{1.31} = 1.478 \times 10^9 \text{ bbl}$$

Using best recovery technology, which typically allows for 2/3 recovery over 20 to 30 years

$$STB = \left(\frac{2}{3}\right) 1.478 \times 10^9 bbl = 9.851 \times 10^8 bbl \text{ (total recovered oil)}$$

Assuming the price of a barrel of oil is \$100/bbl, then the current dollar value of the oil in the reservoir is

$$(\$)value = \$9.85 \times 10^{10}$$

or \$ 98.51 billion. The challenge is to recover this asset through production drilling.

1.5 MOTIVATION FOR EDWIN DRAKE'S OIL WELL

One commodity that was in great demand among the settlers during the 1800s was a fuel for lighting the evening and night time hours. Candles had been used for many years worldwide, but they only produced light over a short distance. Whale oil was also used but was only affordable by the wealthy. In addition, whale population was dwindling and on the verge of extinction. Tree resin was tried as an illuminate but was explosive and unsafe. On the other hand, kerosene was slow burning, safe and cheap. The ability to extract kerosene (coal oil) from coal was discovered by a Canadian chemist and was fast becoming a preferred lighting fuel by the settlers.

In 1858, George H. Bissell and Jonathan G. Eveleth, seeing a possible market, hired Professor Benjamin Silliman, Jr, a chemist at Yale University to see if kerosene could be refined from rock oil, a crude oil that had been seeping to the surface in western Pennsylvania for years [4]. This oil was a nuisance and was occasionally used for curing mange on animals. Some early settlers used it for medicine as well. The results from this study showed that kerosene could, in fact, be produced from rock oil. Based on this finding, the investors bought a farm near Titusville, PA and formed the Pennsylvania Rock Oil Company.

On August 27, 1859, Edwin Drake successfully reached oil at 69.5 ft, set pipe in the ground, and began pumping oil to the surface using a hand pump typically used for pumping water. The production from this well was about 8 barrels per day.

This crude oil was put in 50-gal wine barrels and transported in horse drawn wagons. By the time the barrels got to the refinery, some of the oil had spilled out. It was agreed that each barrel contained 42 gal and today the standard conversion factor is 42 gal/bbl. Since the wine barrels were blue, the abbreviation for barrels became "bbl". One gallon was measured in terms of "Queen Ann's gallon or a cylindrical volume having a height of 6 in. and a diameter of 7 in. (1 gal = 231 in.^{3}). The crude oil was sometimes stored in stock tanks; water tanks used to store water for cattle. The term "stock tank barrel" refers to the volume of oil at standard atmospheric pressure and temperature (14.7 psi and 60°F).

The demand for kerosene expanded and soon many drilling rigs were erected around Titusville. Some of these oil wells produced as much as 3000 bbl/day.

A few miles away in Cleveland, Ohio, John D. Rockefeller, along with his partner, M. B. Clark, had a thriving business of selling produce to settlers moving west. His produce included, seed grain, meats, hand tools, gardening equipment, linens, etc. When the market for kerosene opened up, he and his partner built a refinery and began selling kerosene as well. By the end of the Civil War, Cleveland was a major refining center for kerosene.

Rockefeller concluded there was a greater market potential for kerosene than all of the other produce combined. He sold his share in the produce business and started his own refining company to produce kerosene. In 1870, he formed the Standard Oil Company. The word *standard* implied quality and safety (my interpretation). Kerosene burned slowly and was safe. His company not only supplied kerosene to the American market, but he quickly created markets in Europe and Asia.

Commercial oil production spread rapidly throughout the world. Italy produced oil in 1860 followed by Canada, Poland, Peru, Germany, Russia, Venezuela, India, Indonesia, Japan, Mexico, and Argentina.

The cable-tool method was used, from 1860 to 1900. Following this method, a heavy steel chisel is fixed to the end of a long cable and dropped against the bottom, smashing the rock. At regular intervals, drilling stopped while the cuttings were removed from the wellbore. Holes were drilled to a few hundred feet. However, several pre-1900 wells exceeded 2000 ft and in 1953 one well was drilled in excess of 11,000 ft using the cable tool method.

1.6 SPINDLE TOP OIL FIELD

The first recorded "gusher" or uncontrolled production occurred at a well 500 miles north of Baku, Russia in 1893. This well produced 100,000 barrels of oil in eleven hours before the flow rate begins to drop off.

At the beginning of the 1900s, major oil activity began to spring up near Beaumont, Texas. Patillo Higgins, a local resident, observed gas bubbling through standing water near a high spot outside of Beaumont [5,6]. He would sometimes strike a match to the gas bubbles and watch flames dance about the water. Higgins felt that oil and gas might be found by drilling into the top of the dome. It was later shown that the high spot was caused by a subsurface salt dome pushing upward. In order to check out his theory, Higgins organized the Gladys City Oil, Gas and Manufacturing Company (1892). He was unsuccessful for several years in his attempts to drill and locate oil in and around the dome.

These high spots or domes are common across the Louisiana and Texas Gulf coast shoreline. Over geologic time salt domes formed as a result of the buoyancy of salt. Salt is lighter than rock and over time (millions of years) was forced upward by the heavier rock pressure below it. The significance of the dome is that salt is impermeable and provides a perfect trap for migrating oil.

Eight years later, a well near Beaumont, Texas marked the start of the modern petroleum industry. The Spindletop Gusher spouted 100,000 barrels of oil a day for nine days before it was controlled in January of 1901. This well alone produced more than 120,000,000 barrels of oil during the next 30 years. Spindletop is noted for its use of the rotary drilling method replacing the cable-tool method. This well was the first gusher to be brought under control using a series of valves now known as a Christmas tree.

The rotary drilling process involves placing a drill bit at the end of a long drill pipe, and then rotating the pipe to enable the bit to grind, abrade, shear, or pulverize and fracture the rock formation. Drilling fluid, called drilling mud, is pumped down the inside of the pipe and through nozzles located within a drill bit to cool and lubricate the bit. Drilling fluids also clean the well bore by washing the cuttings from around the drill bit and back to the surface.

As the hole gets deeper, additional lengths of pipe are added to make the drilling pipe longer. The amount of weight applied to the drill bit is regulated by a pulley arrangement supported by a derrick. Rotary rigs were not new in 1900: the first had been patented in 1833 and by 1901 more than 100 wells had been drilled in Texas using this method.

All elements of rotary drilling, as known today, were first brought together at Spindletop in 1901. This initial drilling system included: steam engines, drawworks, rotary table, mud pumps, swivel, traveling block and crown block, and a drill bit.

The first drill bits used in rotary drilling were fish tail drag bits. They were successful in soft formations, such as found along the Gulf coast. In harder formations, this kind of drill bit was not effective. Fish tail drag bits have been used along the Gulf as late as the 1960s. In 1906 Howard Hughes, Sr., a drilling contractor, began experimenting with a drill bit containing rolling cones with spiked cutters. After obtaining a patent in 1909, he quit contract drilling and started the Sharp-Hughes Tool Company [7].

When a drill bit wears out, the entire drillstring is pulled out of the well bore. A new drill bit is attached to the bottom end and the entire string of pipe is lowered back into the hole. Until around 1950, a typical well depth was around 10,000 ft. Today, well depths are much deeper and traverse a prescribed three-dimensional path to a reservoir target.

The rock-cutting tool or drill bit is an important component in the drilling system. Since the early drill bits by Hughes Tool, improvements have been made to metallurgy, cutter hardness and bearing seals. Each improvement increased drill bit life and rate of penetration. Other drill bit types have also been developed such as man-made diamond drag bits and natural or industrial diamond drill bits.

Several oil field tool and service companies were spawned during this time. Erle P. Halliburton established the Halliburton Oil Well Cementing Company in 1919. He developed a process and equipment for cementing large pipe (casing) in an open well bore. His process mixed dry powered cement with the right amount of water so that the slurry could be pumped. This mixture was pumped down the inside of the casing and back up around the outside to secure casing. The ratio of water to cement was important for strength, too. The casing stabilizes an open hole for drilling and production. It is also needed for well control.

One of the key developments by Halliburton is a high pressure, high-volume pump that is required to force the cement slurry downhole and up the annular space. These pumps also proved to be useful in hydraulic fracturing for secondary recovery.

1.7 EAST TEXAS OIL FIELD

The East Texas Oil Field is the largest oilfield in the lower 48 states (Figure 1-8). It extends across Gregg and Rusk Counties with some overlapping into Upshur,

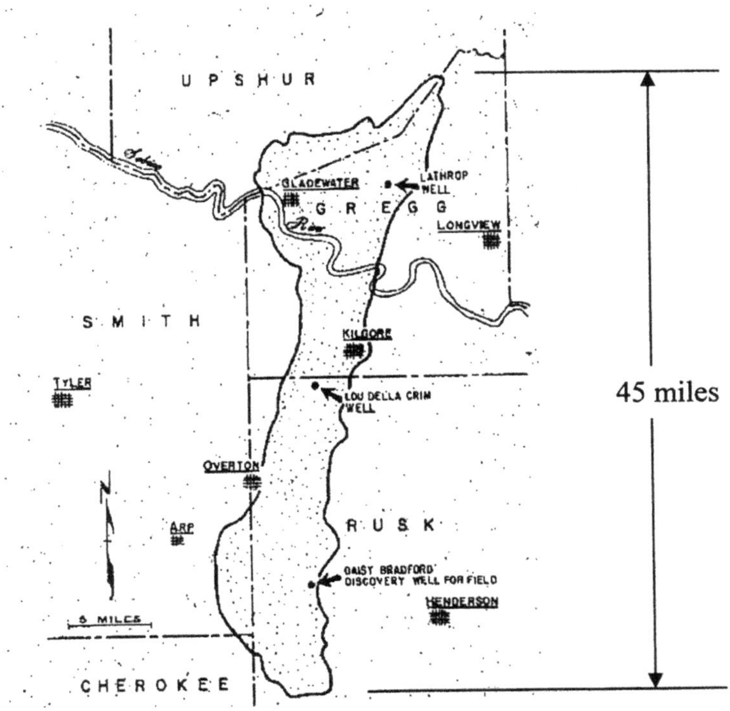

Figure 1-8. East Texas oil field.

Smith and Cherokee Counties. The north/south dimension is about 45 miles and the east/west dimension is about 5 miles; it covers some 140,000 acres [8]. The field is an extension of the Woodbine formation, which extends to the West of Dallas. This sandstone formation was formed when East Texas was a shallow sea, approximately 100 million years ago. Subsequently it was uplifted, eroded and covered by an impermeable layer of chalk, creating a stratigraphic trap (Figure 1-6).

The Woodbine formation to the west supplies drinking water to the Dallas area. To the east, oil migrated until trapped by a sedimentary layer of chalk.

Since its discovery in 1930 this reservoir has produced some 5.42 billion barrels (STB) of oil from the Woodbine formation. Calculations indicate there are about 70 million barrels (STB) that can still be recovered under current production practices. It is estimated that as much as 550 million barrels of oil still remain in place. The field covers approximately 134,000 acres (210 mi^2). More than 31,200 wells have been drilled. The average well spacing is 4.3 acres.

Despite the 75-year history of this field, it has never been unitized nor have production plans been developed to extract maximum recovery from this field.

The East Texas Field is the result of a stratigraphic trap caused by the Sabine Uplift, which extends across the Texas-Louisiana border. The Woodbine formation is approximated 800 ft thick within the basin and thins gradually toward the Sabine Uplift. The west edge of the Woodbine formation is about 250 ft thick.

Oil companies had drilled 17 dry holes in this area during the 1920s and had condemned it. In the early 1920s Columbus Marion (Dad) Joiner from Ardmore, Oklahoma, bought a lease on 320 acres and planned to sell leases at a profit. Instead he became a wildcatter and bought leases on 4000 more acres at little or no cost [9]. The country was in an economic depression and farmers were desperate for some type of income. On August 11, 1925, Joiner obtained a lease from Mrs. Daisy Bradford for her 975.5-acre farm.

During this time, Joiner met A. D. (Doc) Lloyd, a self-proclaimed geologist without any formal training. He was a super salesman having peddled patented medicines he made from oil. Joiner began to use Lloyd's sales skills to sell leases and referred to him as a nationally known geologist, which of course was not true. In 1927, Joiner mailed hundreds of copies of a report along with a letter and geologic map prepared by Doc Lloyd stating the potential of the Daisy Bradford acreage. His maps showed a salt dome, four anticlines and a fault line running through the Daisy Bradford farm all of which was completely incorrect. At this point, there had been no experience with stratigraphic type oil traps. On the date of the letter, June 15, 1927, Joiner was not drilling a well as the letter implied.

Later in the year, oil drilling started on Daisy Bradford #1. After drilling 1098 ft, the pipe got stuck and the well was abandoned on February 1928. A second well, Daisy Bradford #2 was drilled some 100 ft from the first. This well too was abandoned on March 1929 at 2518 ft because of stuck pipe.

On May 8, 1929, the drilling rig was being skidded to a new location when it suffered a structural failure. Since the rig could not be moved further, the driller, Ed Laster, decided to start drilling Daisy Bradford #3 on this spot. At 3456 ft core samples showed signs of oil. Laster reamed the well to 3486 ft and a definitive test took place on September 5, 1930. By October 1930, more than 8000 people had visited the well site.

Oil tests from other wells drilled in East Texas, confirmed that oil from a wide range of well sites was from the same reservoir. Ironically, had Joiner's well been drilled a quarter mile to the east, he would have missed the field. Had Joiner drilled 2 miles to the west, where Doc Lloyd had recommended, he would have brought in a much bigger well. This oil field became known as the Black Giant or the East Texas Oil Field. It is still the largest oil field in the U.S. This oil field fueled the U. S. military during World War II. A pipeline was built connecting this oil field to Philadelphia where the oil was refined and dispatched to the military.

1.8 UNITIZING OIL RESERVOIRS

During the oil boom days wells were drilled without regard to a reservoir production plan. Anyone could drill on their property; old pictures show wooden derricks side by side. This practice followed the "Law of Capture", which allowed oil wells to be located almost at random.

Wells were drilled into a reservoir without adequate well control equipment or procedures. Oil and gas often came to the surface out of control. In addition to being unsafe and environmentally damaging, reservoirs were damaged to the point where oil recovery was low.

Little thought was given to the environmental impact of oil drilling and production. Salt water was produced with the oil and dumped into streams killing fish and effecting wild life. Contaminated rock cuttings were left in reserve pits. Oil from gushers was retained in large ponds, until the natural reservoir pressure reduced to the point that the well could be capped.

Today the oil industry is a leader in environmental protection. Costly steps are taken to discard chemically treated drilling fluids and downhole cuttings. Production equipment separates oil, gas and water and all are discharged appropriately.

Following current practice, reservoirs are mapped and delineated for the benefit of each lease holding company. Oil and gas reservoirs are unitized so that each leaseholder gets its proper proportion of oil production. One company (called the Operator) is chosen to manage the development of the entire field. All of this is especially important in developing large offshore reservioirs.

1.9 PROBLEMS

1. Given: API gravity = 39°
   Find: Specific Gravity of the Crude Oil (S.G.)
   Density (γ) of the Oil in lbs/ft^3

2. Given: γ = 56.0 lb/ft^3
   Find: SG =
   API gravity =

3. The figure shows three 1 in.^{3} boxes containing: a) one solid sphere, b) eight solid spheres, and c) 64 solid spheres. Show that the porosity of the three boxes is the same.

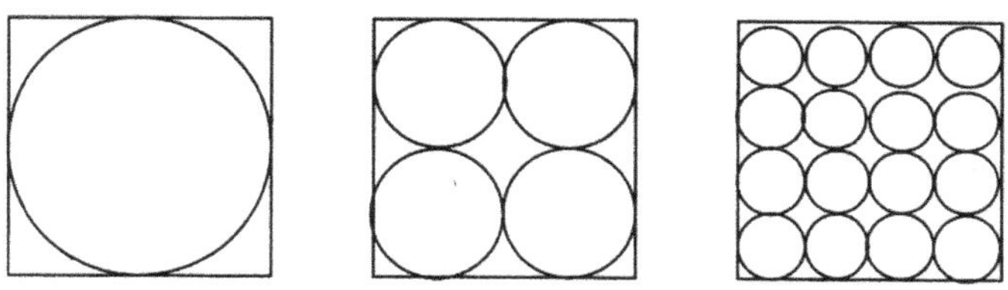

(a) (b) (c)

4. Given: Size of an oil reservoir is 12 miles by 4 miles by 200 ft (thick)
   Formation volume factor, B_0 = 1.4
   Reservoir rock porosity, ϕ = 0.15

   Find: Volume of crude oil in place or Reservoir Barrels (bbl)
   Volume of stock tank barrels (STB)
   Asset value of oil if oil is priced at \$90/bbl (assume 2/3 recovery)

1.10 LIST OF REFERENCES

1. Edward J. Tarbuck and Frederrick K. Lutgens: *The Earth: An Introduction to Physical Geology*; Charles E. Merrill Publishing Company, 1984.

2. Kirk E. Boatright; *Basic Petroleulm Engineering Practices*, 1993.
3. Craft, B. C. and M. F. Hawkins, *Applied Petroleum Reservoir Engineering*, Prentice-Hall, Englewood Cliff, N.J., 1959.
4. Paul H. Giddnes; *The Birth of the Oil Industry*, Macmillan, New York, 1938.
5. James A. Clark and Michael T. Halbouty; *Spindle Top*, Random House, NY, 1952.
6. T. A. Rickard; *Anthony F. Lucus and the Beaumont Gusher*, Mining and Scientific Press, Dec 22, 1917.
7. Hughes Two-Cone Drill Bit; Designated a Historical Mechanical Engineering Landmark, ASME – Baker Hughes Special Publication, The Woodlands, Texas; August 10, 2009.
8. William A. Ambrose, Tucker F. Hentz, Florence Bonnaffe, Robert G. Loucks, L. Frank Brown, Jr., Fred P. Wang and Eric C. Potter; "Sequence-stratigraphic controls on complex reservoir architecture of highstand fluvial-dominated deltaic and lowstand valley-fill deposits in the Upper Cretaceous (Cenomanian) Woodbine Group, East Texas field; Regional and local perspectives"; The American Association of Petroleum Geologists (AAPG) Bulletin, V. 93, NO. 2 (February 2009), pp. 231–269.
9. James A. Clark and Michael T. Halbouty, *The Last Boom*, Shearer Publishing, Texas, 1984.

Chapter 2

OPTIMUM DRILLING PRACTICES

Early drilling rigs were built on site and left in place. Mobile drilling rigs, introduced in the 1950s (Figure 2-1), are transported by truck and set-up on site and then relocated for the next well. The most visible part of a drilling rig is its derrick, a structure that contains a hoisting arrangement, which includes a crown block and a traveling block. The pulley arrangement is designed to create a large mechanical advantage for hoisting equipment (drill pipe, casing, tubing, blow out control equipment, etc.) in and out of the hole.

2.1 SUBSYSTEMS OF DRILLING RIGS

There is much activity during drilling operations. It is helpful to understand these activities by viewing the total rig is being made up of several subsystems.

- Hoisting — The hoisting system includes the derrick, crown block, traveling block and a power-driven drum. The crown block and traveling block provide a huge mechanical advantage for lifting tubulars (~300,000 lb) in and out of bore holes. This pulley arrangement reduces the force on the fast end (drum) of the cable to around 30,000 lbs.
- Rotary Drive and Drillstring — The drillstring extends from the rig floor to the drill bit. Its two basic tubular components are drill pipe and drill collars. This very long pipe transmits torque from the surface to the drill bit and provides a conduit for drill mud. The rotary table transmits torque to drill pipe by means of a square or hexagonal pipe called the Kelly, which allows the drillstring to advance during the drilling process while maintaining rotary torque. A large amount of energy is lost to friction along the drillstring.
- Hydraulic or Circulation — The hydraulic system is central to the rotary drilling method. Surface mud pumps circulate drilling fluid down the drillstring to the drill bit for hole cleaning purposes. Cuttings are carried back to the surface for inspection and analysis. Drill fluid is continually monitored and maintained before being recirculated down hole for bit cleaning. A significant amount of energy is lost to fluid friction over several thousand feet of pipe.
- Well Control — The progressions of tools available to defend against a well blowout are: drilling mud, annular preventer, hydraulic rams. The annular preventer and hydraulic rams make up the Blowout Preventer (BOP) stack. Flow rate returns are continually monitored for possible

Figure 2-1. Drilling site near Lafayette, Louisiana (1962).

formation fluid invasion into the well bore. When this happens, the well is taking a kick from the formation and bad things can happen if this is allowed to continue. Well control measures are taken to control formation fluid invasion.

- Mechanical/Electrical Power (~8000 to 10,000 hp) – Diesel is the prime source of power around a drilling rig. These engines power electric generators, which supply power to motors around the rig for lifting pipe and tools in and out of the well bore, drive the rotary table, and power mud pumps. While this level of power is needed to around a drilling rig, ironically, only about 50 hp is used to break rock formations to advance the well bore.

Drilling muds have greatly improved since Spindletop where, cattle pond water was muddied to add weight for well bore stability and well control. Modern drilling fluids (still called drilling mud) provide: well bore stability, mud weight refinement for well control and drilling pressure balance, and environmentally friendly properties.

The growth in drilling technology over the 20th century was enormous [1,2]. Technology has advanced from crude drill bits and drilling equipment to current tools that drill along specified well paths and with great accuracy. Navigation drilling is accomplished through Measurement While Drilling (MWD), Rotary Steerable Tools and advanced drill bits. Yergin [3] gives an excellent overview of the evolution of the petroleum industry discussing key milestones in history and their political impact on international affairs.

2.2 ELEMENTS OF OPTIMUM DRILLING

There have been major advances in drilling technology since 1900, each contributing to the ability to drill faster, deeper, at lower footage cost and in more difficult regions. These advances have resulted from ongoing efforts by the industry to find new ways to shear, fracture or destroy rock.

There is no single equation that brings all of the drilling parameters together to achieve a mathematical optimum. However, if attention is given to key aspects of drilling, the result is the very best that can be achieved with current drilling technology. Good drilling practice emphasizes

- drill bit selection
  - Roller Cone
  - PDC
  - Diamond
- weight on bit (WOB)
- rotational speed
- hydraulics
- pressure balance
- mud properties
- bottom hole assembly

Paying attention to these factors can lead to maximum rate of penetration (ROP) and minimum footage cost. This practice is commonly called *optimum drilling*.

2.2.1 Drill Bit Selection

Bit selection is based on statistical data giving the performance of drill bits used in a given area. Lowest cost per foot is the basis of bit selection. Footage cost is calculated from

$$C = \frac{C_{rig}(t_t + t_d) + C_{bit}}{ROP * t_d} \tag{2-1}$$

where

C = footage cost (\$/ft)
C_{rig} = rig rate (\$/hour)
C_{bit} = bit cost (\$)
t_T = trip time (hours)
t_d = drilling time (hours)
ROP = average rate of penetration (ft/hour)

This formula does not include costs related to hole problems, casing, etc. It is mainly used to evaluate the cost effectiveness of drill bits. Equation (2-1) is displayed graphically in Figure 2-2 showing that for a given rate of penetration, footage cost steadily reduces with on-bottom drilling time.

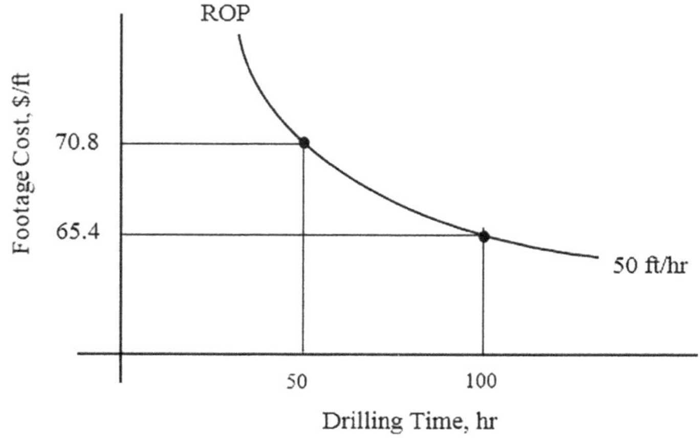

Figure 2-2. Footage cost versus drilling time and ROP.

Consider the following drilling condition.

ROP = 50 ft/hour
Rig Rate = $3000/hour
Trip Time = 6 hours

The life of the drill bit is, say, 50 hours. According to Eq. (2-1) footage cost is $71/ft. Using the assumed ROP, the footage drilled over 50 hours is 2500 ft. The cost to drill 5000 ft

$$\text{Cost} = [(70.8)\ 2500]\ 2 = \$354{,}000$$

This price includes a round trip for the second drill bit and the cost of the second drill bit.

Next, assume the drill bit last (or stays on bottom) for 100 hours. The footage cost in this case is $64/ft. In this case the cost to drill 1000 ft is

$$\text{Cost} = (65.4)5000 = \$327{,}000$$

Doubling the life of the drill bit results in a savings of $27,000.

2.2.2 Weight on Bit and Rotary Speed

Bit weight and rotary speed directly affect rate of penetration (ROP) and bit life [4,5]. In the early 1960s Maurer [6,7] developed a perfect cleaning theory and showed that if chips or cuttings are completely removed from their craters, ROP increased with the square of bit weight. This theory has been verified in drilling situation that has nearly perfect cleaning such as raised boring in open mines and horizontal shaft boring. In reality, bit cleaning is not perfect because actual rates of penetration fall short of the perfect cleaning bench mark.

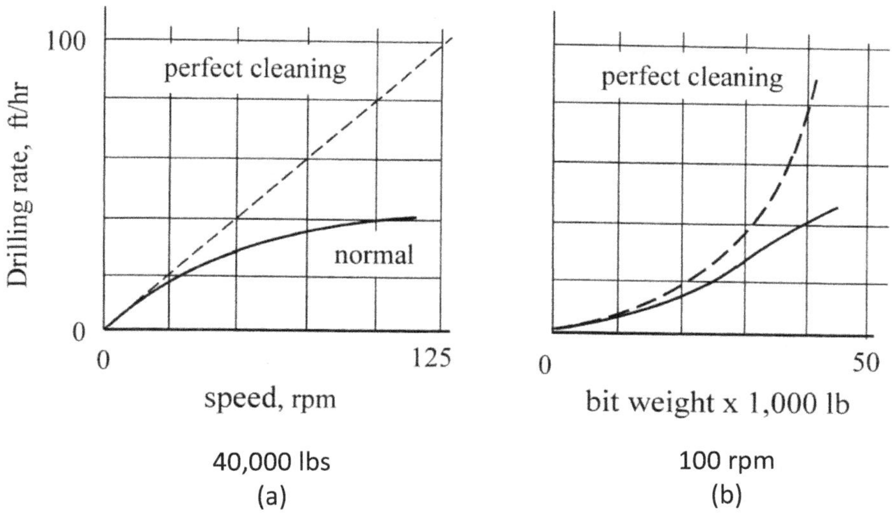

Figure 2-3. Effects of weight on bit and rotary speed on drilling rate. (Courtesy of AIME.)

According to the perfect cleaning theory, rate of penetration should increase directly with rotary speed (Figure 2-3). In practice this does not occur because of the reasons mentioned above.

The difference in rate of penetration between perfect cleaning and normal cleaning is even greater when downhole motors and turbines are used because each operate at much higher speeds. There is much incentive to improve bit cleaning when PDMs and turbines are used. Many attempts have been made to formulate the effects of bit weight and rotary speed on rate of penetration. A common formula is

$$ROP = WOB^{a} RPM^{b} \tag{2-2}$$

where a = 2 and b = 1 for perfect cleaning while a ≤ 2 and b ≤ 1 for imperfect cleaning.

2.2.3 Hydraulics for Best Drill Bit Cleaning

Hydraulics refers to fluid flow and fluid pressures throughout the circulating system. Fluid energy and impact forces are required to clean drill bits and to remove cuttings from underneath the bit. Good drilling rate depends on it. Hydraulic power that gets to the drill bit is the difference between hydraulic power delivered by the mud pump and parasitic losses due to fluid friction. This difference is call available hydraulic horsepower and is a drilling parameter that can be mathematically optimized [8].

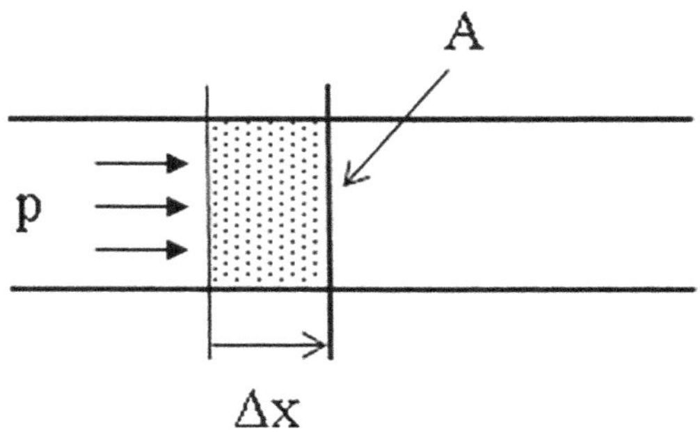

Figure 2-4. Hydraulic horsepower.

Hydraulic power is expressed in terms of local pressure and flow rate according to (see Figure 2-4)

$$\Delta Work = pA\Delta x$$

$$Hydraulic\ Power = pA\frac{\Delta x}{\Delta t} = pAv = pQ \tag{2-3}$$

The units are important.

$$Power = p(psi)\left|\frac{144\ \text{in}^2}{\text{ft}^2}\right|Q(gpm)\left|\frac{1\ \text{ft}^3}{7.48\ \text{gal}}\right|\left|\frac{1\ \text{min}}{60\ \text{sec}}\right|$$

$$Power = 0.3209pQ \quad \text{ft-lb/second}$$

Since 1 hp = 550 ft-lb/second,

$$HHP = \frac{pQ}{1714}\ \text{hp} \tag{2-4}$$

where the units are: p(psi) and Q(gpm). This is a useful formula and is used throughout the text.

2.2.3.1 Maximum Available Hydraulic Horsepower

Consider a long pipe with a pump at one end and a drill bit at the other. The pump delivers power and friction along pipe removes power. The difference (called available HHP) is what gets to the drill bit for cleaning. If system pressure losses (parasitic) due to fluid flow are defined by

$$p_s = CQ^{1.8} \tag{2-5}$$

where

p_s = system pressure loss due to fluid friction, psi
C = friction loss constant
Q = flow rate, gpm

then the hydraulic horsepower available to clean the drill bit is defined by

$$HHP_a = \frac{(p_p - p_s)Q}{1714} \tag{2-6}$$

$$HHP_a = \frac{p_pQ - CQ^{2.8}}{1714} \tag{2-7}$$

where

p_p = pump pressure, psi

The first term on the right side of Eq. (2-6) is the power supplied by surface pumps. The second term represents the system or parasitic losses inside the pipe. The available hydraulic horsepower delivered to the drill bit for bit cleaning can be maximized by calculus as follows.

Mathematically the available hydraulic horsepower reaches a maximum at a particular flow rate [8]. The condition for maximum available hydraulic power is predicted as follows.

$$\frac{d(HHP_a)}{dQ} = p_p - 2.8CQ^{1.8} = 0 \tag{2-8}$$

By substitution of Eq. (2-5)

$$p_p = 2.8p_s \tag{2-9}$$

which means that HHP_a is maximum when

$$p_s = \frac{1}{2.8}p_p \tag{2-10}$$

or when system pressure losses are about ⅓ maximum allowable pump pressure. The flow rate at which this condition occurs depends on the hydraulics of the circulating system defined by Eq. (2-5). By substitution

$$\frac{p_p}{2.8} = CQ_{opt}^{1.8} \tag{2-11}$$

$$Q_{opt} = \left[\frac{p_p}{2.8C}\right]^{\frac{1}{1.8}} \tag{2-12}$$

Assuming C = 0.0142 (its determination is discussed later) and assuming

P_p = 3,000 psi (maximum pump pressure)

then

$$Q_{opt} = \left[\frac{3{,}000}{2.8(0.0142)}\right]^{\frac{1}{1.8}} = 513 \text{ gpm}$$

$$p_s = \frac{1}{2.8}3{,}000 = 1{,}071 \text{ psi}$$

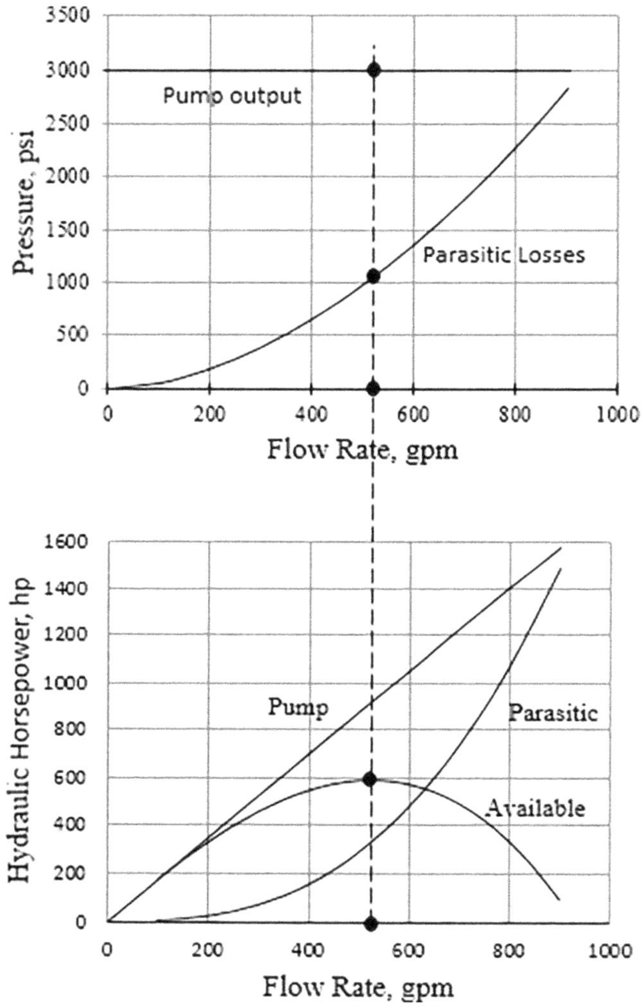

Figure 2-5. System hydraulics.

$$HHP_a = \frac{1,929(513)}{1714} = 577 \text{ hp}$$

Plots of the hydraulics for this example are displayed in Figure 2-5 emphasizing the parameters for optimum hydraulics.

2.2.3.2 Field Application

A practical application of this information is explained below. Assume you go to the rig and find the following drilling data is

- Standpipe pressure is 2500 psi
- Mud density is 12 ppg
- Bit nozzle sizes are 11,11,12 (TFA = 0.2961 in.^{2})
- Flow rate 360 gpm
- Maximum allowable pump pressure, however, is 2800 psi
- A bit change is imminent

The question is — What would be your recommendation for changes in these drilling parameters, if any? The first step is to determine the value of "C" for the hydraulic system at current depth.

Using Table 2-1 we see that the pressure drop across the drill bit is 1622 psi. Knowing the standpipe pressure, then the parasitic losses are $p_s = 2500 - 1622 = 878$ psi. Using Eq. (2-5)

Table 2-1. Pressure drop across bit nozzles (12 ppg drilling fluid density).

	11,11,11	11,11,12	11,12,12	12,12,12
	0.2784	0.2961	0.3137	0.3313
Q (gpm)	Δp (psi)	Δp (psi)	Δp (psi)	Δp (psi)
300	1274	1126	1003	900
310	1360	1203	1071	961
320	1449	1281	1142	1024
330	1541	1363	1214	1088
340	1636	1447	1289	1155
350	1734	1533	1366	1224
360	1834	1622	1445	1295
370	1938	1713	1526	1368
380	2044	1807	1610	1443
390	2153	1903	1696	1520
400	2265	2002	1784	1599
410	2379	2103	1874	1680
420	2497	2207	1967	1763
430	2617	2314	2061	1848
440	2740	2423	2158	1935
450	2866	2534	2258	2024
460	2995	2648	2359	2115
470	3127	2764	2463	2208
480	3261	2883	2569	2303
490	3399	3004	2677	2400
500	3539	3128	2787	2499

$$C = \frac{p_s}{Q^{1.8}} = \frac{878}{360^{1.8}} = 0.022$$

This number defines the total losses throughout the circulating system. Turn now to the drilling parameters that will optimize bit cleaning.

With a maximum allowable pump pressure of 2800 psi, then system losses at maximum hydraulics is

$$p_s = \frac{2,800}{2.8} = 1,000 \text{ psi}$$

So, bit pressure drop is 2800 – 1000 = 1800 psi. Optimum flow rate is

$$Q_{opt} = \left[\frac{2,800}{2.8(0.022)}\right]^{\frac{1}{1.8}} = 387 \text{ pgm}$$

Using Table 2-1 and select nozzle sizes.

Stay with same nozzle size but increase flow rate to 387 gpm. Standpipe pressure should read 2800 psi.

2.2.4 Pressure Balance

The density of drilling fluid is a key to controlling formation fluid pressure while drilling. It is, therefore, important to maintain adequate hydrostatic pressure at the drill bit. Hydrostatic pressure increases linearly with the vertical depth according to

$$p = \gamma x \tag{2-13}$$

where γ is fluid density, x is vertical distance and p is local fluid pressure. Units of the variables are important.

$$p\left(\frac{lb}{ft^2}\right) = \gamma\left(\frac{lb}{ft^3}\right) x\left(ft\right) \tag{2-14}$$

Typically mud density is expressed in units of ppg, vertical depth in ft and pressure in psi. To accommodate these units, we make the following adjustments.

$$p\left(psi\right) = \gamma\left(ppg\right)\left|\frac{7.48gal}{ft^3}\right| x(ft)\left|\frac{ft^2}{144in^2}\right| \tag{2-15}$$

The magnitude of hydrostatic pressure at any vertical depth can be determined by

$$p_m = 0.0519\gamma_m x \tag{2-16}$$

where

p_m = hydrostatic pressure of drilling mud, psi
γ_m = fluid density of drilling mud, ppg
x = vertical depth, ft
0.0519 = conversion factor ~ 0.052

Pressure gradient or rate at which pressure increases with depth is

$$\frac{dp}{dx} = 0.052\gamma_m \text{ psi/ft} \tag{2-17}$$

The pressure gradient of fresh water

$$\frac{dp}{dx} = (0.052)62.4\,\text{lb/ft}^3 \left| \frac{1\,\text{ft}^3}{7.48\,\text{gal}} \right| = 0.433 \text{ psi/ft}$$

Fluids in the formation may also contain pressure associated with depth according to Eq. (2-16). In this case, formation is pressure is considered to be normal. Formation pressures that do not follow this equation are considered to be abnormal. The difference between drilling mud fluid pressure and pore or formation pressure is

$$\Delta p = p_m - p_f \tag{2-18}$$

Another way to express this pressure difference is

$$p_m = 0.052(\gamma_f + \Delta\gamma)L \tag{2-19}$$

$$p_f = 0.052\gamma_f L \tag{2-20}$$

$$\Delta p = 0.052\Delta\gamma L \tag{2-21}$$

where $\Delta\gamma$ is the difference between drilling mud density and formation fluid density, assuming normal formation pressure conditions, and L is well depth.

Typically, normal formation pressure in hard rock areas is equivalent to a mud weight of 8.329 ppg (density of fresh water at 70°F). Normal formation pressure in coastal drilling is equivalent of about 9.0 ppg (density of salt water is $1.025 \times 8.329 = 8.537$ ppg; 1.025 is specific gravity of salt water).

For well control purposes it is important to keep drilling mud pressure slightly greater than formation pressure. However, the pressure difference ($\Delta p = p_m - p_f$) between mud column and formation pressure significantly affects drilling rate of penetration. Tests show that drilling rate may vary by a factor of 10 depending on magnitude of this differential pressure [9,10] (Figure 2-6). Differential pressure is one of the most important factors affecting drilling rate.

Since wellbore pressure is the primary means of controlling formation pressures, the challenge is to seek a safe balance between well control and drilling rate. Balanced pressure drilling means that the hydrostatic pressure is adjusted to match formation pressure. Underbalanced drilling occurs when hydrostatic pressure is less than formation pressure. Under this condition, drilling rate is the highest. Normally drilling mud is overbalanced by about ½ ppg, which means that at 10,000 ft, the magnitude of the pressure overbalance is 260 psi [Eq. (2-21)]. This is ample pressure overbalance pressure for controlling formation pressures.

Typical pressure lines for both drilling mud and formation pressures (assuming normal formation pressure) are illustrated in Figure 2-7. Once the cap rock is reached, drilling rate decreases because fluid formation pressure is zero. However, after drilling through the cap rock and into the reservoir rock,

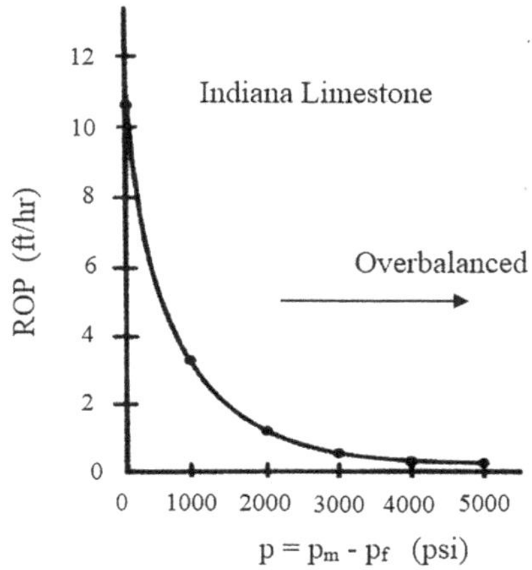

Figure 2-6. Effect of overbalanced pressure on drilling rate. (Courtesy of AIME.)

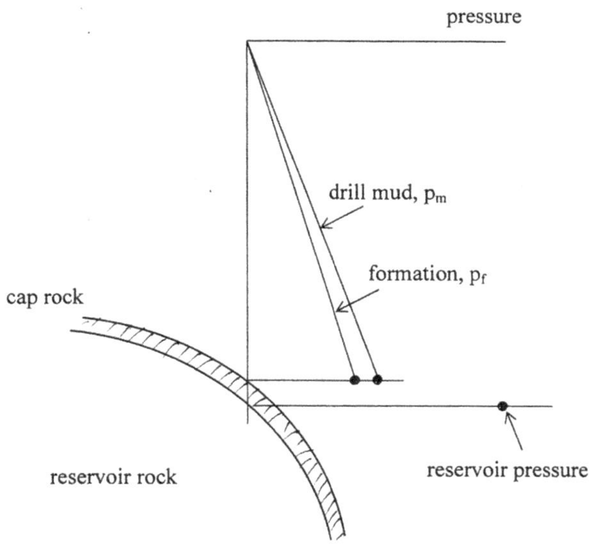

Figure 2-7. Difference between drilling mud and formation pressure.

formation pressure increases to a relatively large value causing a significant increase in drilling rate. The increase in drilling rate is a sure sign the well bore has penetrated the oil reservoir. The sudden increase in ROP is easily detected at the surface.

The explanation for this effect is explained by chip hold down mentioned earlier. Chips are formed under drill bit teeth by shear. Once fracture occurs, the chip is held in place and reground instead of being removed. This regrinding lowers the drill rate.

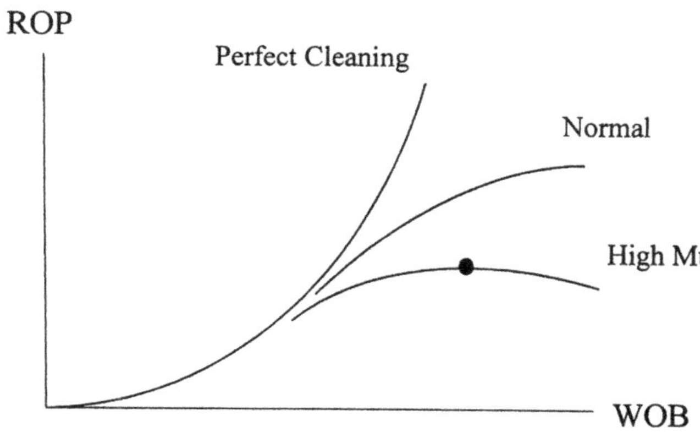

Figure 2-8. Effect of mud weight on ROP.

Figure 2-8 shows how mud weight affects ROP in comparison with the perfect cleaning theory.

The bottom curve shows that if mud weight is too high, there is poor cleaning and ROP is not responsive to increases in WOB. For example, if WOB is increased by, say, 10,000 lbs without an increase in ROP, then mud weight is probably too high. The bold point on the lower curve represents the flounder point; ROP does not increase with weight-on-bit.

2.2.5 Drilling Mud Properties

Drilling mud is the lifeblood of the rotary drilling operation. Its composition and physical properties directly affect rate of penetration. Its density (mud weight), rheological properties (viscosity and gel strength) and chemistry also affect penetration rate.

Since the early days of Spindletop, when muddy water from a cattle pond was used to hold back the formation fluids, the chemistry of current drilling muds is much more complicated. Two basic components are: bentonite, which provides gel strength for lifting cuttings up the annular space and barite, which increases fluid density for well control purposes.

Two rheological fluid models common used to describe the physical properties of drilling mud are Bingham and Power Law models.

$$\tau_{Bingham} = \tau_o + K_0\gamma \qquad \text{(Bingham Model)} \tag{2-22}$$

$$\tau_{PowerLaw} = K\gamma^n \qquad \text{(Power Law Model)} \tag{2-23}$$

where

τ = shear stress
τ_0 = yield point (YP)
γ = shear rate
K_0 = plastic viscosity (PV); slope of straight-line in the Bingham model; its magnitude increases with solids content.

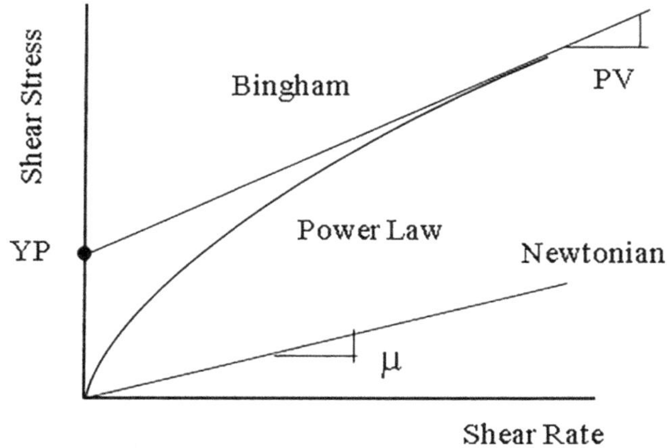

Figure 2-9. Rheology models of drilling fluids.

These mathematical models are useful for predicting pressure losses under laminar flow conditions. Rheological behavior of both models is compared with that of Newtonian fluids in Figure 2-9. By comparison Newtonian fluids behave according to $\tau = \mu\gamma$.

At high shear rates the shear stress-strain rate curves for Bingham and Power Law models converge and both models predict approximately the same pressure drop. Therefore, under turbulent flow conditions, only the plastic viscosity (PV) parameter is needed to predict pressure losses through drillstrings.

Laboratory testing shows that plastic viscosity inhibits cleaning under drill bits and reduces ROP. Studies also show that drilling rate decreases inversely with the square of viscosity [11]. If mud viscosity doubles, drilling rate reduces by one-fourth.

2.2.6 Bottom Hole Assembly

Bottom hole assembly (BHA) refers to the portion of the drill collars and tools directly above the drill bit [12]. This section is typically less than 150 ft. This short section of tools controls and affects everything; ROP, footage cost ($/ft) hole direction, and vibrations. Under straight vertical drilling, the BHA refers to drill bit, stabilizers and drill collars and how they are arranged. Under directional drilling, the BHA refers to drill bit, stabilizers, drilling motor and MWD tools an arrangement that is much more complex, costly and vulnerable. In recent years, drilling tools in the BHA have become more and more complex increasing the probability of failure. Reliability of each tool component becomes increasing important, especially offshore where rig rates are highest.

Overall drill collar length must be chosen to avoid drill pipe buckling and frequency tuning with the drill bit. Controlling lateral vibrations, which are rotary speed related, of drill collars and tool assemblies between stabilizers become increasingly important. The bottom 150 ft of the drill collar section also affects the path of directional wells as well as vibrations throughout the drillstring.

The objective in designing any BHA is to achieve the required goals at minimum footage cost and at low risk as well as directional control.

2.3 DEVELOPING THE DRILLING PROGRAM

Drilling Programs involve drilling teams from both operator and contractor. There is a special process that brings these two teams together to develop and implement a drilling program.

Well planning deals with the selection and application of drilling equipment to drill safely and economically. Safety and cost override everything. Safety includes everything from rig personnel to well control. Cost implies optimum drilling practices, including best rate of penetration, plus lowest footage cost. Drilling economics depends on: skills of rig personnel, logistics, and well planning. Assuming top side operations are conducted efficiently, drilling cost is most affected by the selection and operation of tools used in the bottom hole assembly.

Well planning actually starts with the reservoir development plan, which defines the most economical well spacing and entry point into a reservoir. The plan sets the production tubing size, which in turn sets the casing size and this dictate drill bit size. Well depth and well path affect drill pipe size. Drill bit size changes with each casing interval. The length of the casing depends on formation fracture gradient and pressure gradient of the mud weight.

Other discussions on these topics can be found in books by Robello and Azar [13] and Mitchell and Miska [14].

2.3.1 Operational Requirements

Operational requirements describe what is expected. They are set by management and represent a decision to move ahead with field development. This decision is based on an economic study of reservoir performance and production costs in consideration of market demand and timing. This may include expected production rate, well pattern and points of entry into the reservoir, and other relevant information. This declaration initiates and gives a clear direction for the project.

2.3.2 Design Specifications

Design specifications are developed from the project description as stated in the operational requirements. These specifications detail engineering constraints. They specify expected outcomes.

2.3.3 Creating Design Alternatives

Well planning, like all projects, is an open-ended design activity. There are many possible solutions, and each has to be created and evaluated. In creating design alternatives remember that each has to satisfy the specifications, be technical feasible, cost effective and safe. Past experience in a given production area is useful in this regard.

2.3.4 Evaluating Alternatives

Drillstring design is typically based on the selection of off-the-shelf components, such as drill pipe, drill collars, drill bits, motors, MWD, etc. For each possible component there are many brands, sizes and performance capabilities. The designer must choose among many possible tools to configure drillstrings for each well bore section. In essence there are many possible drillstring configurations and the problem is to select the best tool combination to meet drilling specifications.

2.3.5 Summary of Equations

$$C(\$/ft) = \frac{C_{rig}(t_t + t_d) + C_{bit}}{ROP * t_d}$$

$p_s = CQ^{1.8}$ (C is a constant related to fluid losses in the hydraulic system)

$p_s = \frac{1}{2.8} p_p$, $Q_{opt} = \left[\frac{p_p}{2.8C}\right]^{\frac{1}{1.8}}$ (maximum HHP_a for drill bit)

$p_m = 0.052\gamma_m x$, $\frac{dp}{dx} = 0.052\gamma_m$ psi/ft (hydrostatic pressure, pressure gradient)

$\Delta p = p_m - p_f$ (over balanced drilling)

2.4 PROBLEMS

1. Calculate footage cost for three rates of penetration (ROP = 10, 20, and 30 ft/hour). Assume operating costs are:
   - C_{rig} = \$500/hour
   - C_{bit} = \$12,000
   - t_T = 10 hours

   Plot *footage cost* verses *drilling time* for each ROP in one figure. Vary time from 10 to 100 hours. in increments of 5 hours.

2. A drilling system uses a mud pump having maximum output pressure of 3000 psi. Under current drilling conditions (hole depth, mud properties, pipe size, etc.), parasitic pressure losses vary with flow rate according

$$p_s = 0.015Q^{1.8}$$

   Calculate and plot the available horsepower $(HHP)_a$ verses flow rate (Q) at the drill bit. Let Q vary between 100 to 800 gpm in increments of 50 gpm.

3. A vertical well contains drilling mud having a density of 14.3 ppg. Find the hydrostatic pressure at a hole depth of 7800 ft.

4. A mud column of 3000 ft produces a hydrostatic pressure of 1560 psi at a drill bit. Express mud weight in ppg.

5. As a general rule, bottom hole pressure is overbalanced by ½ ppg. What is the difference between well bore hydrostatic pressure and formation pressure at 10,000 ft.

6. Stabilized shut-in drill pipe pressure is 400 psi, depth is 8000 ft, and original mud weight is 12 ppg. How much mud weight increase is required to balance the formation pressure.

7. Formation pressure in a normal pressure area of the Rocky Mountains is 8.34 ppg. What is the formation pressure gradient?

8. Drilling conditions prior to replacing a drill bit are:
   - Standpipe pressure = 2100 psi
   - Flow rate = 400 gpm
   - Mud weight = 11 ppg
   - Bit nozzle sizes 12-12-12

   If maximum allowable operating pump pressure is 3000 psi
   1. Current hydraulic horsepower across drill bit
   2. Select nozzles for the next bit run
   3. Recommended flow rate
   4. New hydraulic horsepower across drill bit

9. If current drilling conditions are:
   - Standpipe pressure = 2500 psi
   - Maximum allowable pump pressure = 3000 psi
   - Flow rate = 330 gpm
   - Bit nozzle are 11-11-11
   - Mud weight is 12 ppg

   Determine optimum flow rate and nozzle sizes for next bit.
   If the roller cone bit is replaced by a PDC drill bit is to be replaced by a PDC bit having 4 nozzles, determine the nozzles sizes.

2.5 LIST OF REFERENCES

1. Bernt S. Andnoy, Iain Cooper, Stefan Z. Miska, Robert F. Mitchell, Michael L. Payne; *Advanced Drilling and Well Technology*, Society of Petroleum Engineers, 2009.
2. Dareing, D. W.; *Mechanics of Drillstrings and Marine Risers*, ASME Press, 2012.
3. Yergin, Daniel, *The Prize*, Simon & Schuster, NY, 1991.
4. Woods, H. B. and Galle, E. M., "Constant Bit Weight and Rotary Speed", Oil and Gas Journal, Oct. 6, 1958.
5. Woods, H. B. and Galle, E. M., "Bit Weight and Rotary Speed", Oil and Gas Journal, Nov 14 & 21, 1963.
6. Maurer, W. C., "The Perfect-Cleaning Theory of rotary Drilling", J. of Petroleum Technology, AIME, **225**, November 1962, pp. 1270–1274.
7. Maurer, W. C., "Bit Tooth-Penetration Under Simulated Borehole Conditions", J. of Petroleum Technology, AIME **234**, December 1965, pp. 1433–1442.
8. Kendall, H. A. and Goins, W. C. "Design and Operation of Jet-bit Programs for Maximum Horsepower, Maximum Impact Force and Maximum Jet Velocity", Trans., AIME, 1960.

9. Murray, A. S. and Cunningham, R. A., "Effect of Mud Column Pressure on Drilling Rates", Trans. AIME **204**, 1955, pp. 196–204.
10. Cunningham, R. A. and Eenink, J. G., "Laboratory Study of Effect of Overburden, Formation and Mud Column Pressure on Drilling Rate of Permeable Formations", Trans. AIME, **216**, 1959, pp. 9–17.
11. Eckel, J. R., "Microbit Studies of the Effect of Fluid Properties and Hydraulics on Drilling Rate", paper SPE 2244, SPE Annual Conference, 1968.
12. Bruce H. Walker; "Some Technical and Economic Aspects of Stabilizer Placement", Journal of Petroleum Technology, June 1973.
13. G. Robello Samuel and Jamel J. Azar; *Drilling Engineering*, Penn Well Corp. 2007.
14. Robert F. Mitchell and Stefan Z. Miska, *Fundamentals of Drilling Engineering*, SPE Textbook Series, 2011.

Chapter 3

DRILL BIT TYPES AND FEATURES

Since the early days of rotary drilling, numerous drill bit types have been developed. The first rotary drill bits were drag bits with two cutters forming the shape of a fish tail. These drill bits were limited by design and metallurgy and used primarily in soft formations. Fish-tail drag bits were followed by roller cone bits having two cones with non-intermeshing teeth. The two cone bits gave low rate of penetration and it was not until a three-roller cone bit with intermeshing teeth was developed that roller cone bits were widely accepted.

During the late 1950s there was much interest in novel methods for drilling through rock. Ledgerwood [1] gives an overview of international efforts to find new methods (lasers, pellets, grinding, spark, explosives) for oil well drilling. Studies by Maurer [2] on novel drilling methods show that the rotary drilling method is still the most economical method for boring through rock.

Three basic drill bit types commonly used today are:

1. Roller Cone
   a. Milled tooth
   b. Insert tooth
2. Polycrystalline Diamond Compact (PDC)
3. Natural Diamond
   a. Radial flow
   b. Feeder collector

Roller cone bits contain three cones, which rotate relative to the bit body. Each cone contains cutters or teeth. Bit forces are transmitted through bearings mounted within the cones. PDC and Natural diamond bits are rigid body bits and are made with various profile shapes and cutter densities. Recent advances in drill bits are due to improvements in metallurgy, design and manufacturing. Drill bit selection is made from a wide range of drill bits allowing for a best choice for a given formation. The objective in bit selection is minimum footage cost.

3.1 ROCK MECHANICS

3.1.1 Porosity

Porosity, ϕ, is a measure of the void space within a given volume of rock.

$$\phi = \frac{Vol_{void}}{Vol_{total}}$$

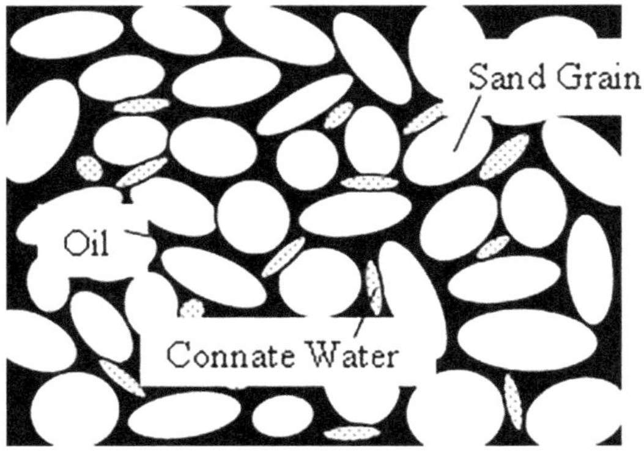

Figure 3-1. Microscopic view of sandstone.

Oil and gas occupy this void space in a reservoir. This volume represents the quantity of oil or gas (depending on water saturation) within a given reservoir. It is a factor in calculating revenue potential. This void space is typically filled with oil (liquid), gas or water or a combination of the three. Connate water is always present with oil within the pores of reservoir rock as illustrated in Figure 3-1. Water also exits separately directly below the oil zone as shown earlier.

3.1.2 Permeability

Permeability is a rock property that is used to predict volume flow rate through the pores of rock. It is a property that was identified by the French Engineer, Henri Darcy in 1856 while studying flow of sewage through sand compacted pipe. By testing several liquids, he concluded that rock permeability is *independent of fluid viscosity* [3]. The Darcy equation [Eq. (3-1)] is basic to recovery predictions from oil and gas reservoirs.

$$Q = \frac{kA\left(p_1 - p_2\right)}{\mu L} \tag{3-1}$$

where

Q = volume flow rate, cc/second
A = cross section area of flow, cm^2
μ = viscosity, centipoise
Δp = pressure drop, atm
L = length of flow, cm
k = permeability, darcy (note: 1 darcy = 1000 millidarcys)

Darcy's equation is also expressed in differential form as

$$Q = \frac{kA}{\mu}\frac{dp}{dx}$$

which has the same form as the Hagen-Poiseuille equation

$$Q = \frac{\pi R^4}{8\mu}\frac{dp}{dx}$$

Permeability, k, is determined experimentally by measuring the following variables. The liquid is usually water.

$$k = \frac{Q\mu L}{A\Delta p} \tag{3-2}$$

A permeability of one darcy is much higher than commonly found in reservoir rock, therefore, rock permeability is usually expressed in millidarcy's. Porosity is not an indicator of permeability, but permeability is a good indicator of porosity.

3.1.3 Flow Through Porous Media

Equation (3-1) is fundamental to reservoir engineering and is used to prediction oil flow from oil and gas reservoirs. The pressure profile throughout a reservoir determines flow rate (directional derivative) and flow direction (vector normal to constant pressure contour lines). Steady flow of an incompressible fluid, assuming a 2-D model, is illustrated in Figure 3-2.

From continuity of flow across the control volume

$$Q_x + Q_y = Q_x + \frac{\partial Q_x}{\partial x}dx + Q_y + \frac{\partial Q_y}{\partial y}dy \tag{3-3}$$

but

$$Q_x = \frac{kA_x}{\mu}\frac{\partial p}{\partial x} = \frac{kdya}{\mu}\frac{\partial p}{\partial x} \tag{3-4}$$

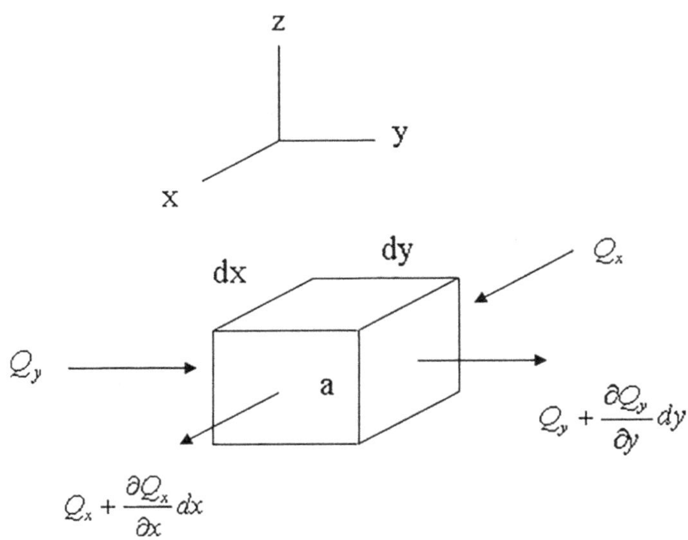

Figure 3-2. Control volume within a reservoir.

$$Q_y = \frac{kA_y}{\mu}\frac{\partial p}{\partial y} = \frac{kdxa}{\mu}\frac{\partial p}{\partial y} \tag{3-5}$$

By substitution into Eq. (3-3)

$$\frac{\partial^2 p}{\partial x^2} + \frac{\partial^2 p}{\partial y^2} = 0 \tag{3-6}$$

The solution to the Laplace equation, along with boundary conditions, establishes the pressure distribution throughout a porous media assuming steady flow conditions. Flow rates are then determined from Eqs. (3-4) and (3-5). Actual flow direction is normal to constant pressure lines and is determined from directional derivatives,

$$Q_n = \frac{kA}{\mu}\frac{dp}{dn} \qquad \text{where A = ds a.} \tag{3-7}$$

The polar form of the Laplace equation is most useful in predicting flow into a well bore.

$$\frac{1}{r}\frac{\partial}{\partial r}\left(r\frac{\partial p}{\partial r}\right) + \frac{1}{r^2}\frac{\partial^2 p}{\partial \theta^2} = 0 \tag{3-8}$$

For an idealized case for pressure not a function of θ, then Eq. (3-8) simplifies to

$$\frac{d}{dr}\left(r\frac{dp}{dr}\right) = 0 \tag{3-9}$$

which is solved by direct integration giving

$$p(r) = C_1 \ln r + C_2 \tag{3-10}$$

Assuming an idealized reservoir having a boundary pressure of p_2 at r_2 flowing into a well bore pressure of p_1 at r_1, then pressure within this idealized reservoir would be defined by

$$p(r) = p_2 - \frac{p_2 - p_1}{\ln\frac{r_2}{r_1}}\ln\frac{r_2}{r} \tag{3-11}$$

Flow rate into the well bore from

$$Q = \frac{kA}{\mu}\frac{dp}{dr} \tag{3-12}$$

where

$$\frac{dp}{dr} = \frac{1}{r}\frac{p_2 - p_1}{\ln\frac{r_2}{r_1}}$$

In reality, the pressure distribution in a producing petroleum reservoir is a function of time and its solution establishes the time history and rate of oil depletion. In its simplest form the flow equation is [4,5]

$$\frac{\partial^2 p}{\partial x^2} + \frac{\partial^2 p}{\partial y^2} = \frac{\phi \mu c}{k} \frac{\partial p}{\partial t} \tag{3-13}$$

or in polar coordinates

$$\frac{\partial^2 p}{\partial r^2} + \frac{1}{r} \frac{\partial p}{\partial r} = \frac{\phi \mu c}{k} \frac{\partial p}{\partial t} \tag{3-14}$$

where

ϕ = porosity
μ = viscosity
c = compressibility of fluid
k = permeability

Equation (3-14) is the diffusivity equation for radial flow through porous media. Historically, this equation first arose out of the study of heat conduction. It also appears in diffusion and electrical potential distribution problems. The direct solution to Eq. (3-14) is found by separation of variable method and can be found in ref [6].

In reality the time history of fluid flow from actual reservoirs is complex and typically require numerical analysis conducted on large computers. Reservoir analysis are complicated by complex boundary conditions and initial conditions. A complete reservoir analysis predicts production history and expected recovery volume. Drilling and production planning are based on a reservoir analysis, which defines the best drilling plan for maximum recovery.

3.1.4 Mechanical Strength of Rocks

Rocks are known to have relative high strength in compression, while being weak in tension. This characteristic is true of rocks encountered in oil well drilling (Table 3-1). The state of stress in deep formations is believed to be near

Table 3-1. Strength properties of various rocks.

	Compression (psi)	Tension (psi)	Shear (psi)
Granite	10,000–40,000	600–1000	2000–4300
Syenite	14,000–28,000		
Diorite	16,000–35,000		
Quartzite	16,000–45,000		
Limestone	2,500–28,000	280–890	1200–3000
Marble	8,000–27,400	400–2300	1300–6500
Basalt	28,000–67,000		
Serpentine	11,000–28,000	800–1600	2600–5000
Sandstone	5,000–20,000	280–500	300–3000
Slate	7,000–31,000	3000–4300	2000–3600

hydrostatic with horizontal and vertical components equal to overburden pressure. This was the condition during the lithification process.

Overburden pressure is the stress in the formation due to the weight of rock. In hard rock areas, overburden pressure is about 1 psi/ft of depth (equivalent to about 19 ppg). In coastal areas, the overburden pressure varies from 0.5 psi/ft at 1000 ft to 0.9 psi/ft (equivalent to about 18 ppg) at depths greater than 12,000 ft. Note that overburden pressure is different from formation pressure. The stress magnitude in rock, say at 16,000 ft, would be about 14,400 psi in both vertical and horizontal directions; (0.9) 16,000 = 14,400 psi.

Rocks do not fail under a hydrostatic state of stress. However, if horizontal forces are generated due to tectonic plate movement, shear failure can cause sliding along shear planes, creating formation shifting and faults.

The structural strength of rock is defined in terms of: compressive, tension, and shear stress limits. As in all brittle materials they are characterized by:

1. Failure occurs by fracture, i.e., there is no yield prior to failure.
2. Compressive strength is typically many times tension strength
3. Shear strength is approximately the same as the tension strength

Most rocks follow this trend as shown in Table 3-1. These strengths are established from uniaxial stress tests.

3.1.4.1 Criteria of Failure for Brittle Materials

In reality, rock formations in the earth are subjected to complex bi-axial or tri-axial states of stress. Formation rocks are subjected to overburden as well as horizontal loads. When a well bore penetrates through rock formations, well bore pressure is applied to rock as well as overburden. The analysis of rock stresses is further complicated by formation fluid pressures.

Commonly used criteria of failure for brittle materials are:

- Maximum normal stress theory
- Coulomb-Mohr theory
- Modified Mohr theory

Each of these theories is illustrated in Figure 3-3.

Note

S_{ut} = ultimate tension strength as determined from uniaxial tests

S_{uc} = ultimate compressive strength as determined from uniaxial tests

The Maximum Normal Stress Criteria of Failure theory is adequate for normal stress conditions in the first quadrant. In cases where there is a combination of tension and compression (second and fourth quadrant) the other two criteria are more appropriate. Consider the biaxial stress state for the fourth quadrant. The Coulomb-Mohr criteria are shown as a line between S_{ut}, ultimate strength in tension and S_{uc}, the ultimate strength in compression. This line recognizes that failure is affected by the Law of Coulomb-Mohr.

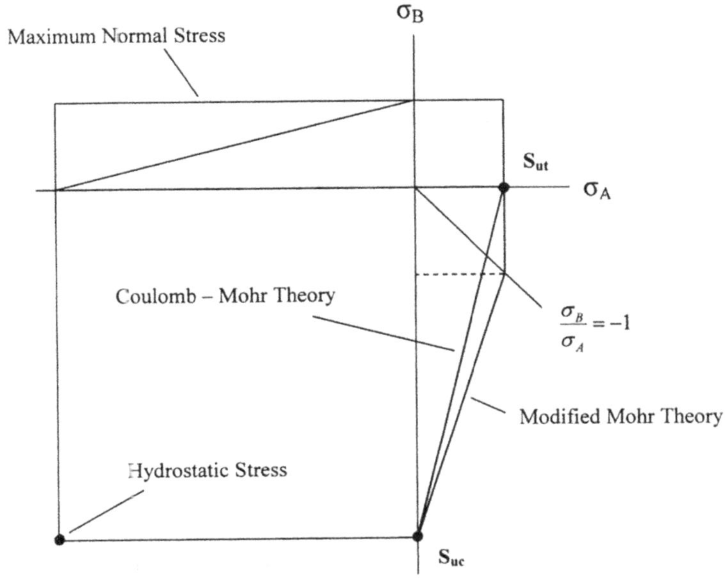

Figure 3-3. Criteria of failure of brittle materials.

$$\tau = c + \sigma \tan\psi \tag{3-15}$$

where

- τ = maximum shear stress at rupture
- c = cohesive strength of rock
- σ = compressive stress normal to the plane of rupture (compression is a + number)
- ψ = angle of internal friction

If principal stresses are arranged so that $\sigma_1 \succ \sigma_2 \succ \sigma_3$, then the critical stress limits for σ_1 and σ_3 are σ_A and σ_B. These two stress limits are related to S_{ut} and S_{uc} by

$$\frac{\sigma_A}{S_{ut}} - \frac{\sigma_B}{S_{uc}} = 1 \tag{3-16}$$

In this equation, S_{uc} is a positive number. For example, (using numbers for sandstone in Table 3-1); S_{uc} = 20,000 psi, S_{ut} = 500 psi

$$\sigma_B = S_{uc}\left(\frac{\sigma_A}{S_{ut}} - 1\right) \tag{3-17}$$

This equation predicts failure in the 4th quadrant.

$$\sigma_B = 20{,}000\left(\frac{\sigma_A}{500} - 1\right) \tag{3-18}$$

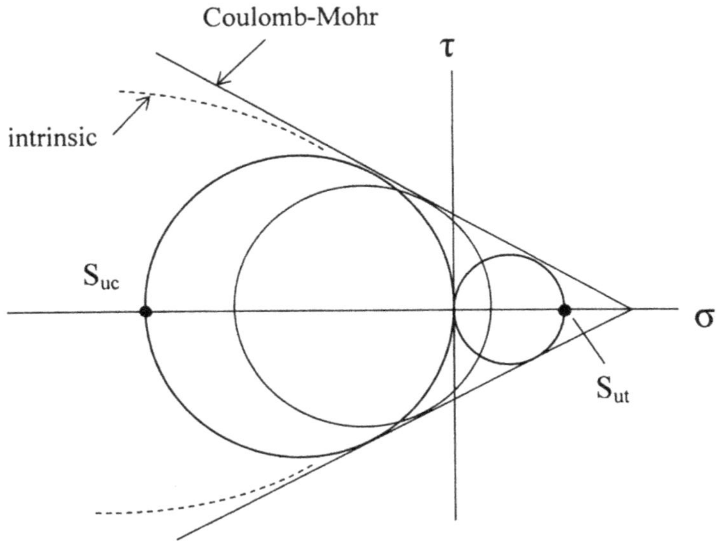

Figure 3-4. Mohr diagram for brittle fracture.

At $\sigma_A = 500$, $\sigma_B = 0$ and at $\sigma_B = -20{,}000$, $\sigma_A = 0$. A biaxial state of stress (σ_1 and σ_3) located within this line does not produce failure, however, a biaxial state of stress outside this line does produce failure according to this criterion.

The Coulomb-Mohr line in Figure 3-3 is also illustrated by the Mohr diagram shown in Figure 3-4 [7]. Two circles are first drawn based on S_{ut} and S_{uc}. The tangent line to the two circles represents Eq. (3-16). If the principal stress circles are within this envelope, failure will not occur. Failure occurs when any stress circle touches or extends outside of the envelope.

3.1.5 Modified Mohr Theory

Experimental data show that actual points of failure fall outside of the Coulomb-Mohr criteria. One point in particular is the case where $\frac{\sigma_B}{\sigma_A} = -1$. An example is the twisting of classroom chalk, which causes failure on a 45 deg. plane. Under this torsion loading, the principal normal stresses are equal but opposite and oriented on planes 45 deg. from the axis of twist (Figure 3-5). The Modified Mohr line is drawn from this point. The data of Coffin [8] and Grassi and Cornet [9] fall along this line. The Modified Mohr criteria, then, appears to be more accurate than the Coulomb-Mohr criteria.

3.1.5.1 Mode of Failure in 3rd Quadrant

The mode of rock fracture due to compressive principal stresses (3rd quadrant) is different. The intrinsic line of failure is indicated in Figure 3-4 [10]. The Maximum Normal Stress Criteria of Failure does not apply in this case.

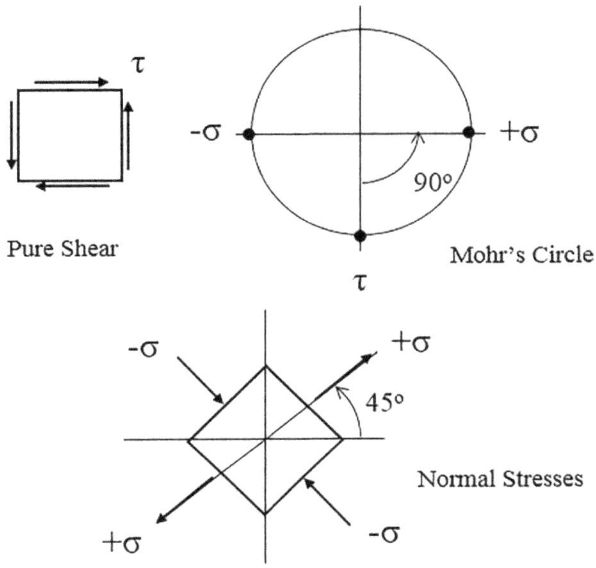

Figure 3-5. Brittle failure by pure shear.

Consider the case where all principal normal stresses are equal. Such a condition is indicated by the corner point identified by "hydrostatic stress". Experience shows that hydrostatic stress alone does not cause failure. Living creatures in the deepest oceans are good examples.

3.1.5.2 Chips Formed Under a Bit Cutter

Laboratory tests [11] show that chips under a drill bit tooth are created by shear fracture according to the Coulomb-Mohr criteria. Figure 3-6 illustrates how chips are formed when a bit tooth penetrates rock. The wedging effect of

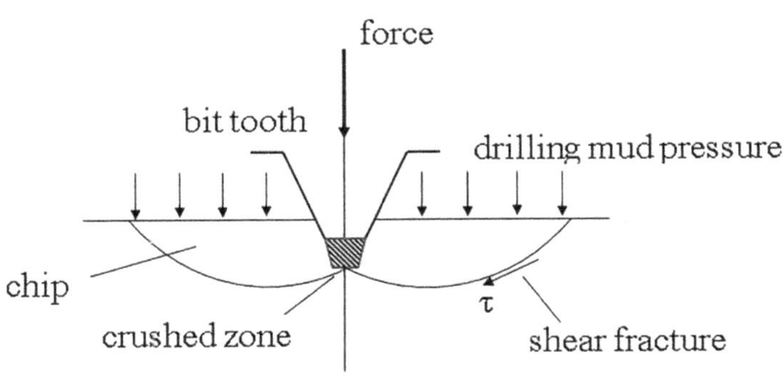

Figure 3-6. Rock fracture and chip hold-down.

the cutter creates side forces and stresses on both sides of the cutter. This side forces eventually causes a crack to form as shown.

PDC cutters form chips by shearing layer of the formation directly. Depth of cutter depends on applied bit force. Natural diamonds tend to plow through layers of rock.

3.2 HYDRAULIC FRACTURING

Hydraulic fracturing (or fracking) is a technique used to create cracks in an oil bearing formation for the purpose of increasing rock permeability and production. Formation fracturing is accomplished by forcing special fluids into a formation under high pressure and at high flow rates.

High pump pressure is needed to overcome overburden stresses and breakdown the formation. Fracture openings propagate and are filled with the fracturing fluid containing a propping agent (small solid particles). When pump pressure is removed, the propants keep the cracks open and create flow channels, greatly enhancing permeability and production. Basic to creating these formation cracks is the ability to pump fracturing fluids (with suspended

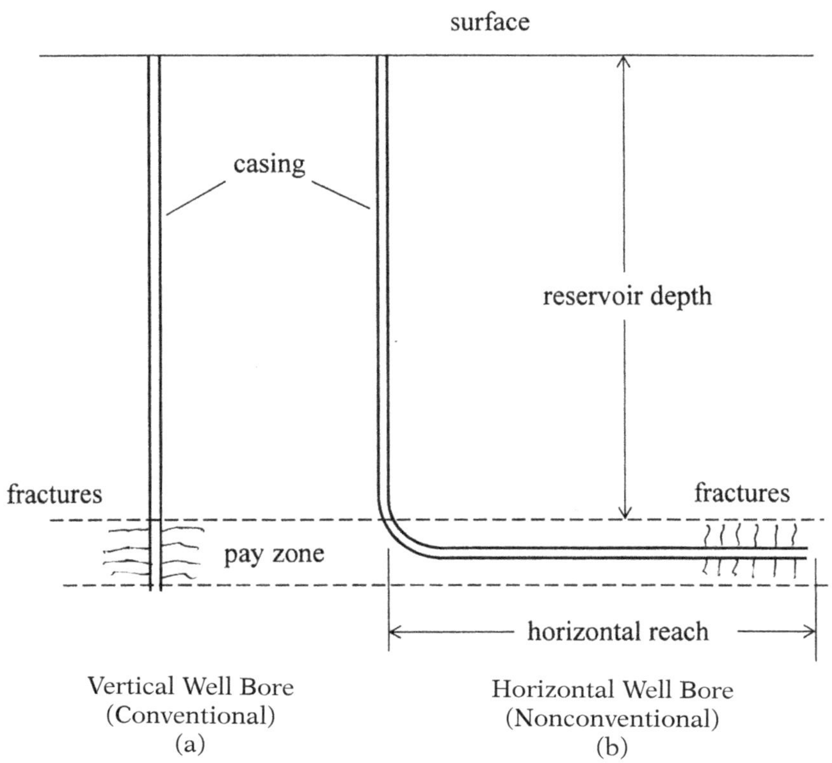

Figure 3-7. Two types of hydraulic fracturing.

propants) at high pressures and high flow rates. This requires special pumps with high hydraulic horsepower capability.

Two types of hydraulic fracturing (Figure 3-7) are *conventional* and *nonconventional*. Conventional fracking is conducted from a single vertical well bore (Figure 3-7a) and creates crack volumes of about 500,000 gal. Nonconventional fracking is conducted from a horizontal bore hole as shown in Figure 3-7b and creates crack volumes of up to 7 million gals/well. Fluid volume is directly related to crack volume.

Conventional fracking is conducted from vertical wells. Fracking fluid is pumped down casing and/or production tubing and enters the formation through casing perforations. Cracks can be either horizontal or vertical or randomly oriented depending on hole depth, formation type and preexisting planes of weakness, left in place by faulting and formation movement over time. In general, cracks formed by fracking are horizontal at depth above 1000 to 2000 ft and vertical at depths below these levels.

Nonconventional fracking is conducted from horizontal well bores, which are drilled off of a vertical well bore and can reach out as much as a mile or more. Because of the length of the horizontal well bore, fracking is typically conducted in segments ranging in lengths of 100 ft to 1000 ft. Fracking from horizontal bore holes is especially effective in tight formations, such shale.

3.2.1 Historical Background

Conventional hydraulic fracturing evolved from earlier (1932) efforts to stimulate oil production by injecting acids into oil reservoirs. Acidizing became a common practice by 1934 for enhancing oil production from limestone formations [12].

The phenomenon of formation fracturing has been explained [13] as one of lifting the overburden load through the use of fluid pressure within fractured cracks. Early field experience showed that there is a critical break-through pressure, which initiates cracks near the well bore. Once a crack is initiated, fluid pressure required to propagate the crack is fairly constant. In other words, the rate of fluid injection (or crack growth) can be increased with little or no increase in injection pressure [14,15]. Critical pressure varies with depth depending on weight of the overlaying rock. Assuming a specific gravity of water saturated rock to be 2.2, the overburden pressure gradient would be about 1.0 psi per foot of depth. In general, critical fracture pressure gradients range from 0.8 to 1.4 psi/ft.

Hydraulic fracturing was commonly used in the 1950s as a means of secondary recovery, i.e., enhanced recovery beyond the ability of the reservoir to drive oil into the well bore. It was implemented after natural reservoir pressures had been depleted. Its effectiveness depends on the reach of the cracks produced by the fracturing process.

Fracking pumps are required to deliver flow rates in the order of 110 bbl/minute (4620 gpm) and pump pressures in the order of 10,000 psi. Volume flow rate is accomplished by multiple pumps working in parallel. Pressure requirements are much higher than normal drilling mud pump capability so special pumps had to be developed.

Drainhole drilling was under development during the early 1960s as a means of enhancing oil recovery. Problems related to: high angle curvature, pipe friction, bit thruster and bit navigation. Each limited the length of the horizontal bore hole. This technology was greatly enhanced by directional drilling technology allowing the drill bit to navigate several thousand feet through narrow layers of oil bearing formations. Fracking from these horizontal bore holes has greatly enhance oil and gas recovery.

3.2.2 Formation Fracture Pressure

Farris [16] studied bottom hole pressures required to break down formations in the Texas-Louisiana Gulf Coast by reviewing the records of cement squeeze jobs. In each case, the pressure required to break down each formation at the depth of the casing perforations, was calculated. The results are summarized in Figure 3-8. These data are bounded by the maximum and minimum lines. The maximum depth for this data was ~12,000 ft. The trend with depth is indicated by the dashed lines.

The data shows an average fracture pressure gradient of ~0.7 psi/ft beyond a depth of 4000 ft depth. The formation fracture curves do not pass through the origin of the plot indicating formation fracture pressure must overcome the cohesive strength of the rock plus the overburden load. This phenomenon is more pronounced at depths less than 4000 ft.

The data shown in Figure 3-8 indicates that at depths greater than 4000 ft, fracture pressure is approximately 80% of overburden pressure.

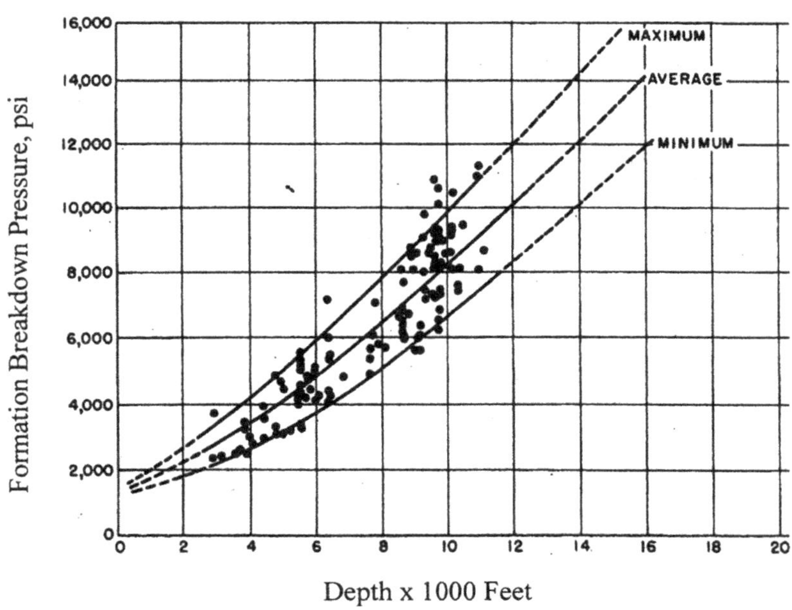

Figure 3-8. Formation breakdown pressures (AIME).

Overburden pressure (or stress) is caused by weight of rock. In hard rock areas, stress gradients are about 1 psi/ft of depth (equivalent to mud weight of about 19 ppg). In areas of low rock strength, such as the Gulf Coast, overburden pressure gradients are about 0.5 psi/ft (at 1000 ft) to 0.9 psi/ft (at 12,000 ft).

3.2.3 Theories of Rock Failure

With regard to in-situ hydraulic fracturing of rock, two events appear to take place. The first event is the initiation of a fracture caused by high levels of fluid pressure at casing perforations. The second is the propagation of a crack with continual fluid flow. Field observations show that crack propagation requires little or no increase in fluid pressure with even a drop in fluid pressure once the crack starts. The second event indicates local stresses at the tip of a crack are great enough to extend the crack, a characteristic of brittle fracture.

Attempts have been made to apply several criteria of failure, including those mentioned earlier. Local stress calculations produced by increased well bore pressure are usually based on some form of Lame's [17] equation for thick wall cylinders. Griffith Theory of Brittle Strength [18] shows that stress concentration at the tip of a crack is great enough to overcome molecular bonds of rock material, thus allowing the crack to propagate with little or no increase in fluid pressure. The Griffith theory is widely accepted.

3.2.4 Drilling Considerations

Formations exposed to drilling mud in an open hole are susceptible to breaking down, too. The concern with an exposed formation is that drilling mud may have to be increased to contain formation fluids later down hole. If this happens it is possible that the exposed formation at the last casing shoe could break down.

A formation breakdown could cause lost circulation or even migration of gas to the surface creating a hazardous condition under the drilling rig. Formation breakdown at a casing shoe is determined by a leak-off test.

Schuh [19] shows that the magnitude of well bore pressure likely to breakdown or fracture formation around a well bore can be approximated by

$$p_F = p_f + K\left(p_O - p_f\right) \tag{3-19}$$

where

p_F = formation fracture pressure; expressed in mud weight, ppg
p_0 = pressure of the overburden expressed in lb/gal; approximately 18.6 ppg (0.9672 psi/ft) in the Gulf Coast and California areas; 19.2 ppg (0.9984 psi/ft) in hard rock areas
p_f = formation or pore pressure; also, expressed in ppg
K = fracture gradient stress ratio;
Nominal values of K are given in ref [20–23]

Data show that K varies from ~0.3 at shallow wells to 1.0 near 20,000 ft.

3.3 ROLLER CONE DRILL BITS

There are many different types of roller cone bits and each has unique features and benefits. They can be divided into milled tooth bits and tungsten carbide insert cutters. Bearing performance has been extended over the years through design and seal improvements. PDC and diamond drill bits are also options and each has been design for special applications and formations.

Three components of roller cone bits are cutters, bearings, and bit body. The *cutters* are fixed to cones, which rotate on bearings relative to the bit body. Cutter elements are either machined directly on the roller cones (Figure 3-9a) or are tungsten carbide inserts, which are pressed into the cone surfaces (Figure 3-9b). Cutters on milled tooth bits are longer and the cones are offset to create dragging action making these bits aggressive in soft formations. Insert bits are designed for hard formations and breakup the rock by crushing action rather than chipping and gouging action.

The typical bearing arrangement within each of the three roller cones is illustrated in Figure 3-10. The bearing arrangement includes a cylindrical roller bearing, a ball bearing and a sleeve bearing. If a total bit force of 45,000 lbs, each cone would carry 15,000 lbs each. Ideally the cylindrical bearing carries the bulk of the applied load. The main purpose of the roller bearing is to retain the cone onto the bit body; they support little or no load. The balls are inserted through a hole from the outside until all of the balls are in place. The purpose of the sleeve bearing is to support part of the side load to the cone and resist any thrust force component.

Normally cylindrical and journal bearings are not paired on the same shaft because clearance in the journal would create uneven loading on the cylindrical rollers. However, because of space limitation within each roller cone, bearing

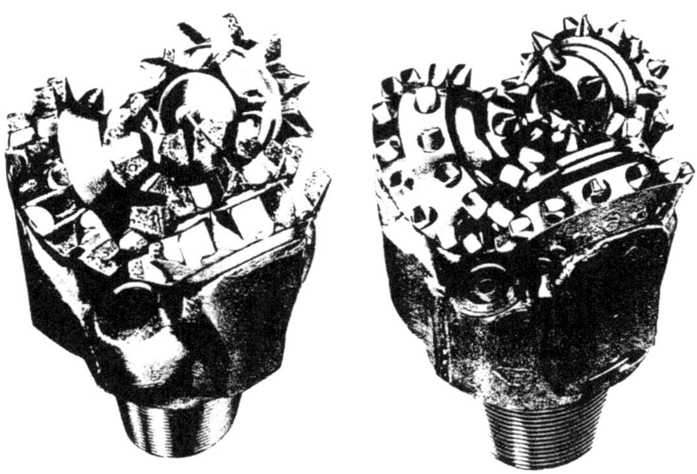

(a) (b)

Figure 3-9. Milled tooth and insert roller drill bits.

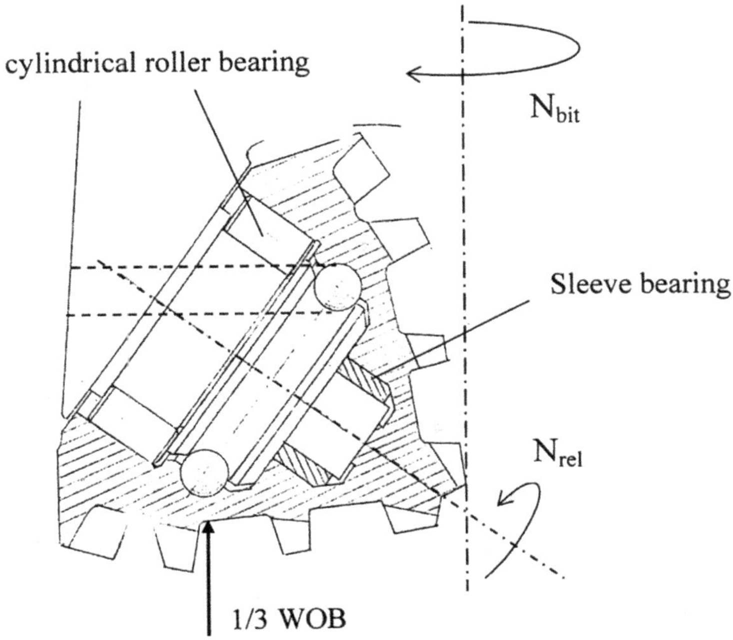

Figure 3-10. Typical bearing cavity in bit roller cones.

design alternatives are limited. This inherent weakness in bearing design, is further compromised by lack of lubrication and an abrasive environment associated with drilling mud.

Bearing cavities in early bit designs were not sealed and bearings wore out within eight to ten hours. Challenges to drill bit bearing design has been recognized for years. Efforts to improve on drill bit bearing design include:

1. Dynamic seal between cone and bit body
2. Lubrication sub directly above the drill bit capable of supplying a large volume of oil to each bit bearing cavity
3. Lubrication reservoir within the legs of each bit body section (three legs)
4. Improvements to dynamic seals
5. Replacing cylindrical roller bearings with journal bearings

The motivation for increasing the life of drill bits is lower footage cost. Large improvements in bearing life have been achieved through these advances.

Roller cone rock bits are made with many design features and for a wide range of formation hardness. The selection of any of these bits depends on bit cost, expected rate of penetration and bit life. Bit performance prediction is based on a data base of bit records and is a statistical prediction.

Cleaning and cuttings removal from underneath drill bits are enhanced by pressure drop across bit nozzles. Bit nozzles transform available energy into

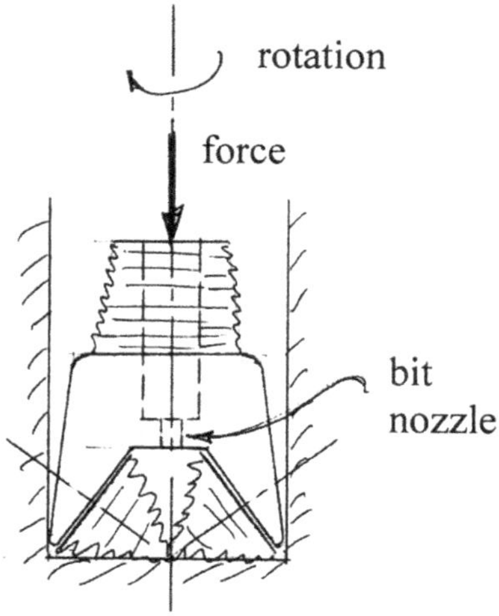

Figure 3-11. Fluid jets for bottom hole cleaning.

kinetic energy, which is discharged under the bit for cleaning and cuttings removal (Figure 3-11). Flow rate selection and nozzle sizes are important for making the best use of available hydraulic horsepower. If, for example, the available HHP is maximum when Q = 400 gpm and p_a = 2100 psi, then according to Eq. (4-71), nozzles having a total flow area (TFA) of 0.2639 in.^{2} (assuming 10 ppg mud) would convert this HHP into high velocity jets for bottom hole cleaning under the drill bit.

3.4 POLYCRYSTALLINE DIAMOND COMPACT (PDC) BITS

Polycrystalline Diamond Compact (PDC) bits are drag bits containing multiple cutters made of synthetic diamonds. The size of synthetic diamonds is about 175 diamonds per carat. These tiny diamonds are bonded to form a disc shape (about the size of a nickel) and backed by a thick layer of tungsten carbide substrate. The synthetic diamond layer is about 0.025 in. thick. The tungsten carbide backing is about 0.115 in. thick (Figure 3-12). This cutting structure is then bonded onto a metal stud, which is pressed into a steel body, which forms the drill bit. These cutters are also embedded directly into the bit body. When used in tungsten carbide matrix bits, they are bonded onto a short disc, which is then bonded into the matrix.

Rock bits of all types fail rock by developing high shear stresses under cutters. PDC cuter develop shear stresses efficiently provided they can be forced into the rock. The application areas for PDC drill bits are soft to medium hard formations. PDC bits were introduced in the early 1970s and became a good companion bit for downhole motors and turbines, which operate at high rotary speeds.

In general, the flatter profiles are:

- Used in hard formations because load is more evenly distributed among cutters (PDC bits)

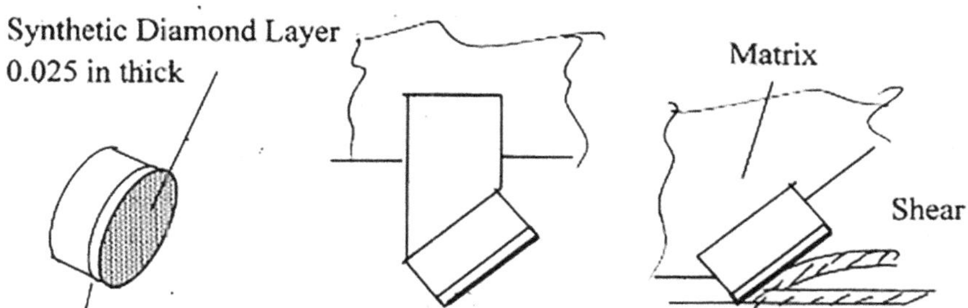

Figure 3-12. Polycrystalline Diamond Compact (PDC) cutter.

- Weak in bit stabilization
- Easier to clean because of shorter distance for fluid to travel across bit face

Parabolic profiles have the following performance characteristics:

- Greater number of cutters in flank area, giving more uniform wear among the cutters
- Good performance at high RPM and commonly used with downhole motors
- Outside cutters generate more heat because of higher velocity and are more difficult to cool because fluid velocity is lowest. High flow rate is desirable with these bit profiles
- Has strong gauge or shoulder protection which is especially needed in motor/turbine applications

3.5 NATURAL DIAMOND DRILL BITS

Natural diamond drill bits were introduced in the late 1940s. These bits allow exploration of deep reservoirs, which usually mean harder and more abrasive formations. Individual diamonds are set in the matrix of the bit so that about 1/3 of the diamond is exposed and 2/3 is buried within the matrix. The average depth of cut is generally 1/3 of the exposure of the stone.

Industrial diamonds are hand-set in a machined graphite mold. A special mixture of matrix granules is placed between the mold and blank, then, heated in a furnace until the matrix material melts, capturing the diamonds onto the blank. A shank is then threaded onto the blank and welded to form the diamond drill bit.

As a general rule, the smaller the diamond bit, the better it will perform (ROP, t_d) relative to roller cone bits of the same size. The life of a roller bit bearing reduces directly with bearing size. Also, rate of penetration with diamond bits is usually higher than rate of penetration with rock bits in small holes and hard formations. Therefore, diamond bits usually lead to a lower footage cost in deep wells where hole size is smallest. On the other hand, diamond bits do not perform as well as rock bits in larger holes and softer formations (see Table 3-2). For example, in drilling areas where rock bits would drill 7 ft/hour to 9 ft/hour, diamond bits out-perform rock bits for the smaller bit sizes.

The flow of drilling mud under a diamond bit is parallel to formation and bit face. Water courses in diamond bits make up part of the "nozzle". The formation makes up the other part.

Pressure drop calculations for diamond bits are based on the TFA of the flow channels at the control diameter normally located at the nose of the bit (Figure 3-13). A diamond bit behaves much like a hydrostatic thrust bearing in this regard. Fluid pressure creates an upward force tending to lift the bit off bottom. Assume, for example, the control diameter is 4 in. and the pressure drop across the control diameter is 1,000 psi. The pump off force would be 12,500 lbs. This pump-off force has to be overcome by additional bit weight.

Fluid pressure under a diamond bit can create an effective overbalance condition between well bore pressure and formation pressure making the formation more difficult to drill. An overbalance of 0.5 ppg is common for rock bits. This is equivalent to about 250 psi at 10,000 ft. A pressure drop of 2100 psi under a *diamond* bit is equivalent to a 4.04 ppg overbalance in regards to ROP.

$$\Delta p = 0.052\Delta\gamma L$$

$$\Delta\gamma = \frac{2100}{0.052(10{,}000)} = 4.04 \text{ ppg}$$

This increase in overbalance pressure can cause a significant reduction in drilling rate, as formations cannot distinguish between hydrostatic and dynamic fluid pressure. Fortunately, deep wells where diamond bits are normally used, bit pressure drop is usually small.

Table 3-2. Expected diamond bit performance (ROP, ft/hr).

Dia. bit size, in	Comparable rock bit performance (ROP, ft/hr)			
	1-3	4-6	7-9	10-12
4	2-6	5-8	8-11	10-12
5	2-6	5-8	7-11	10-12
6	1.5-6	4.5-6.5	7-10	10-12
7	1.5-4.5	4-7	7-9	8-12
8	1.5-4.5	4-7	7-9	8-12
9	1-4.5	4-6	6-9	6.5-11
12	1-3	3-6	5-9	6-10

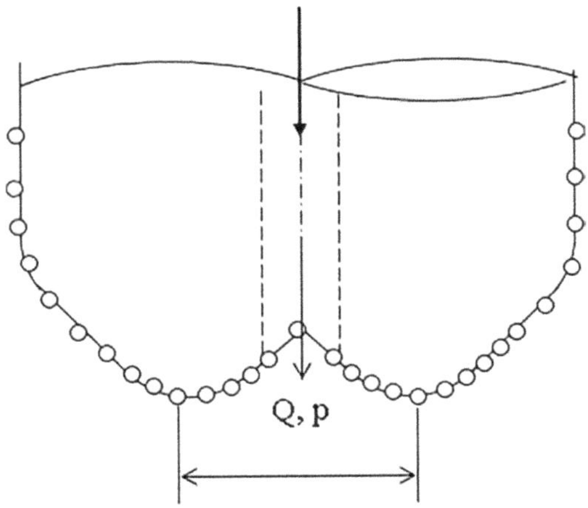

Figure 3-13. Leading face of a natural diamond drill bit.

In some cases, high HSI across a diamond bit may strengthen or harden a *medium-hard* formation and thereby improve the performance of these bits; this mechanism could prevent bit balling. Therefore, hydraulic horsepower or HSI under a diamond bit should be low in comparison with HSI for rock bits. HSI for diamond bits normally ranges from 1.5 to 2.5 for hard formations and from 2 to 3.5 for medium hard formations. Fortunately, in regions where diamond drill bits are normally used (deep formations) flow rate for bit cleaning and cooling is more important than hydraulic horsepower.

3.6 HEAT DISTRIBUTION UNDER RIGID BODY BITS

Rigid body drill bits have no moving parts and cutting elements remove chips with shearing action. Diamond and PDC drill bits do not necessarily drill better than roller bits; they tolerate high speed better. Therefore, they are typically used with downhole drill motors and turbines. Diamonds, whether natural or synthetic, wear due to high levels of heat generated in the cutting action. It is, therefore, important to locate high friction and heat areas so they can be cooled with the drilling mud flow. The purpose of this section is to identify areas of high heat rate per cutter, HQ(r), and cutter wear. Key parameters in the following discussion are:

- Profile
- Cutter density, d(r)
- Average pressure, p(r)
- Force per cutter, q(r)
- Heat rate per cutter, HQ(r)

These variables are interrelated as will be shown.

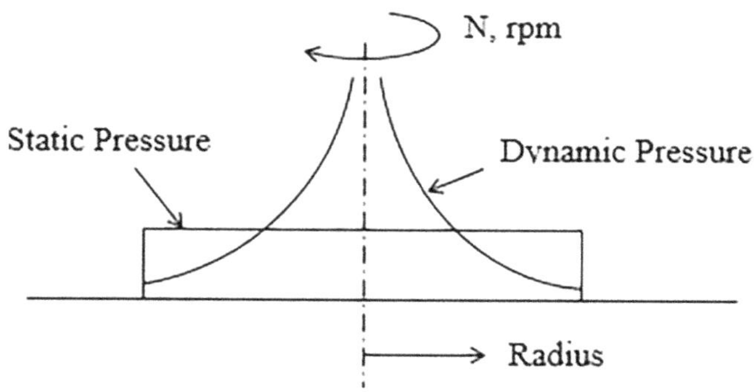

Figure 3-14. Pressure distributions under a flat profile.

Consider the average pressure distribution under a flat circular profile with, say, sandpaper as the cutting element. If this "bit" is pressed onto a perfectly flat wood surface the static pressure would be uniform. Now assume the flat "bit" profile is rotated. The cutter elements near the outer edge work harder than the cutter elements near the center, and thus are able to relieve the contact pressure over the outer edge. Since the total force remains the same for both static and rotating cases, contact pressure increases at the center of the profile. The rotating pressure distribution under this flat rigid body bit is much different from its initial static pressure distribution (Figure 3-14).

Average pressure under all rigid body bit is redistributed in a similar manner. Consider average pressure distribution under a rigid-body drill bit having a flat profile.

Figure 3-15 depicts an area containing a specified number of cutters. A few basic relationships used in our discussion are

- $d = \frac{n}{A}$ (cuter density)
- $p = \frac{F}{A}$ (average pressure)
- $q = \frac{F}{n}$ (force per cutter)

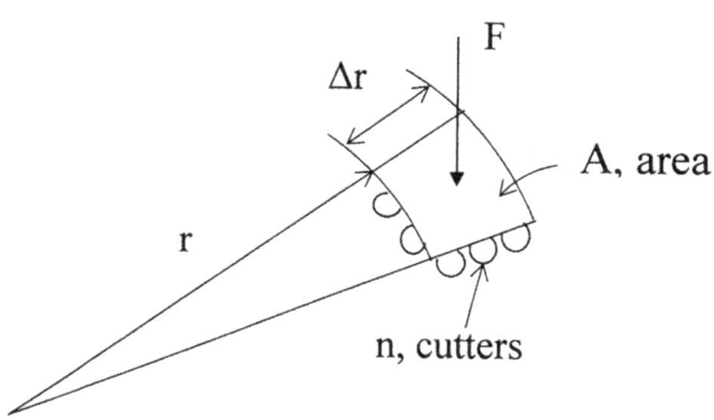

Figure 3-15. Average pressure under a flat profile.

These three relationships combine to give

$$p(r) = q(r)d(r) \tag{3-20}$$

This is a fundamental relationship between average pressure (p), force per cutter (q) and cutter density (d). Each of these parameters may vary with radial distance from the center of a drill bit.

3.6.1 Formulation of Pressure Distribution

In rigid body bits, each cutter progresses forward in the direction of the well bore at the same rate. Let δ represent bit advancement per one-bit rotation. This means that the volume of rock removed by a ring of cutters at radial distance, r, and radial width, *dr*, is $\delta\left(2\pi rdr\right)$.

$$\frac{d(Vol)}{rev} = \delta 2\pi rdr \tag{3-21}$$

Now consider the capability of each cutter to remove rock. Let

v – volume of rock removed by a single cutter when it is loaded by a unit force and travels a unit distance.

Then the volume of rock removed by a ring of cutters is $v\left[2\pi rdrd(r)\right]q(r)2\pi r$. Equating both volumes gives

$$\delta(2\pi rdr) = v\left[2\pi rdrd(r)\right]q(r)2\pi r \tag{3-22}$$

The assumption here is that volume removed by a single cutter is directly proportional to cutter load. This is a reasonable assumption for PDC and diamond cutters. Equation (3-22) reduces to

$$\delta = 2\pi rvd(r)q(r) \tag{3-23}$$

which shows that

$$rd(r)q(r) = \frac{\delta}{2\pi v} \tag{3-24}$$

or since δ and v are constants

$$p(r) = \frac{C}{r} \tag{3-25}$$

where

$$C = \frac{\delta}{2\pi v}, \text{ lb/in}$$

Average pressure under a rigid body drill bit varies inversely with radial distance from the center of drill bit or axis of rotation. Equation (3-25) is consistent with the rotating pressure distribution illustrated in Figure 3-14. The constant, C, can also be determined from equilibrium of the drill bit. Equating cutter forces under a drill bit to weight on bit (WOB)

$$WOB = \int_0^R p(r) 2\pi r dr \tag{3-26}$$

By substitution

$$WOB = \int_0^R C 2\pi dr$$

from which

$$C = \frac{WOB}{2\pi R} = \frac{WOB}{\pi D} \text{ lb/in}$$

Note that Eq. (3-25) is fundamental to all rigid bits with flat profiles regardless of cutter type. These equations will now be applied to two cutter types (PDC and Natural Diamond) and cutter arrangements.

3.6.2 Heat Rate per Cutter

Diamond cutters used in PDC and natural diamond drill bits are the hardest materials and yet they do wear. The basic cause of diamond wear is heat generated in the rock cutting process. The following gives the engineering bases for heat prediction for these two bit types and shows the effects of bit geometry on cutter wear rate.

3.6.2.1 Relation Between Variables

$$p(r) = q(r) d(r) \tag{3-27}$$

The units are homogeneous as shown below

$$\frac{F}{A} = \frac{F}{n}\frac{n}{A} \text{ (units)}$$

From equal advancement of each cutter

$$p(r) = \frac{C}{r}, \quad C = \frac{WOB}{\pi D} \tag{3-28}$$

Force per cutter is given by

$$q(r) = \frac{1}{d(r)} p(r) \tag{3-29}$$

Heat generated per cutter is

$$HQ(r) = q(r) f r \omega$$

$$HQ(r) = q(r) f r \frac{2\pi N}{60}, \text{ in – lb/second} \tag{3-30}$$

$$HQ(r) = q(r)fr\,\frac{2\pi N}{60}\left|\frac{1\text{ ft}}{12\text{ in}}\right|\left|\frac{1\text{ BTU}}{778\text{ ft}-\text{lb}}\right|\text{Btu/sec}$$

3.6.2.2 Flat Bit — Uniform Cutter Density

For this case cutter density is constant (Figure 3-16) and could represent the number of diamonds per square inch on a diamond bit.

$$d(r) = C_1 \tag{3-31}$$

$$q(r) = \frac{p(r)}{d(r)} = \frac{C}{r}\frac{1}{C_1} = \frac{C}{C_1}\frac{1}{r} \tag{3-32}$$

$$HQ(r) = rfF\omega,\quad \text{note F = q (force per cutter)} \tag{3-33}$$

$$HQ(r) = rf\,\frac{C}{C_1}\frac{1}{r}\frac{2\pi N}{60} \tag{3-34}$$

$$HQ(r) = f\,\frac{C}{C_1}\frac{2\pi N}{60}\ \text{(heat rate same for all cutters)} \tag{3-35}$$

The constant, C, depends on bit size and WOB. Cutter density, C_1, is a given. Assuming

$$C = \frac{WOB}{\pi D} = \frac{40{,}000}{\pi 10} = 1{,}273$$

$$C_1 = 50 \text{ cutters per square inch}$$

$$\text{N} = 400 \text{ rpm (41.89 rad/second)}$$

$$f = 0.4$$

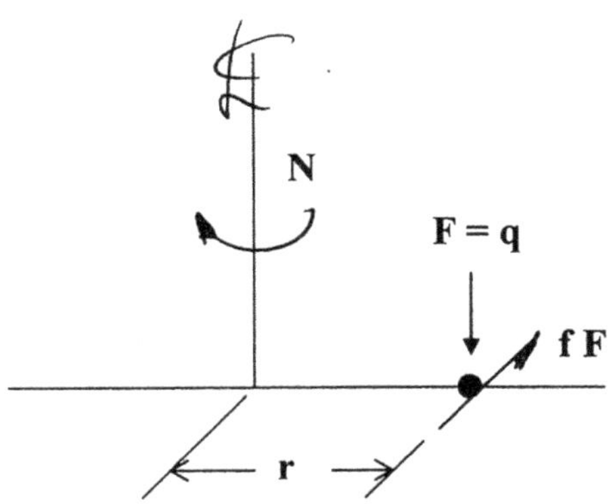

Figure 3-16. Force per cutter in diamond drill bit.

Then

$$HQ(r) = 0.4\frac{1,273}{50}(41.89) = 426.6 \text{ in.-lb/second}$$

$$HQ(r) = \frac{426.6}{12}\left|\frac{1 \text{ Btu}}{778 \text{ ft} - \text{lb}}\right| = 0.0457 \text{ Btu/second (per cutter)}$$

3.6.2.3 Flat Bit — 4 (m) Rows of PDC Cutters

This case represents a PDC bit with four rows of cutters (Figure 3-17). The size of the cutters, Δr, is a given in this case. As before, C, depends on WOB and bit size.

$$d(r) = \frac{4}{2\pi r\,\Delta r} = \frac{2}{\pi\Delta r}\frac{1}{r}, \quad (\Delta r \text{ is cutter spacing}); \text{ Let } C_2 = \frac{2}{\pi\Delta r} \tag{3-36}$$

$$q(r) = \frac{p(r)}{d(r)} = \frac{C}{r}\frac{r}{C_2} = \frac{C}{C_2} \tag{3-37}$$

$$HQ(r) = rfF\omega = rf\frac{C}{C_2}\frac{2\pi N}{60}, \text{ Heat rate is greatest near outer edge.} \tag{3-38}$$

Using the same input numbers as above with $\Delta r = 0.75$ in.

$$C_2 = \frac{2}{\pi 0.75} = 0.849$$

$$HQ(r) = r(0.4)\frac{1,273}{0.849}(41.89) = 25,124r \text{ in-lb/second}$$

$$HQ(r) = r\,\frac{25,124}{12}\left|\frac{1 \text{ Btu}}{778 \text{ ft} - \text{lb}}\right| = 2.69r \text{ Btu/second (per cutter)}$$

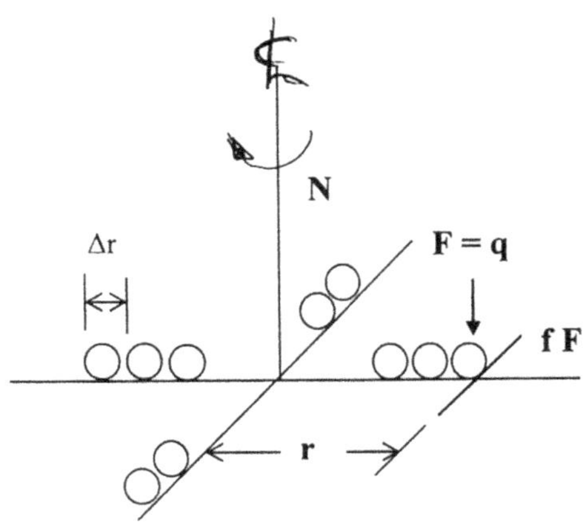

Figure 3-17. Force per cutter in PDC Drill Bit.

The greatest heat rate and wear occurs at r = 5 in.

$$HQ(5) = 13.46 \text{ Btu/second}$$

The outer cutters work harder than the cutters toward the center of the bit and rate of wear is greatest near the outer edge. Worn drill bits having similar profiles and cutter spacing support this prediction.

To combat this type of uneven wear pattern, rigid bodied bits are designed with large sweeping parabolic shaped profiles, which put more cutters near the outer radial position. Another approach is to fill the space near the outer edge with additional cutters.

Drill bit profile and number of cutters directly affect bit cost. However, parabolic profiles have greater dynamic stability and are preferable for high speed motor/turbines applications. On the other hand, shallow profiles are easier to clean and cool because drilling mud has a shorter distance to travel.

3.6.3 Curved Profile

Similar relationships are valid for curved drill bit profiles. Consider a differential section of a surface oriented at an angle θ from the horizontal (Figure 3-18). The notation for this sketch is

r = radial distance to differential element, in.
ds = tangential distance, in.
dr = radial interval, in.
δ = depth of cut, in.
θ = slope of bit surface at distance, r

The circumferential *horizontal* ring area at radius, r is

$$dA = 2\pi r dr \tag{3-39}$$

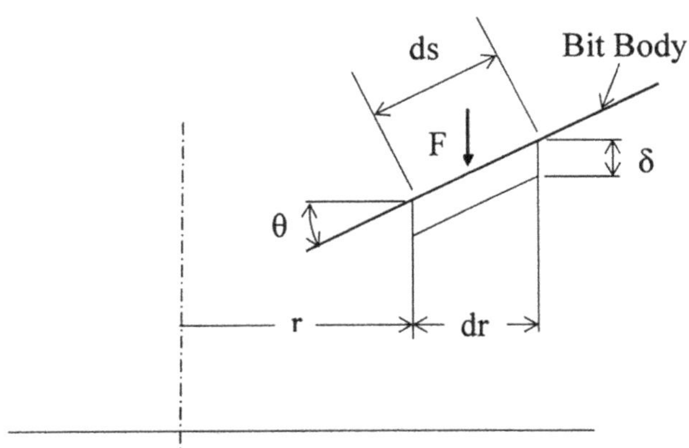

Figure 3-18. Average pressure under curved profile.

The circumferential *surface* area at radius, r is

$$dA_s = 2\pi r ds \tag{3-40}$$

From Figure 3-19

$$dA = dA_s \cos\theta \tag{3-41}$$

By definition

$$p(r) = \frac{F}{A} \tag{3-42}$$

$$d(r) = \frac{n}{A} = \frac{n}{A_s \cos\theta} = \frac{d_s(r)}{\cos\theta} \tag{3-43}$$

$$q(r) = \frac{F}{n} \tag{3-44}$$

where A is circumferential surface area (A_s) projected onto the horizontal plane, n represents number of cutters on the circumferential *surface* area (A_s) and $d_s(r)$ is cutter density on the circumferential *surface* area (Figures 3-18, 3-19, and 3-20). As before

$$p(r) = d(r)q(r) \tag{3-45}$$

Each parameter is referenced to the horizontal plane. The formula expressing the volume of rock removed at a given radius during one revolution is the same as before.

$$rd(r)q(r) = \frac{\delta}{2\pi v} \tag{3-46}$$

which, combined with Eq. (3-45) gives

$$p(r) = \frac{C}{r} \tag{3-47}$$

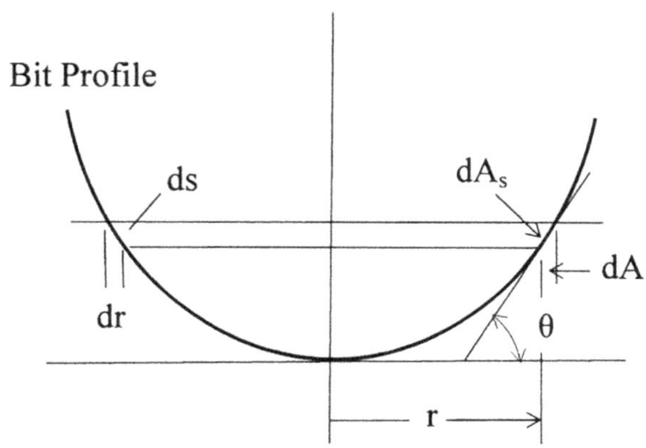

Figure 3-19. Generalized drill bit profile.

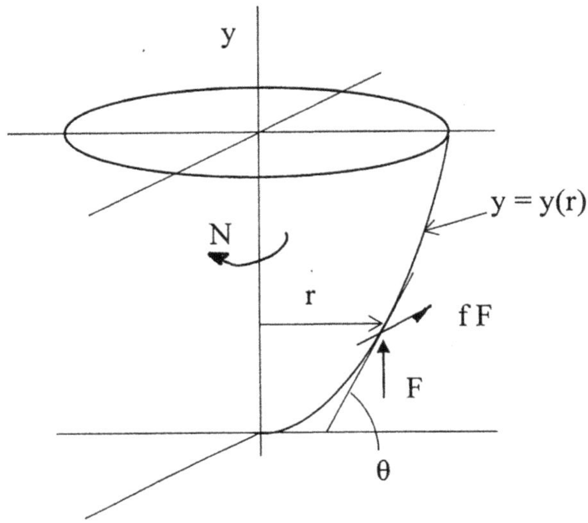

Figure 3-20. Force per cutter in curved profile diamond drill bit.

where

$$C = \frac{\delta}{2\pi v}$$

The constant, C, can also be determined from equilibrium of the drill bit. Summing forces in the axial direction,

$$WOB = \int_0^R p(r) 2\pi r dr \tag{3-48}$$

$$WOB = \int_0^R C 2\pi dr \tag{3-49}$$

showing that

$$C = \frac{WOB}{\pi D}, \text{ lb/in} \tag{3-50}$$

Bit curvature does not affect the value of "C". Bit profile is defined by the outer edges of the cutters.

3.6.3.1 *Uniform Cutter Density*

This formulation applies to diamond drill bits having either a radial flow or feeder/collector diamond pattern.

$$d(r) = \frac{d_s(r)}{\cos\theta} \tag{3-51}$$

$$q(r) = \frac{p(r)}{d(r)} = \frac{C}{r}\frac{\cos\theta}{C_1} \tag{3-52}$$

$$HQ(r) = rfF\frac{2\pi N}{60} \tag{3-53}$$

$$HQ(r) = rf\frac{C}{C_1}\frac{\cos\theta}{r}\frac{2\pi N}{60} \tag{3-54}$$

$$HQ(r) = f\frac{C}{C_1}\cos\theta\frac{2\pi N}{60} \tag{3-55}$$

As a result the heat rate depends on cos θ, which is geometry dependent. The calculation procedure is the same as before except for the cos θ term. The greatest heat rate occurs where the slope of bit profile is lowest.

3.6.3.2 "m" Rows of PDC Cutters

In this case the surface of the bit profile is a line drawn along the outside of the PDC cutters (Figure 3-21).

$$d_s(r) = \frac{m}{2\pi r\,\Delta s} = C_2\frac{1}{r} \tag{3-56}$$

where

m = rows of PDC cutters
Δs = spacing of individual cutters along surface profile

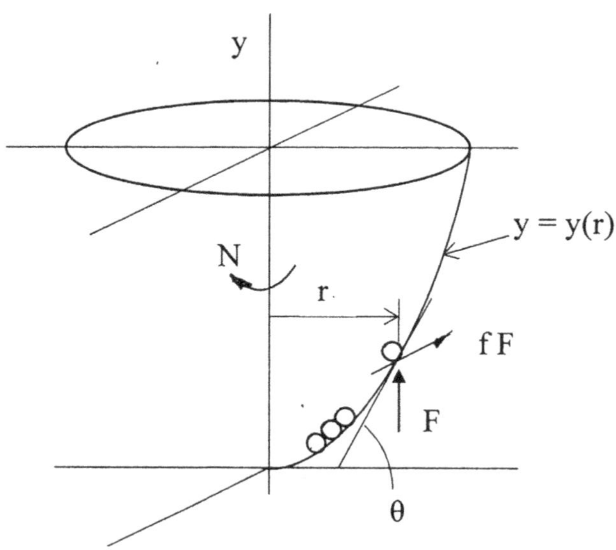

Figure 3-21. Force per cutter in PDC curved profile bit.

$$C_2 = \frac{m}{2\pi\Delta s} \tag{3-57}$$

$$q(r) = \frac{p(r)}{d(r)} \tag{3-58}$$

$$d(r) = \frac{d_s(r)}{\cos\theta} = \frac{C_2}{r\cos\theta} \tag{3-59}$$

$$p(r) = \frac{C}{r} \tag{3-60}$$

$$q(r) = \frac{C}{C_2}\cos\theta \tag{3-61}$$

Heat rate per cutter

$$HQ(r) = rfF\frac{2\pi N}{60} = f\frac{C}{C_2}r\cos\theta\frac{2\pi N}{60} \tag{3-62}$$

$$HQ(r) = f\frac{C}{C_2}\frac{2\pi N}{60}r\cos\theta \quad \text{(heat rate per cutter)} \tag{3-63}$$

Heat rate calculations are the same as before except for the cos θ term.

For example consider a curved profile containing four rows of PDC cutters and cutter spacing is Δs. The average pressure, p(r), profile is still the same as before but now cutter density varies with r.

$$d(r) = \frac{d_s(r)}{\cos\theta} \tag{3-64}$$

$$d(r) = \frac{4}{2\pi r\,\Delta s}\frac{1}{\cos\theta} = C_2\frac{1}{r\cos\theta} \tag{3-65}$$

where

$$C_2 = \frac{2}{\pi\Delta s}$$

Force per cutter is

$$q(r) = \frac{p(r)}{d(r)} = \frac{C}{C_2}\cos\theta \tag{3-66}$$

Cutters near the nose of the bit carry the greater load. Those near the outer edge carry little load but serve to ream the hole and stabilize the bit.

The rate at which heat is being generated per cutter

$$HQ(r) = q(r)fr\omega \tag{3-67}$$

$$HQ(r) = \frac{C}{C_2}fr\cos\theta\frac{2\pi N}{60} \tag{3-68}$$

which shows that heat rate per cutter depends directly on "$r\cos\theta$" or the shape of the bit profile.

3.7 APPLICATION TO GENERAL CURVED PROFILES

3.7.1 Uniform Cutter Density

For uniform cutter density on the surface, $d_s(r) = C_1$. Then

$$d(r) = \frac{d_s(r)}{\cos\theta} = \frac{C_1}{\cos\theta} \tag{3-69}$$

$$p(r) = \frac{C}{r} \tag{3-70}$$

$$q(r) = \frac{p(r)}{d(r)} = \frac{C}{C_1}\frac{\cos\theta}{r} \tag{3-71}$$

$$HQ(r) = q(r)fr\,\frac{2\pi N}{60} \tag{3-72}$$

Everything is straight forward with θ being determined from the slope of the function at r.

$$\theta = \tan^{-1}\left(\frac{dy}{dr}\right) \tag{3-73}$$

3.7.2 Rows of PDC Cutters

For this case

$$d_s(r) = \frac{m}{2\pi r\,\Delta s} = \frac{m}{2\pi\Delta s}\frac{1}{r} = C_2\frac{1}{r} \tag{3-74}$$

where

m = rows of PDC cutters
Δs = spacing of individual cutters along surface profile
$C_2 = \dfrac{m}{2\pi\Delta s}$

It follows that

$$d(r) = \frac{d_s(r)}{\cos\theta} = \frac{C_2}{r\cos\theta} \tag{3-75}$$

$$p(r) = \frac{C}{r} \tag{3-76}$$

$$q(r) = \frac{p(r)}{d(r)} = \frac{C}{C_2}\cos\theta \tag{3-77}$$

$$HQ(r) = q(r)fr\omega \tag{3-78}$$

$$HQ(r) = \frac{C}{C_2} f \frac{2\pi N}{60} r\cos\theta \qquad \text{in.-lb/second per cutter} \tag{3-79}$$

θ is determined from the slope of the function at r. The point of maximum heat rate depends on where $r \cos\theta$ is maximum.

3.7.3 Summary of Equations

$$Q = \frac{kA}{\mu}\frac{dp}{dx} \text{ (Darcy equation)}, \quad k = \frac{Q\mu L}{A\Delta p} \text{ (permeability, darcy units)}$$

$$\frac{\sigma_A}{S_{ut}} - \frac{\sigma_B}{S_{uc}} = 1 \quad \text{(Coulomb-Mohr Criteria of failure)}$$

Idealized flow from oil reservoirs

$$\frac{\partial^2 p}{\partial x^2} + \frac{\partial^2 p}{\partial x^2} = 0 \qquad \frac{d}{dr}\left(r\frac{dp}{dr}\right) = 0 \qquad Q_n = \frac{kA}{\mu}\frac{\partial p}{\partial n}$$

Rate of Heat generated in rigid body cutters

$$p(r) = q(r)d(r)$$

$$p(r) = \frac{C}{r}, \quad C = \frac{WOB}{\pi D}$$

Force per cutter is given by

$$q(r) = \frac{1}{d(r)}p(r)$$

Heat generated per cutter is

$$HQ(r) = q(r)fr\omega$$

$$HQ(r) = q(r)fr\frac{2\pi N}{60}, \text{ in. – lb/second}$$

$$HQ(r) = q(r)fr\frac{2\pi N}{60}\left|\frac{1\text{ ft}}{12\text{ in}}\right|\left|\frac{1\text{ BTU}}{778\text{ ft – lb}}\right|$$

Curve Profile

$$d(r) = \frac{d_s(r)}{\cos\theta}$$

$$HQ(r) = f\frac{C}{C_1}\cos\theta\frac{2\pi N}{60}$$

3.8 PROBLEMS

1. An idealize reservoir is cylindrical with an outer boundary of r_2 = 75 meters. It is being produced through a 20 cm (r_1 = 10 cm) well bore. Reservoir pressure at the outer boundary is p_2 = 800 atm. The

pressure within the well bore is maintained at p_1 = 200 atm. If oil viscosity is 50 cp and reservoir rock permeability is 20 md

a. plot pressure distribution with the reservoir assuming steady state flow
b. determine the volumetric flow rate (bbl/day) of oil into the well

2. Draw the failure lines in the 4th quadrant for:
   a. Coulomb Mohr Criteria
   b. Modified Mohr Criteria

   for limestone rock having tensile and compressive strengths of 1500 psi and 12,000 psi, respectively. Locate the biaxial stress state (+1000 psi and –8000 psi) in the diagram. Under this stress condition, do you expect failure?

3. The Coulomb-Mohr failure criteria is defined as

$$S = c + \sigma \tan \phi \quad \text{(compressive stress is +)}$$

This shear failure line is shown as tangent to the two stress circles.

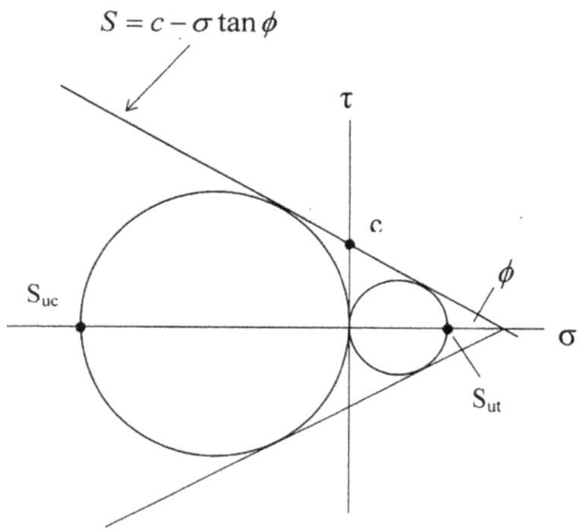

Prove that the two unknown constants, c and ϕ, can be expressed in terms of σ_{ut} and σ_{uc} by

$$\sin \phi = \frac{S_{uc} - S_{ut}}{S_{uc} + S_{ut}}$$

$$c = \frac{1}{2} S_{ut} \left(\tan \phi + \frac{1}{\cos \phi} \right)$$

4. Determine the TFA required to generate 2.5 HSI under an 8 ½ in radial flow diamond bit. Flow Rate is 300 gpm and mud weight is 12 ppg.

5. Determine and plot bit hydraulic horsepower across a bit having a TFA of 0.31 for the following flow rates: Q = 200, 300, 400, and 500 gpm. Assume mud weight is 13 ppg.

6. Current drilling conditions are:
   - Standpipe pressure = 2,800 psi
   - Flow rate = 300 gpm
   - Bit nozzle sizes = 11-11-11
   - Mud weight = 12 ppg

   The rock bit is being replaced with a radial flow diamond bit having a TFA of 0.34 in.^{2}.
   a. Determine system losses based on this information.
   b. Determine the new standpipe pressure assuming flow rate stays the same.

7. Determine the heat rate per cutter distribution under a natural diamond drill bit assuming:
   - Flat profile = 10 in. in diameter
   - Cutter density = 10 diamonds per square inch
   - WOB = 40,000 lbs
   - Rotary speed = 400 rpm
   - Friction factor (f) = 0.4

8. Determine the heat rate per cutter under a PDC drill bit assuming:
   - Flat profile = 10 in. in diameter
   - Four rows of cutters = 0.5-in. cutter spacing
   - WOB = 40,000 lbs
   - Rotary speed = 400 rpm
   - Friction factor (f) = 0.4

9. A PDC drill bit has a flat bottom profile. Cutters are arranged as shown below. Assuming WOB = 40,000 lbs and rotary speed is 400 rpm, determine the maximum heat-rate per cutter for the following two cutter arrangements. Cutter spacing in both cases is 0.5 in. per cutter. Assume a bit radius of R = 10 in. and a friction factor of f = 0.4.

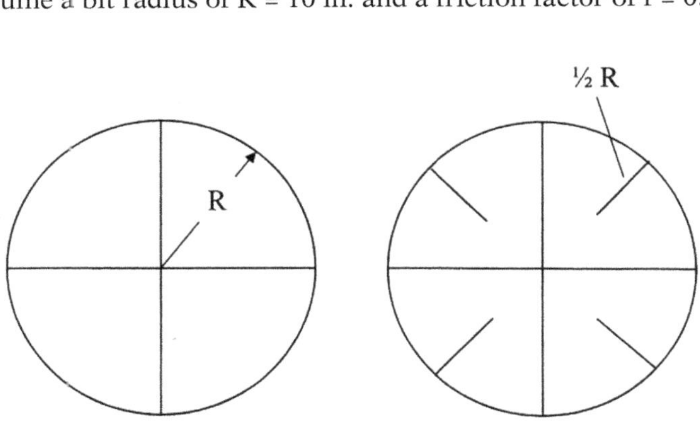

10. Consider a curved profile defined by a parabolic function, $y(r) = \frac{1}{3} r^2$. This function passes through r = 6 in. and y = 12 in. In other words, the bit diameter is 12 in. and bit height is 12 in. Assume there are eight rows of PDC cutters having a Δs = 5/8" spacing. Bit rotation is N = 400 rpm, friction coefficient is 0.4 and WOB = 50,000 lbs. Determine at two locations: r = R and r = R/2.

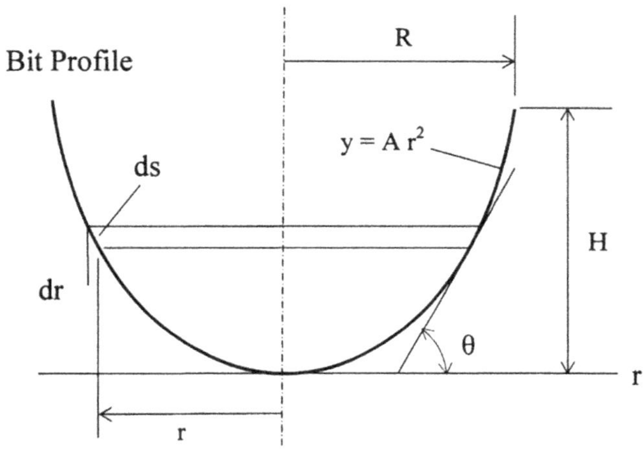

3.9 LIST OF REFERENCES

1. Ledgerwood, Jr. L. W.; "Efforts to Develop Improved Oilwell Drilling Methods", *J. of Petroleum Technology*, 1960.
2. William C. Maurer; *Advanced Drilling Techniques*, Petroleum Publishing Co, Tulsa, OK, 1980.
3. Carl Gatlin; *Petroleum Engineering*, Prentice-Hall, Inc, Englewood Cliffs, New Jersey, 1960.
4. Hubbert, M. King: "Darcy's Law and the Field Equations of the Flow of Underground Fluids", Trans. AIME (1956) **207**, 222–239.
5. C. S. Matthews and D. G. Russell, *Pressure Buildup and flow Tests in Wells*, SPE of AIME, 1967.
6. H. W. Reddick and F. H. Miller: *Advanced Mathematics for Engineers*, 3rd Ed, John Wiley & Sons, Inc. 1955.
7. S. Timoshenko; *Strength of Materials, Part II*, 3rd Ed., D Van Nostrand Co., 1956, pp. 159–162.
8. L. F. Coffin; "The Flow and Fracture of Brittle Materials", Trans. ASME, *J. Appl. Mech.*, Vol. 17, 1950, pp. 233–248.
9. R. C. Grassi and I. Cornet, "Fracture of Gray Cast Iron Tubes under Biaxial Stress", Trans. ASME, Vol. 71, *J. Appl. Mech.*, Vol. 16, 1949, pp. 178–182.

10. Charles Jaeger, *Rock Mechanics and Engineering*, Cambridge University Press, 1972.
11. Maurer, W. C., "The Perfect-Cleaning Theory of Rotary Drilling", *Journal of Petroleum Technology*, AIME, **225**, November 1962, pp. 1270–1274.
12. G. C. Howard and C. R. Fast; *Hydraulic Fracturing*, Society of Petroleum Engineers of AIME, 1970.
13. Clark, J. B.: 'A Hydraulic Process for Increasing the Productivity of Wells", Trans., AIME (1949) **186**, 1–8.
14. Dickey, P. A. and Andersen, K. H.: "Behavior of Water Input Wells – Part 4", *Oil Weekly* (Dec 10, 1945).
15. Grebe, P. A.: "Tools and Aims of Research", *Chem. And Eng. News* (Dec 10, 1943) 21, No. 23, p2004.
16. G. C. Howard and C. R. Fast, Hydraulic Fracturing, SPE of AIME, Dallas, 1970 [Farris, R. F.: unpublished report, Pan American Petroleum Corp., Tulsa, Okla. (1946)].
17. Lame' and Clapeyron; "Memoire sur l'equilibre interieur des corps solides homogenes; *Memoirs presents par divers savans*" (1833) 4.
18. Griffith, A. A.: "The Phenomena of Rupture and Flow in Solids", *Philios. Trans.*, Royal Soc. (1920) **221A**, 163.
19. Frank Schuh; *Drilling Equations*, Petroleum Engineer Publishing Co. 1977.
20. Bell, L. N., "Pressures and Fracture Gradients", Atlantic Richfield Drilling Manual (June 1969).
21. Christman, Stan, "Offshore Fracture Gradients", *Journal Petroleum Technology* (Aug. 1973) pp. 910–914.
22. Eaton, Ben A., "Fracture Gradient Prediction and Its Application in Oilfield Operations", *Journal Petroleum Technology* (Oct. 1969) pp. 1353–1360.
23. Matthews, W. R. and Kelly, John; "How to Predict Formation Pressures and Fracture Gradient", *Oil and Gas Journal* (Feb 20, 1967).
24. Michael J. Economides and Kenneth G. Nolte; *Reservoir Stimulation*, 2nd Ed., Schlumberger Educational Services (1989); TN 871.R467.

Chapter 4

HYDRAULICS OF ROTARY DRILLING

The purpose of the circulating system (see Figure 4-1) is to cool drill bit cutters, clean cuttings from underneath drill bits and carry them back to the surface for analysis and disposal. Downhole motors and turbines also extract power from the hydraulic system.

Drilling mud is mixed in holding tanks and pulled into circulation by mud pumps, which establish fluid pressure and flow rate. Drilling mud passes down the drillstring, losing pressure due to friction, but arrives at the drill bit with enough pressure to clean and remove cuttings from underneath the bit. The cuttings are carried back to the surface and passed across a shale shaker for inspection. Drilling mud is checked for fluid and chemical properties and prepared for recirculation. Hydrodynamic pressure pushes drilling mud down the inside and up around the outside of drillstrings. Total input pressure is party consumed by friction. The remaining pressure is used to clean drill bits and power downhole motors.

4.1 FUNDAMENTALS OF FLUID FLOW

Total fluid pressure can be viewed as the sum of hydrostatic pressure and hydrodynamic pressure. Hydrostatic mud pressure affects many aspects of drilling including well control, drillstring mechanics and drilling rate. Hydrodynamic pressure, produced by the mud pumps, reduces along the drillstring due to fluid friction.

4.1.1 Hydrostatic Pressure

Hydrostatic pressure is the result of gravity acting on a drilling mud column. Hydrostatic pressure increases with true vertical depth and mud density according to

$$p = \gamma y \tag{4-1}$$

In terms of oil field units

$$p = \frac{7.48\gamma y}{144} = 0.052\gamma y \text{ (psi)} \tag{4-2}$$

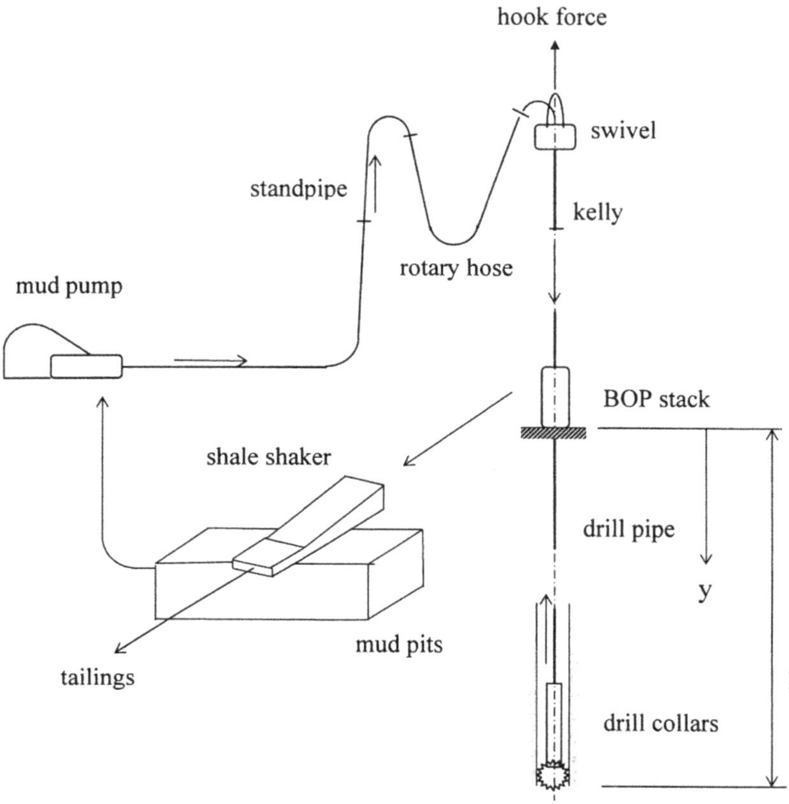

Figure 4-1. Diagram of circulating system used in rotary drilling.

where

γ = mud weight, ppg
y = true vertical depth, ft

This equation applies to all fluids regardless of rheology.

4.1.2 Continuity of Fluid Flow

Continuity of flow is based on the conservation of mass, which requires that mass entering a control volume is equal to mass leaving the control volume.

$$(Q\rho)_{in} = (Q\rho)_{out} \tag{4-3}$$

where

Q = flow rate (volume/time)
ρ = mass per volume

If fluid is assumed incompressible, the volume flow rates are the same and

$$(Q)_{in} = (Q)_{out} = \text{constant} \tag{4-4}$$

or

$$(VA)_{in} = (VA)_{out} \tag{4-5}$$

where

V = average velocity
A = cross sectional area of flow

4.1.3 Bernoulli Equation (1st Law)

The energy equation of fluid flow, assuming frictionless flow, is defined by the Bernoulli equation

$$\frac{V_1^2}{2g} + \frac{p_1}{\gamma} + y_1 + E_{in} = \frac{V_2^2}{2g} + \frac{p_2}{\gamma} + y_2 \tag{4-6}$$

where each term has units of feet. We view each term as having units of ft-lb/lb. E_{in} is energy per lb (ft-lb/lb) being put into the control volume between location 1 and 2. This formula is used to determine pump power required to move fluid from one location to another, assuming friction losses are negligible.

A more general form of Bernoulli's equation accounts for fluid frictional and mechanical energy taken out.

$$\frac{V_1^2}{2g} + \frac{p_1}{\gamma} + y_1 + E_{in} = \frac{V_2^2}{2g} + \frac{p_2}{\gamma} + y_2 + E_{out} + h_f \tag{4-7}$$

where E_{out} (ft-lb/lb) is mechanical energy taken out (bit cleaning, drilling motors) of the control volume and h_f is the friction head in units of feet. One way to view the circulation system is thru a U-tube analogy (Figure 4-2).

Applying Bernouli's equation [Eq. (4-7)] between points 1 and 2 gives

$$E_m = E_{out} + h_f \tag{4-8}$$

which is expressed in units of "ft" or "ft-lb/lb". In terms of hydraulic horsepower

$$\frac{p_p Q}{1714} = \frac{\Delta p_{bit} Q}{1714} + \frac{p_s Q}{1714} \tag{4-9}$$

The application of this equation will be discussed later.

4.2 FRICTIONAL FLOW OF NEWTONIAN FLUIDS

Flow through pipe can be divided into laminar flow and turbulent flow. When flow is laminar, fluid flows along streamlines in a predictable manner. When

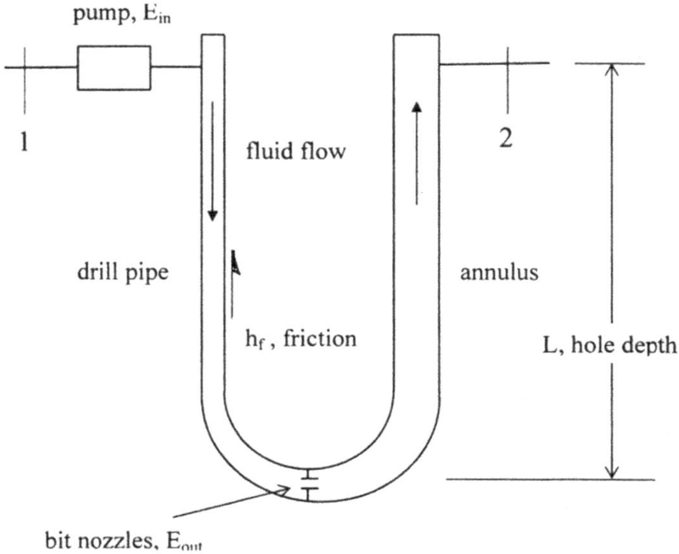

Figure 4-2. U-Tube analogy of circulating system.

flow is turbulent, fluid advances along a chaotic path, which is difficult to model mathematically.

4.2.1 Laminar Flow

Summing forces on the fluid plug (Figure 4-3) and assuming a Newtonian fluid gives

$$\frac{r}{2}\frac{dp}{dy} = \tau = -\mu \frac{du}{dr} \tag{4-10}$$

$$\frac{du}{dr} = -\frac{1}{2\mu}\frac{dp}{dy} r$$

The velocity distribution of fluid moving within pipe follows.

$$u(r) = \frac{1}{4\mu}\frac{dp}{dy}\left(R^2 - r^2\right) \tag{4-11}$$

which shows that fluid velocity is maximum in the center (r = 0) and zero at the inside wall of the pipe (r = R). The derivation assumes Newtonian fluid having viscosity of μ. Using this expression for the velocity profile and integrating across the inside pipe area gives the Hagen-Poiseuille equation.

$$Q = \frac{\pi}{8\mu}\frac{dp}{dy} R^4 \tag{4-12}$$

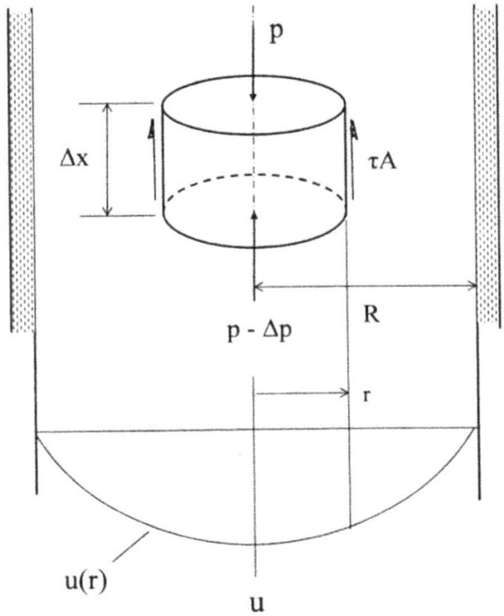

Figure 4-3. Freebody diagram of fluid plug inside pipe.

Rearranging and letting Q = V A

$$\frac{dp}{dy} = \frac{8\mu}{\pi R^4} AV \tag{4-13}$$

The pressure drop in pipe for laminar flow conditions is

$$\frac{\Delta p}{L} = \frac{32\mu}{D^2} V \tag{4-14}$$

4.2.2 Reynolds Number

Reynolds number is a dimensionless number defined by

$$N_R = \frac{VD\rho}{\mu} = \frac{VD\gamma}{\mu g} \tag{4-15}$$

It can be viewed as the ratio of fluid kinetic energy divided by viscous forces at the tubular surface. This number dictates whether flow is laminar or turbulent. If kinetic energy is high and wall friction is low, flow is turbulent. On the other hand, if kinetic energy is low and wall friction is high, flow is laminar. Experiments on Newtonian fluids show that flow ceases to be laminar when Reynolds number is around 2000 and turbulent flow is fully developed when Reynolds number is around 4000.

4.2.3 Friction Head for Laminar Flow

The friction head for laminar flow is determined from the Bernoulli equation [Eq. (4-7)] assuming $V_1 = V_2$. The term, $(y_1 - y_2)\gamma$, represents hydrostatic pressure. Since we are interested only in dynamic pressure changes, this term is dropped.

$$h_f = \frac{\Delta p}{\gamma} \tag{4-16}$$

But from Hagen-Poiseuille

$$\Delta p = L\frac{32\mu}{D^2}V \tag{4-17}$$

By substitution

$$h_f = \frac{L}{D}\frac{32\mu}{\gamma D}V \tag{4-18}$$

The friction head in terms of Reynolds number is

$$h_f = \left(\frac{64}{N_R}\right)\left(\frac{L}{D}\right)\frac{V^2}{2g} \tag{4-19}$$

4.2.4 Turbulent flow

The friction head under turbulent conditions is usually expressed by the Darcy-Weisbach empirical equation

$$h_f = f\left(\frac{L}{D}\right)\left(\frac{V^2}{2g}\right) \tag{4-20}$$

The factor f depends on the Reynolds number and pipe surface roughness.

$$f = f\left(N_R, \frac{e}{D}\right) \tag{4-21}$$

This factor is given in the Moody diagram [1] (Figure 4-4), which shows that the friction factor follows laminar low theory up to a Reynolds number of 2000. There is a transition region between 2000 and 4000. Above 4000 surface roughness is a factor. However, at high values of e/D, friction factor becomes less dependent on the Reynolds number.

As an example, assume the following flow conditions.

Pipe Size = 4 ½ in. (16.6 lb/ft) (drill pipe)
ID = 3.826- in. (Area = 11.497 in.^{2})
Flow Rate = 400 gpm
Viscosity = 1 cp (water)
Density (γ) = 62.4 lb/ft^3 (water)

Converting to uniform units:

$$Q\left(\frac{\text{ft}^3}{\text{min}}\right) = 400\left(\frac{\text{gal}}{\text{min}}\right)\left|\frac{1\,\text{ft}^3}{7.48\,\text{gal}}\right| = 53.48\,\text{ft}^3/\text{min}$$

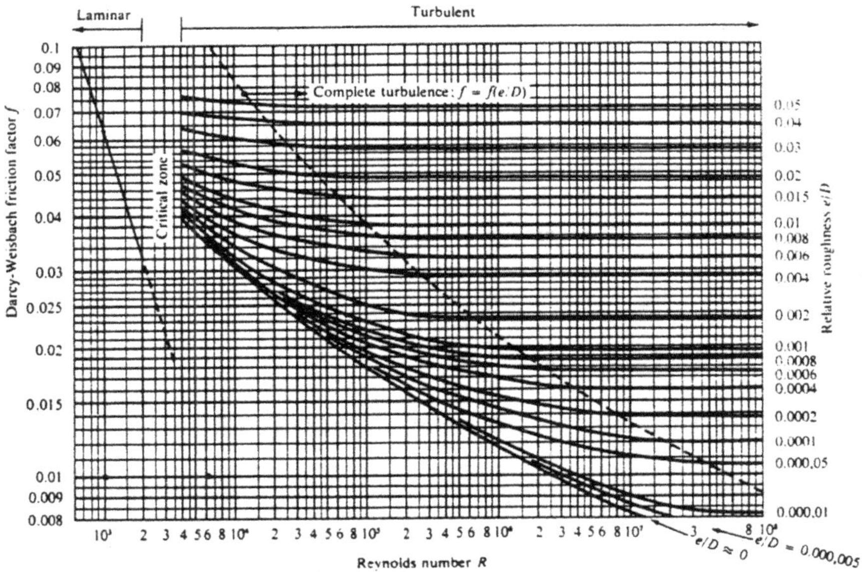

Figure 4-4. Moody diagram.

$$V\left(\frac{\text{ft}}{\text{sec}}\right) = 53.48\left(\frac{\text{ft}^3}{\text{min}}\right)\left|\frac{1\,\text{min}}{60\,\text{sec}}\right|\frac{1}{11.5\,\text{in}^2}\left|\frac{144\,\text{in}^2}{1\,\text{ft}^2}\right| = 11.16\,fps$$

$$\mu(\text{reyns}) = 1\;cp = \left|\frac{1.45\times10^{-7}\,reyns}{1cp}\right| = 1.45\;10^{-7}\,\text{lb second/in.}^2$$

$$\mu = 1.45\times10^{-7}\left(\frac{\text{lb sec}}{\text{in}^2}\right)\left|\frac{144\,\text{in}^2}{\text{ft}^2}\right| = 208.8\,\frac{\text{lb sec}}{\text{ft}^2}\times10^{-7}$$

Pressure drop predictions for turbulent flow (Newtonian fluid) are determined from the following steps.

$$N_R = \frac{VD\rho}{\mu} = \frac{VD\gamma}{\mu g} = 3.43\times10^5$$

f = 0.038 (from Moody diagram, assuming e/D = 0.01)

$$h_f = f\left(\frac{L}{D}\right)\left(\frac{V^2}{2g}\right) \qquad \text{(Darcy-Weisbach)}$$

$$\frac{p_1}{\gamma} - \frac{p_2}{\gamma} = h_f \qquad \text{(Bernoulli equation)}$$

Combining last two equations gives

$$\frac{\Delta p}{L} = f\left(\frac{\gamma}{D}\right)\left(\frac{V^2}{2g}\right)$$

Substitution of all input data gives

$$\frac{\Delta p}{L} = 0.038\left(\frac{62.4}{3.826}12\right)\left[\frac{11.16^2}{2(32.2)}\right] = 14.383\,\frac{\text{lb/ft}^2}{\text{ft}}$$

$$\frac{\Delta p}{L} = 14.383\,\frac{\text{lb/ft}^2}{\text{ft}}\left|\frac{1\,\text{ft}^2}{144\,\text{in}^2}\right| = 0.1\ \text{psi/ft}$$

For a 10,000 ft pipe length the pressure loss due to fluid flow would be 1000 psi.

Note that over the laminar range, the friction factor is defined by

$$f = \frac{64}{N_R} \tag{4-22}$$

4.3 FRICTIONAL FLOW OF NON-NEWTONIAN FLUIDS

The velocity flow profile for both Bingham [Eq. (2-22)] and Power Law [Eq. (2-23)] models can be determined using their respective rheological mathematical formulations [2]. In each case the derivation starts with the freebody diagram in Figure 4-3.

4.3.1 Laminar Flow — Bingham Fluids

From Figure 4-3 and Eq. (2-22)

$$\frac{dp}{dy}\frac{r}{2} = \tau = \tau_o + K_o\left(-\frac{du}{dr}\right) \tag{4-23}$$

$$\frac{du}{dr} = \frac{1}{K_o}\left[\tau_o - \frac{dp}{dy}\frac{r}{2}\right] \tag{4-24}$$

Integrating and satisfying boundary conditions give

$$u(r) = \frac{1}{K_o}\left[\frac{dp}{dy}\frac{1}{4}\left(R^2 - r^2\right) - \tau_o\left(R - r\right)\right] \tag{4-25}$$

Integrating this velocity profile across the pipe area gives

$$Q = \frac{2\pi}{K_o}\left[\frac{dp}{dy}\frac{1}{4}\frac{R^4}{4} - \tau_o\frac{R^3}{6}\right] \tag{4-26}$$

The pressure gradient is expressed as

$$\frac{dp}{dy} = \frac{32K_oV}{D^2} + \frac{16}{3}\frac{\tau_o}{D} \tag{4-27}$$

4.3.2 Laminar Flow — Power Law Model

In this case, the velocity profile is determined as follows. From Figure 4-3 and Eq. (2-23)

$$\frac{dp}{dy}\frac{r}{2} = \tau = K\left(-\frac{du}{dr}\right)^n$$

$$\frac{du}{dr} = -\left[\frac{1}{2K}\frac{dp}{dy}\right]^{\frac{1}{n}} r^{\frac{1}{n}}$$

$$u(r) = \left(\frac{1}{2K}\frac{dp}{dy}\right)^{\frac{1}{n}} \frac{n}{1+n}\left(R^{\frac{n+1}{n}} - r^{\frac{n+1}{n}}\right) \tag{4-28}$$

Values of "n" depend on rheology of fluid. Integrating the velocity profile across the pipe area gives

$$Q = \pi\left(\frac{1}{2K}\frac{dp}{dy}\right)^{\frac{1}{n}} \frac{n}{3n+1} R^{\frac{3n+1}{n}} \tag{4-29}$$

Applying the continuity equation, Q = VA, gives

$$\frac{dp}{dy} = \frac{2K}{R}\left[\frac{V}{R}\frac{3n+1}{n}\right]^n \tag{4-30}$$

The formulations of these laminar flow equations are rather straight forward. However, in practice, fluid flow inside drillstrings is usually turbulent requiring an empirical approach.

4.3.3 Turbulent Flow — Non-Newtonian Fluids

At high shear rates both Bingham and Power Law models converge (Figure 2-9) and both models predict the same pressure losses under turbulent flow. As before, the Darcy-Weisbach equation is used to predict parasitic losses [Eq. (4-20)]. In this case, the friction factor is predicted using

$$f = \frac{a}{N_R^{\,b}} \tag{4-31}$$

which is commonly called the Blasius approximation for non-Newtonian fluids under turbulent flow conditions.

For weighted drilling muds flowing through drill pipe, Eq. (4-31) is taken as

$$f = \frac{0.046}{N_R^{\,0.2}} \tag{4-32}$$

In terms of oil field units, Reynolds number is given as [3]

$$N_R = \frac{15.47\gamma VD}{\mu} \tag{4-33}$$

where

γ = density, ppg
V = average velocity, ft/minute

D = inside diameter, in.
μ = viscosity, cp

In addition, to accommodate turbulent flow of non-Newtonian drilling fluids, the viscosity term in Eq. (4-33) is assumed to be [3]

$$\mu = \frac{PV}{3.2}\ \text{cp} \tag{4-34}$$

Combining Eqs. (4-33) and (4-34) gives

$$N_R = \frac{49.5\gamma VD}{PV} \tag{4-35}$$

4.3.3.1 Pressure Loss Inside Pipe

Substituting these equations into the Darcy-Weisbach equation [Eq. (4-36)] and the Bernoulli equation [Eq. (4-37)]

$$h_f = f\left(\frac{L}{D}\right)\left(\frac{V^2}{2g}\right) \quad \text{(Darcy-Weisbach)} \tag{4-36}$$

$$\frac{\Delta p}{\gamma} = h_f \quad \text{(Bernoulli)} \tag{4-37}$$

gives

$$\frac{\Delta p}{L} = f\left(\frac{\gamma}{D}\right)\left(\frac{V^2}{2g}\right) \tag{4-38}$$

In *oil field* terms [3]

$$\frac{\Delta p}{L} = f\,\frac{\gamma V^2}{93{,}000D} \tag{4-39}$$

where

Δp = pressure drop, psi
L = pipe length, ft
f = friction factor
V = average fluid velocity, ft/minute
D = diameter of flow area, in.
γ = density of drilling mud, ppg

When the friction factor, f, [Eq. (4-32)] is incorporated into Eq. (4-39)

$$\frac{\Delta p}{L} = 2.27(10^{-7})\frac{PV^{0.2}\gamma^{0.8}V^{1.8}}{D^{1.2}} \tag{4-40}$$

Expressing V in terms of Q gives

$$V = \frac{24.51Q}{D^2} \text{ ft/minute} \tag{4-41}$$

By substitution, pressure loss inside pipe [4,5] becomes

$$\frac{\Delta p}{L} = 7.7(10^{-5})\frac{PV^{0.2}\gamma^{0.8}Q^{1.8}}{D^{4.8}} \tag{4-42}$$

where

D = ID of pipe, in.
PV = plastic viscosity, cp
γ = mud weight, ppg
Q = flow rate, gpm
L = pipe length, ft
Δp = pressure drop, psi

For example, assume the following

Mud weight = 11 ppg
Plastic Viscosity = 10 cp
Drill Pipe Size = 4½ in. (16.6 lb/ft) (ID = 3.826 in.)
Flow rate, Q = 400 gpm

Substituting directly into Eq. (4-42) gives

$$\frac{\Delta p}{L} = 0.0634 \text{ psi/ft}$$

Pressure loss over 10,000 ft of pipe is predicted to be 634 psi.

4.3.3.2 Comparison With Moody Diagram Predictions

Figure 4-5 gives a comparison between Newtonian and Non-Newtonian theories for the following input:

Pipe inside diameter = 4 in.
Pipe length = 10,000 ft
Plastic viscosity = 10 cp (Non-Newtonian)
Fluid density = 11 ppg

Flow rate, Q, is an independent variable.

Frictional pressure losses for three different surface roughness factors, $\frac{\varepsilon}{D}$ = 0.005, 0.010, 0.015, following the Moody diagram (Figure 4-4), were generated assuming a Newtonian viscosity of μ = 10 cp. The Newtonian model predicts higher pressure losses than does the non-Newtonian model (Figure 4-5).

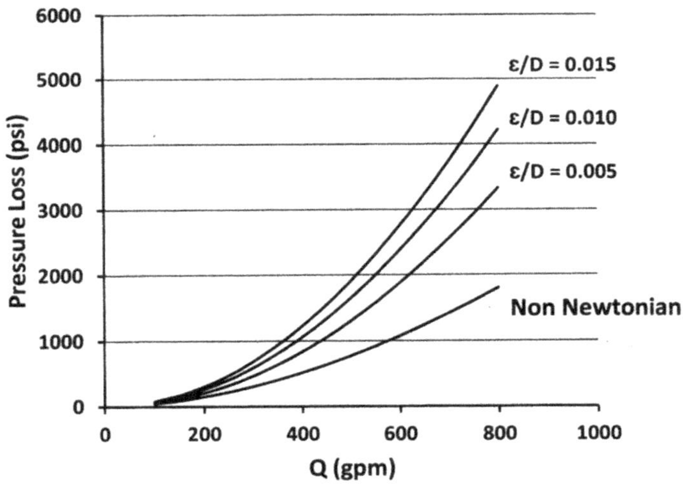

Figure 4-5. Comparison of non-Newtonian and Newtonian pressure loss predictions.

This exercise shows that approximations of this type are inaccurate. The fourth curve is based on Eq. (4-42) with PV = 10 cp.

4.3.3.3 Pressure Loss in Annulus

Pressure loss in the annulus is predicted by [4,5]

$$\frac{\Delta p}{L} = 7.7(10^{-5})\frac{PV^{0.2}\gamma^{0.8}Q^{1.8}}{\left(D_h - D_p\right)^3\left(D_h + D_p\right)^{1.8}} \tag{4-43}$$

4.3.3.4 Pressure Loss in Surface Equipment

Pressure losses also occur through surface equipment. Four categories of surface equipment have been defined by the industry.

One formula commonly used to estimate pressure losses in surface equipment is [6]

$$\Delta p = C\gamma\left(\frac{Q}{100}\right)^{1.8} \tag{4-44}$$

where

γ = mud weight, ppg
Q = flow rate, gpm
C = constant depending on surface equipment type as defined in Table 4-1.

Table 4-2 lists recommended values for each of the four Types.

Table 4-1. Types of surface equipment.

	Stand Pipe		Hose		Swivel		Kelly	
Type	length, ft	ID, in	length, ft	ID, in	length, ft	ID, in	length, ft	ID, in
1	40	3	45	2	4	2	40	21/4
2	40	31/2	55	21/2	5	21/4	40	31/4
3	45	4	55	3	5	21/4	40	31/4
4	45	4	55	3	5	3	40	4

Table 4-2. Losses through surface equipment.

Type	C
1	1.0
2	0.36
3	0.22
4	0.15

For example, assuming a mud weight of 10 ppg, Type 4 surface equipment, and a flow rate of 400 gpm

$$\Delta p = 0.15(10)\left(\frac{400}{100}\right)^{1.8} = 18.2 \text{ psi}$$

4.4 MUD PUMP POWER SUPPLY

Two types of oil field mud pumps are triplex and duplex. Triplex pumps are single acting and have three pistons. Duplex pumps are double acting and have two pistons. Pumping units contain both pump and power source and are usually skid mounted. The power end is driven by electric motors. Older versions were driven by diesel engines by means of a direct drive.

4.4.1 Mud Pump Performance

Mud pumps have two distinct parts: 1) fluid end and 2) power end (Figure 4-6). The fluid end contains: fluid manifold, intake and exhaust valves, pistons, and liner. The power end is a mechanical drive containing a slider crank mechanism, which converts rotary motion and torque into linear piston motion and force.

Mud pumps are positive displacement pumps. Volume output is directly related to the number of strokes of the crank. Output capacity is normally expressed in terms of volume per stroke. Volume output depends on liner size and stroke length. Actual flow rate output depends on volumetric efficiency of the pump and suction intake.

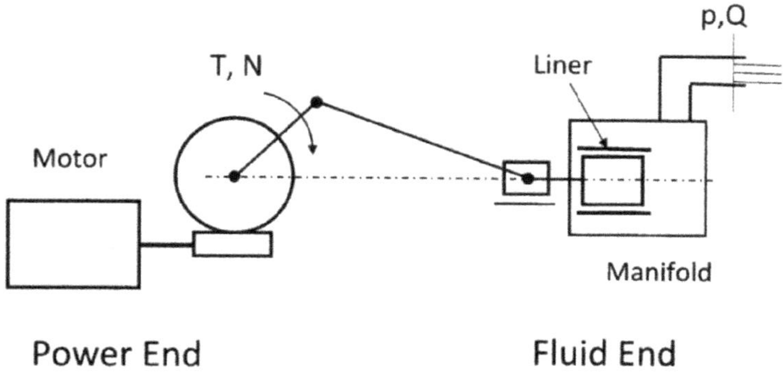

Figure 4-6. Schematic of mud pumps.

Fluid ends are limited by fluid pressure in the manifold. Triplex pumps are limited to a maximum discharge pressure of approximately 5500 psi. Higher manifold pressures can cause fatigue cracks and premature fatigue failure.

Power ends are limited by force transmitted from the piston through the slider crank mechanism. Bearings limit the maximum allowable force on the piston and this force depends on fluid pressure and piston area. Allowable fluid pressure in large liners is low (high volume). Allowable pressure in small liners is typically high (small volume).

While pressure and volume output change with liner size, power input remains the same. Hydraulic horsepower is related to pressure and flow rate by

$$Power = pQ$$

which is adjusted to horsepower units by

$$HHP = pQ\left|\frac{144\ \text{in}^2}{1\ \text{ft}^2}\right|\left|\frac{1\ \text{ft}^3}{7.48\ \text{gal}}\right|\left|\frac{1\ \text{min}}{60\ \text{sec}}\right|\left|\frac{1\ \text{hp}}{550\ \text{ft}-\text{lb/sec}}\right|,\ \text{hp}$$

$$HHP = \frac{pQ}{1714},\ \text{hp} \tag{4-45}$$

where

HHP = hydraulic horsepower, hp
p = fluid pressure, psi
Q = volume flow rate, gpm

Performance of mud pumps is measured in terms of hydraulic horsepower output. Mud pumps produce maximum hydraulic horsepower with different combinations of output pressure and flow rate (Table 4-3). The constant horsepower curve is shown in Figure 4-7.

The rectangle that touches the horsepower curve represents pressure-flow rate capability the pump. The horizontal line indicates the maximum operating pressure set by the manufacturer. Drilling contractors usually operate below the maximum pressure. Flow rates listed in Table 4-3 are limits for each of the liner sizes.

Table 4-3. Typical triplex mud pump performance features.

Liner (in.)	p_{max} (psi)	F_{max} (lb)	Flow Rate (gpm)	HHP (hp)
5 ½	5558	132,048	481	1560
6	4665	131,879	573	1560
6 ½	3981	132,090	681	1571
7	3423	131,717	778	1554
7 ½	2988	132,010	894	1559

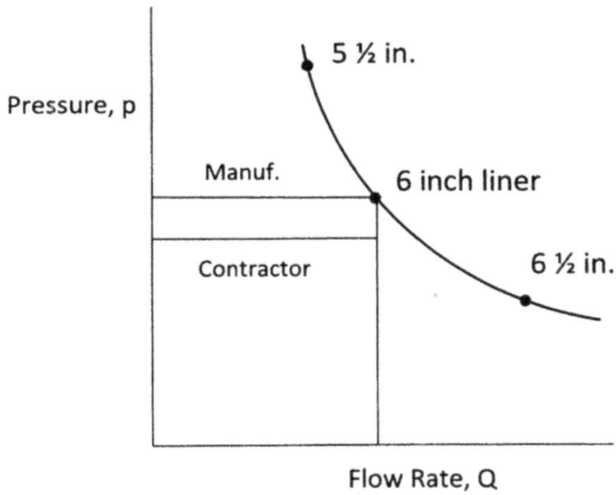

Figure 4-7. Pump performance versus liner size.

4.5 OPTIMIZATING AVAILABLE HYDRAULIC HORSEPOWER

When drilling mud is pumped through a drilling system, energy is lost due to fluid friction. This friction loss is taken from the hydraulic horsepower and can be viewed as parasitic. Losses occur in surface equipment, inside and out of the drillstring. Energy loss across drill bit nozzles is not considered parasitic. Parasitic losses are substantial and must be considered in designing the overall circulating system. A major portion of the total hydraulic horsepower input by the mud pumps is lost to friction. The remaining part is used for bottom hole cleaning and power for downhole motors or turbines. Parasitic losses occur in

a. Surface equipment
b. Inside drill pipe
c. Inside drill collars
d. Annular space around drill collars
e. Annular space around drill pipe

The previous section gives the analytical tools to predict pressure losses throughout the circulating system. Bit cleaning, however, depends upon hydraulic horsepower that is created underneath the bit; $HHP_{bit} = \Delta p_{bit} Q$. This section develops the rationale for maximizing HHP for best bottom hole cleaning.

4.5.1 Maximizing Hydraulic Horsepower

Available bottom hole pressure, by itself, is not enough to judge the cleaning capability of drilling mud. Fluid work capability at the bottom of the hole depends on fluid pressure and flow rate.

The difference between the power supplied by the pump and the power consumed by friction is what is left over at the bottom of the drillstring. Similar calculations at other flow rates define a curve showing how the available bottom hole hydraulic horsepower varies with flow rate.

System pressure loss equations [Eqs. (4-42), (4-43), (4-44)] show that

$$p_s = CQ^{1.8} \tag{4-46}$$

This formula can be used to determine system losses corresponding to any flow rate provided system losses at one flow rate are known. The coefficient "C" can be determined *analytically* using previous equations or *numerically* from rig data.

The value of "C" can be established at the rig as follows:

$$p_s = p_{sp} - \Delta p_{bit} \tag{4-47}$$

where

p_s = system pressure loss at flow rate Q
p_{sp} = standpipe pressure
Δp_{bit} = pressure drop across bit

Pressure drop across the bit is calculated by knowing nozzle sizes (TFA) and flow rate. Standpipe pressure is observed at the rig floor.

For example, if rig data is

P_{sp} = 1925 psi
γ = 10 ppg
TFA = 0.2961 in.^{2}, (11,11,12)
Q = 350 gpm

then

Δp_{bit} = 1287 psi
ps = 638 psi

$$C = \frac{638}{350^{1.8}} = 0.0168$$

System losses can now be determined for all flow rates using Eq. (4-46).

Mathematically, the available hydraulic horsepower reaches a maximum at a particular flow rate [7]. The condition for maximum available hydraulic power is predicted as follows.

$$HHP_a = \frac{1}{1714}(p_p Q - CQ^{2.8}) \tag{4-48}$$

$$\frac{d(HHP_a)}{dQ} = p_p - 2.8CQ^{1.8} = 0 \tag{4-49}$$

By substitution of Eq. (4-46)

$$p_p = 2.8\, p_s \tag{4-50}$$

which means that HHP_a is maximum when

$$p_s = \frac{1}{2.8}\, p_p \tag{4-51}$$

or when system pressure losses are about 1/3 maximum allowable pump pressure. The flow rate at which this condition occurs depends on the hydraulics of the circulating system defined by Eq. (4-46). By substitution

$$\frac{p_p}{2.8} = CQ_{opt}^{1.8} \tag{4-52}$$

$$Q_{opt} = \left[\frac{p_p}{2.8C}\right]^{\frac{1}{1.8}} \tag{4-53}$$

Using the previous drilling data, which gave C = 0.0168 and assuming

Pp = 2800 psi (maximum pump pressure)

then

$$Q_{opt} = \left[\frac{2{,}800}{2.8(0.0168)}\right]^{\frac{1}{1.8}} = 450 \text{ gpm}$$

$$p_s = \frac{1}{2.8} 2{,}800 = 1000 \text{ psi}$$

$$HHP_a = \frac{1{,}800(450)}{1714} = 473 \text{ hp}$$

This was the drilling condition under which Figure 4-9 was developed.

4.5.2 Maximizing Impact Force

The impact force of drilling mud striking the hole bottom is a measure of the fluid's ability to blast chips or cutting off of the formation directly under a bit [7]. The magnitude of this fluid impact depends on fluid velocity, v, leaving the bit nozzles and mud weight. Impact force is the result of fluid momentum change. The fluid momentum vector is turned 90 deg. from the axis of nozzles. An impact force is required to turn the fluid 90 deg. as the fluid strikes the formation.

From impulse-momentum (see Figure 4-8)

$$F = Q_m v = \frac{\gamma}{g} Q v \tag{4-54}$$

From Bernoulli's energy equation

$$\frac{1}{2}\frac{v^2}{g} = \frac{\Delta p}{\gamma} \tag{4-55}$$

$$v = \sqrt{\frac{2\Delta p g}{\gamma}} \tag{4-56}$$

Combining Eqs. (4-54) and (4-56) gives

$$F = \frac{\gamma}{g} Q \sqrt{\frac{2\Delta p g}{\gamma}} \tag{4-57}$$

or

$$F = Q\sqrt{\frac{2\gamma\Delta p}{g}} \tag{4-58}$$

In terms of oil field units

$$F = Q\sqrt{\frac{\gamma\Delta p}{3008}} \tag{4-59}$$

where

F = hydraulic impact force, lbs
Q = circulation rate, gpm
γ = mud weight, ppg
Δp = bit pressure drop (Δp_{bit}), psi

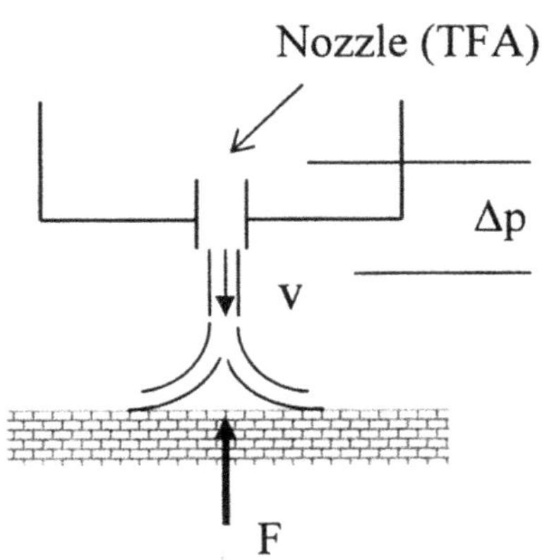

Figure 4-8. Impact fluid force on rock.

Both impact force and hydraulic horsepower are shown in Figure 4-9, assuming parasitic pressure loss is equal to $p_s = 0.0166\, Q^{1.8}$ and all available pressure is absorbed across the bit; there is no downhole motor. Note that while bit hydraulic horsepower reaches a maximum (472 hp) at a flow rate of Q = 450 gpm (corresponding tc a bit pressure drop of 1800 psi), the hydraulic impact force reached a maximum of 1400 lb at a flow rate of Q = 600 gpm (corresponding to a bit pressure drop of 1327 psi); mud weight was taken as 12.35 ppg.

The condition for which impact force is maximized is determined as follows. From Eq. (4-57)

$$F = Q\left[\frac{2\gamma}{g}\left(p_p - p_s\right)\right]^{\frac{1}{2}} \tag{4-60}$$

$$F = \left[\frac{2\gamma}{g}\left(p_p Q^2 - CQ^{3.8}\right)\right]^{\frac{1}{2}} \tag{4-61}$$

Differentiating F with respect to Q and setting the expression equal to zero gives

$$p_s = \frac{2}{3.8}\, p_p \tag{4-62}$$

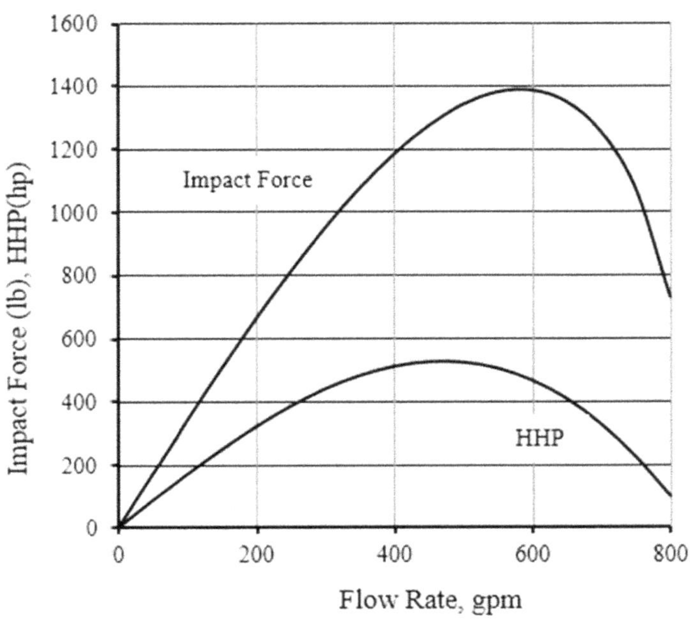

Figure 4-9. Comparison of impact force and hydraulic horsepower.

which states that impact force is maximum when system pressure loss is about ½ the allowable pump pressure. By Eq. (4-46), the flow rate with maximized the impact force is

$$\frac{2}{3.8}p_p = CQ_{opt}^{1.8}$$

$$Q_{\max} = \left(\frac{2p_p}{3.8C}\right)^{\frac{1}{1.8}} \tag{4-63}$$

The maximum impact force is determined from

$$F_{\max} = Q\left[\frac{2\gamma}{g}(0.474)p_p\right]^{\frac{1}{2}} \tag{4-64}$$

Whether it is best to maximize hydraulic or impact force is debatable. It does appear that hydraulic horsepower is preferable where bit cleaning and formation cleaning are needed simultaneously, and impact force is preferable where bottom hole formation cleaning is needed such as in excessive chip hold down cases. Fortunately, when hydraulic horsepower is maximized, impact force is 95 percent of its maximum value; and when impact force is maximized, hydraulic horsepower is 95 percent of its maximum value. It is, therefore, difficult to distinguish which cleaning mechanism is more effective.

4.5.3 Nozzle Selection

Cleaning and removing cuttings from underneath drill bits is enhanced by pressure drop across bit nozzles. Bit nozzles transform available energy into kinetic energy, which is discharged under the bit for cleaning and cuttings removal. Flow rate selection and nozzle size are important for making the best use of available hydraulic horsepower.

Bit pressure drop depends on total flow area (TFA) of all nozzles in a drill bit. Applying Bernoulli's energy equation between sections 1 and 2 (see Figure 4-10) gives

$$\frac{p_1}{\gamma} + \frac{V_1^2}{2g} = \frac{p_2}{\gamma} + \frac{V_2^2}{2g} \tag{4-65}$$

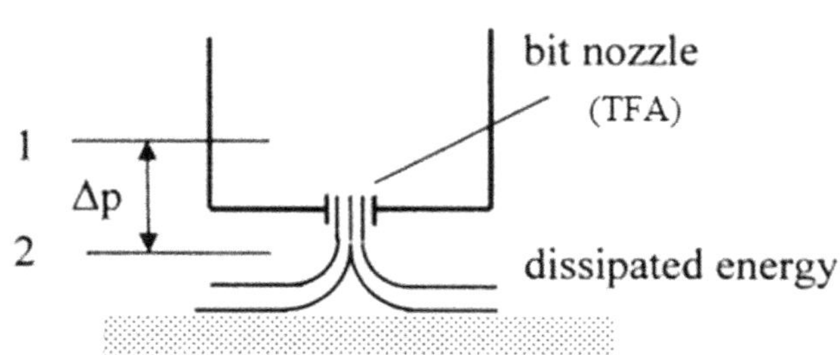

Figure 4-10. Schematic of fluid flow through bit nozzles.

Since V_1 is small by comparison with V_2

$$p_1 - p_2 = \Delta p = \frac{\gamma}{2g} V^2 \tag{4-66}$$

In oil field units

$$\Delta p = \frac{7.48\gamma}{2(32.2)(144)} V^2 \tag{4-67}$$

$$\Delta p = \frac{\gamma V^2}{1240} \tag{4-68}$$

where

Δp = psi
V = fps (note the unit change)
γ = ppg

Applying a nozzle coefficient of 0.95 [3] gives

$$\Delta p = \frac{\gamma V^2}{1120} \tag{4-69}$$

The continuity flow equation is

$$V = \frac{0.32Q}{A_n} \text{ fps} \tag{4-70}$$

By substitution

$$\Delta p = \frac{\gamma Q^2}{10{,}938 A_n^2} \tag{4-71}$$

where

Δp = psi (scales directly to mud weight)
Q = gpm
γ = ppg (pressure scales directly with γ)
A_n = total nozzle area (TFA), sq in

which predicts pressure drop (Δp_{bit}) across bit nozzles.

Consider, for example, the pressure drop across three (3), $\frac{11}{32}$ nozzles (TFA = 0.2784 in.^{2}) is Δp_{bit} = 1887 psi assuming mud weight of γ = 10 ppg and a flow rate of 400 gpm. It is desirable to use up available horsepower for best bottom hole cleaning. This is accomplished by selecting the best flow rate and then solving for total nozzle flow area.

Dividing bit hydraulic horsepower $\left(HHP = \frac{\Delta p Q}{1714}\right)$ by the cross-sectional area of the hole gives hydraulic horsepower per square inch of hole or HSI. An HSI of 5 to 7 is considered a high level of hydraulic horsepower across roller cone drill bits.

4.5.4 Annular Velocity Requirements

Fluid flow in the annulus carries bit cuttings back to the surface for geological evaluation and disposal. While the upward movement of the drilling mud is lifting the cuttings, gravity tends to pull the cuttings back down the hole and cause them to slip relative to the fluid flow. A minimum annular velocity is, therefore, required to adequately clean the wellbore. If annular velocity is too high, fluid turbulence can develop and possibly erode the wellbore.

Formulas for calculating slip velocity and minimum flow rate have developed. Zamora [8] presents the annular velocity required to limit the concentration of cuttings to 4% in the annulus.

A general guide commonly used to establish appropriate annular velocities is

$$V_{an} = \frac{11800}{D_h \gamma} \tag{4-72}$$

where

V_{an} = average annular velocity, fpm
D_h = hole diameter, in.
γ = drilling mud density, ppg

As an example, assume a hole diameter of 10 in. and drill mud of 12 ppg, then Eq. (4-72) gives

$$V_{an} = \frac{11800}{10(12)} = 98 \text{ ft/min}$$

Typical annular flow rates are around 120 ft/minute.

Actual average annular velocity is determined from

$$V_{an} = 19.25 \frac{Q}{A_{an}}, \text{ fpm}$$

where

Q = flow rate, gpm
A_{an} = cross sectional area of annulus, in.^{2}

4.6 EXAMPLE OF DRILLING HYDRAULICS

The purpose of a hydraulic analysis is to establish available hydraulic horsepower at the bottom of a drillstring in order to make best use of what is available to drive motors (or turbines), clean drill bits and carry cutting out of the well bore. For the class exercise consider the follow operating conditions:

Hole Size = $9\frac{7}{8}$ in.
Hole depth = 8500 ft
Mud weight = 11 ppg
Plastic Viscosity = 10 cp
Drill Pipe Size = 4½ in. (16.6 lb/ft); ID = 3.826 in.

Drill Pipe length = 7900 ft
Drill collar size = 8 × 3 in.
Drill collar length = 600 ft
Pump data:
  6-in. liner
  12-in. stroke
  Liner pressure rating is 4665 psi
  Operating pressure limit is 3000 psi
Surface Equipment—Type 4

Step # 1 Calculate parasitic pressure losses for a flow rate of 400 gpm.
Step # 2 Calculate parasitic pressure losses for a range of flow rates (Figure 4-11).
Step # 3 Calculate and plot the hydraulic horsepower loss curve for each flow rate (Table 4-4).
Step # 4 Superimpose the three hydraulic horsepower curves to obtain the available power curve (Figure 4-12).

The straight line shows how the hydraulic horsepower supplied by the pump varies with flow rate. This line is based on the maximum operating pump pressure of 3000 psi. For example, the fluid power supplied by the pump at 400 gpm is

$$HHP_P = \frac{3000(400)}{1714} = 700 \text{ hp}$$

while the fluid power consumed by friction at 400 gpm is

$$HHP_f = \frac{686(400)}{1714} = 160 \text{ hp}$$

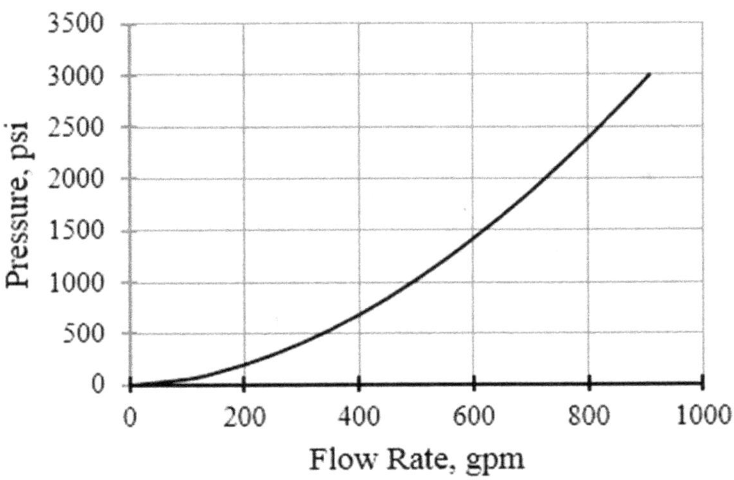

Figure 4-11. Effect of flow rate on parasitic loss.

Table 4-4. Pressure losses and available hydraulic horsepower.

Flow rate (gpm)	Drill pipe, ID (psi)	Collar ID (psi)	Pipe annulus (psi)	Collar annulus (psi)	Surface equip. (psi)	Total system (psi)	Input pump (HHP)	System losses (HHP)	Available (HHP)
0	0	0	0	0	0	0	0	0	0
100	42	10	1.4	1.7	1.7	57	175	3	172
150	87	21	2.9	3.5	3.4	117	263	10	252
200	145	35	4.8	5.8	5.7	197	350	23	327
250	217	53	7.2	8.7	8.6	294	438	43	395
300	301	74	10.0	12.1	11.9	409	525	72	454
350	398	97	13.2	16.0	15.7	540	613	110	502
400	506	123	16.8	20.3	20.0	686	700	160	540
450	625	153	20.8	25.1	24.7	848	788	223	565
500	755	184	25.2	30.4	29.9	1025	875	299	576
550	897	219	29.9	36.1	35.5	1217	963	391	572
600	1049	256	34.9	42.2	41.5	1424	1050	498	552
650	1211	296	40.3	48.7	47.9	1644	1138	624	514
700	1384	338	46.1	55.7	54.8	1879	1225	767	458
750	1567	383	52.2	63.1	62.0	2127	1313	931	382
800	1760	430	58.6	70.8	69.7	2389	1400	1115	285
850	1963	479	65.4	79.0	77.7	2665	1488	1321	166
900	2176	531	72.5	87.6	86.1	2953	1575	1551	24
908	2211	540	73.6	89.0	87.5	3001	1589	1590	0

Note the difference between the maximum operating pump pressure (3000 psi) and system losses at a given flow rate represents the pressure available at the bottom of the drill bit.

As a check note that

$$C = \frac{p_s}{Q^{1.8}} = \frac{1879}{700^{1.8}} = 0.0142 \quad \text{(using numbers from Table 4-4)}$$

System loss and optimum flow rate are

$$p_s = \frac{3,000}{2.8} = 1,071 \text{ psi}$$

$$Q_{opt} = \left[\frac{p_p}{2.8C} \right]^{\frac{1}{1.8}} = 513 \text{ gpm}$$

$$HHP_{\max} = \frac{3000 - 1071}{1714} 513 = 577.3 \text{ hp}$$

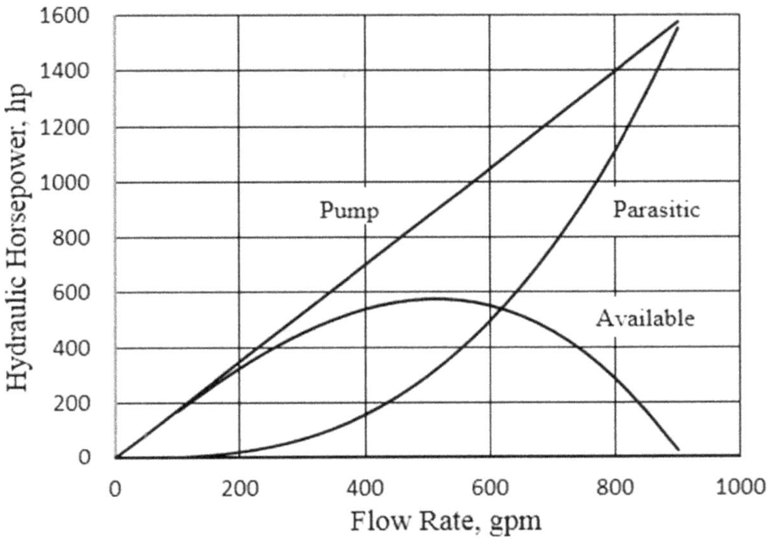

Figure 4-12. Available hydraulic horsepower at bottom of drillstring.

which agrees with the numbers in Table 4-4. Note that the flow rate limit of one pump having a 6-in. liner is 573 gpm, which is slightly higher than the optimum flow rate.

The example shows that the maximum available hydraulic power of 577.3 hp is achieved at a flow rate of 513 gpm. The system pressure loss at 513 gpm is 1071 psi, which is $\frac{1}{2.8}$ of the maximum allowable operating pressure of 3000 psi. This also means that 577.3 HHP (1929 psi at 513 gpm) is available for motor, turbine or bit.

The annular velocity between drill pipe and well bore is determined from

$$V_{an} = \frac{Q}{\frac{\pi}{4}\left(D_h^2 - D_{dp}^2\right)} \left|\frac{144}{7.48}\right|$$

$$V_{an} = 24.51 \frac{Q}{\left(D_h^2 - D_{dp}^2\right)} \text{ fpm} \tag{4-73}$$

where

Q = flow rate, gpm
D_h = well bore diameter, in.
D_{dp} = OD of drill pipe, in.

Both HHP_a and V_{an} are shown in a composite diagram (Figure 4-13). The selection of the best flow rate is a compromise between highest HHP_a and annular velocity requirements.

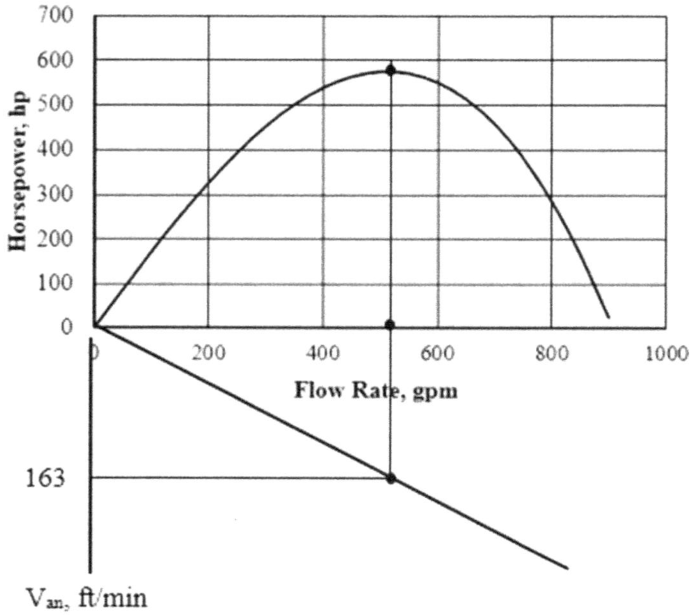

Figure 4-13. Composite diagram of HHP and annular velocity.

4.6.1 Change in Available Horsepower with Well Depth

The available hydraulic power curve changes with depth. Assuming a bit drills the next 1000 ft before being removed, the available power curve changes as shown in Figure 4-14.

It is not always possible or necessary to optimize hydraulics. Consider the three cases shown in Figure 4-15.

High flow rates (5000 ft) may be necessary for bit and hole cleaning drilling the early part of the hole (5000 ft). In this case two pumps may be necessary as well as large pump liners (Table 4-3). If two pumps are used, the maximum pressure limit stays the same, but flow rate can double. Even with high flow rates parasitic losses will be low and it is not necessary to operate at peak of the hydraulic curve.

4.6.2 Effect of Liner Size on Available HHP_a

Optimum hydraulics may be applicable to the intermediate portion (10,000 ft) of the hole. As the well depth increases, parasitic losses consume more and more of available horsepower. In this case, smaller liners (Table 4-3) may be necessary. Fortunately, bit cooling (flow rate) is often more important that bit cleaning (HHP). Annular velocity required to lift cuttings should also be considered.

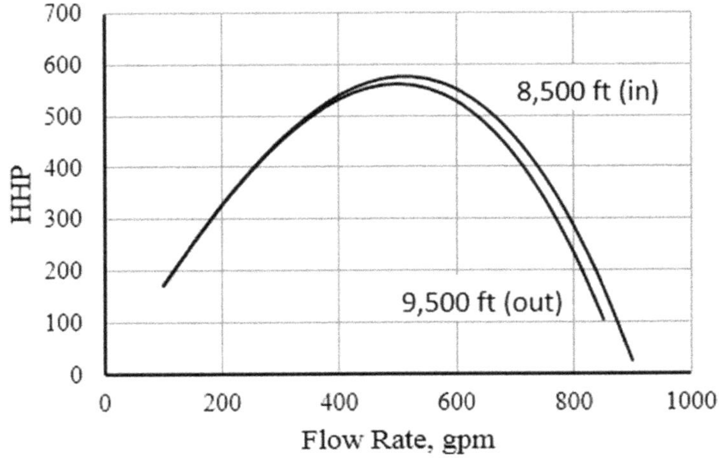

Figure 4-14. Effect of 1000 ft drilled interval on available HHP.

The vertical lines (Figure 4-16) indicate the flow rate limits for each of the three liner sizes

- 481 gpm (5.5 in. liner). P_p = 4000 psi
- 681 gpm (6.5 in. liner), p_p = 3000 psi
- 894 gpm (7.5 in. liner), p_p = 2000 psi

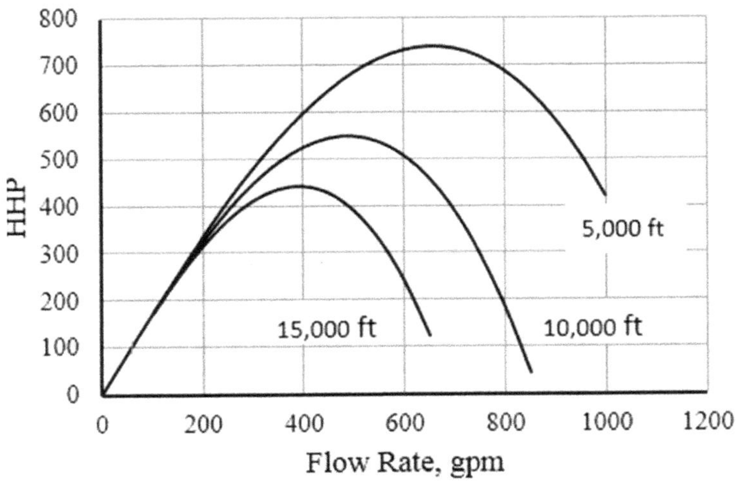

Figure 4-15. Hydraulics for surface, intermediate, and deep holes (drilling data from the previous example are used here).

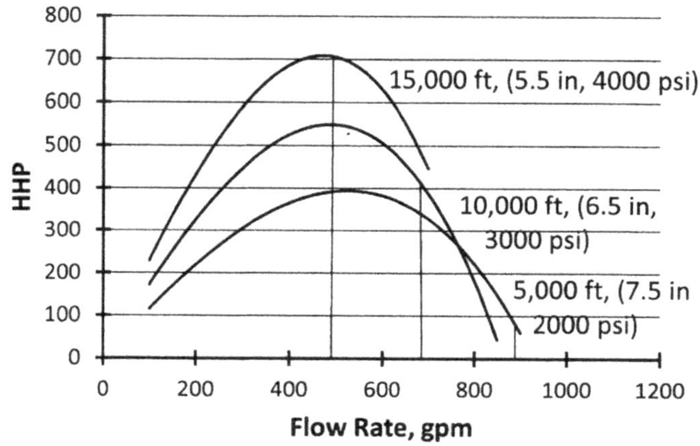

Figure 4-16. Effect of liner size on optimum flow rate (drilling data same, except for liner sizes).

4.7 POWER DEMANDS OF DOWNHOLE MOTORS

Power consumption of positive displacement motors (PDMs) has to be considered in designing drilling hydraulics. Part of the available horsepower (see Figure 4-12) is consumed by the motor and reduces the amount of power available to clean the drill bit.

4.7.1 Power Conversions

Downhole motors and turbines transform hydraulic power back into mechanical power at the drill bit. The power in the circulating system goes through several transformations starting with mechanical power delivered to the mud pumps. The power chain includes:

- Diesel engine — mechanical
- Generator — electrical
- Electric motor — electrical
- Mud pump — mechanical to hydraulic
- PDM/turbine — hydraulic to mechanical

The mechanical power output of downhole motors/turbines starts with mechanical power delivered by diesel engines.

$$P_M = \frac{2\pi TN}{33{,}000} = \frac{TN}{5{,}252}\ \text{hp} \tag{4-74}$$

where

P_M = mechanical power, hp

T = torque, ft-lbs
N = rotary speed, rpm

Hydraulic power, delivered by mud pumps, is related to output fluid pressure and flow rate by

$$P_H = \frac{pQ}{1714}\ \text{hp} \tag{4-75}$$

where

P_H = hydraulic power, hp
p = manifold fluid pressure, psi
Q = flow rate, gpm

The exchange of hydraulic power into mechanical power is illustrated in Figure 4-17.

Applying Bernoulli's energy equation across the motor gives

$$\frac{p_1}{\gamma} = \frac{p_2}{\gamma} + E_{out},\ \frac{ft - lb}{lb} \tag{4-76}$$

The unit of each term is "ft" or ft-lbs per lb of fluid flowing thought the control volume. To convert this equation to units of power

$$\frac{\Delta p}{\gamma}(\gamma Q) = E_{out}(\gamma Q),\ \text{ft-lb/second} \tag{4-77}$$

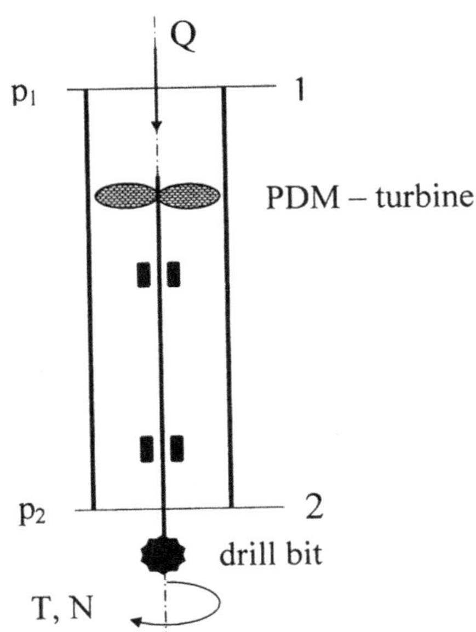

Figure 4-17. Hydraulic power converted to mechanical power.

In terms of hydraulic and mechanical power

$$\left(P_H\right)_{in} = \left(P_M\right)_{out} \tag{4-78}$$

$$\frac{\Delta pQ}{1714} = \frac{TN}{5252} \tag{4-79}$$

assuming no friction losses.

The efficiency of a motor is determined by

$$\eta = \frac{\left(P_M\right)_{out}}{\left(P_H\right)_{in}} \tag{4-80}$$

The performance of PDMs is monitored through rig data. Speed (N) is directly related to flow rate (Q) and torque (T) is directly related to pressure drop (Δp) as will be shown below. This data is available to the driller by means a flow meter and the standpipe pressure gauge.

4.7.2 Performance Testing

Analytical tools are useful for configuring a motor for a given set of operational specifications. However, after a tool is built its true performance is determined by laboratory testing. Test results establish how a given motor will performance in the field. A typical test arrangement is shown schematically in Figure 4-18.

Fluid, usually water, is pumped through the tool and each of the above parameters are measured. The dynamometer generates torque on the drive shaft and sensors measures each parameter for a number of incrementally increased torque settings. Performance tests are made for each size and model.

4.7.3 Performance of Positive Displacement Motors (PDM)

Typical performance data for PDMs is shown in Figure 4-19. Pressure drop and output speed are recorded for various dynamometer torque settings. Flow rate, Q, is held constant. The test is then repeated at other flow rates.

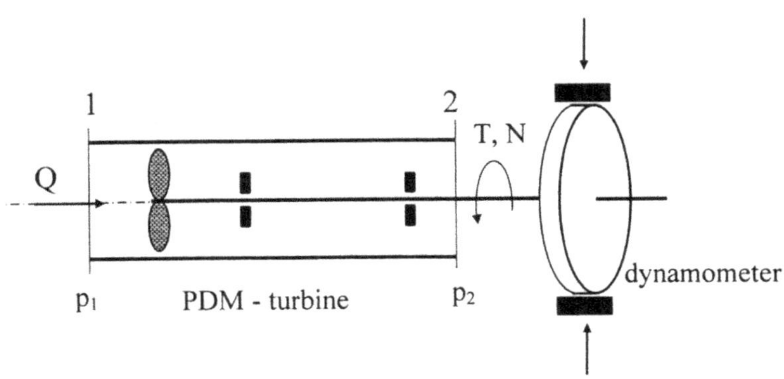

Figure 4-18. Performance testing parameters.

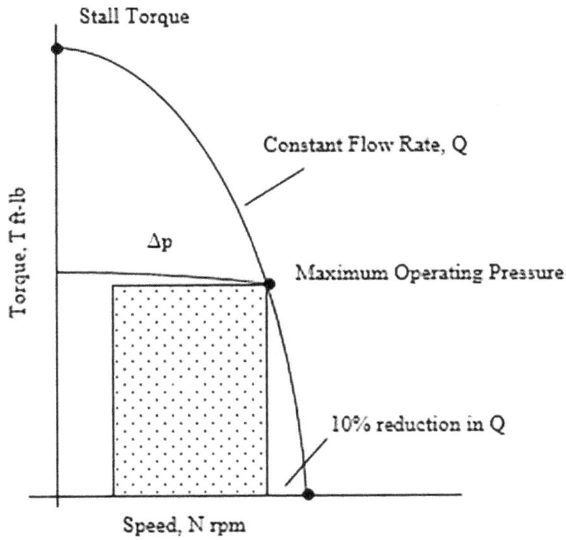

Figure 4-19. Typical performance diagram for Moineau type motors.

Pressure differential is required to overcome internal friction but is lowest under no torque conditions. When torque is increased, rotational speed drops off slightly as a result of fluid leakage through each stage. Higher torque requires higher differential pressures and causes a reduction in rotational speed due to leakage.

When applied torque is taken to the extreme, PDMs stall out. This means that the pressure level is great enough to force fluid through motors without rotation. In each case, fluid leakage is 100 percent. Even though torque is a maximum under stall conditions, power output is zero because the output shaft does not turn. Therefore, operating a PDM near stall conditions produces little or no power to the drill bit. PDMs are normally rated at torque and pressure levels corresponding to about a 10% leakage through the motor.

PDMs operate between recommended maximum and minimum flow rates. The maximum flow rate is based on a maximum allowable planetary angular velocity of the rotor. The rotor is constantly undergoing accelerations, which generate forces on rubber undulations in stators. This could cause premature failure of the rubber. Minimum flow rate is based on inefficiency, which is typical at low flow rates. Based on these guidelines, laboratory performance data can be approximated as shown in Figure 4-20. The numbers are typical of a 1:2 PDM. The performance of multiple lobed PDMs deviates somewhat from this model. Performance data for a variety of motor sizes are given in Table 4-6.

It is important to note that maximum recommended operating torque and pressure levels are not stall conditions. Stall torque is approximately 50% higher than recommended operating torque.

The capability to monitor performance of PDMs from the surface is a distinct advantage. Pressure drop across these motors can be monitored by standpipe pressure and output speed can be monitored by pump flow rate.

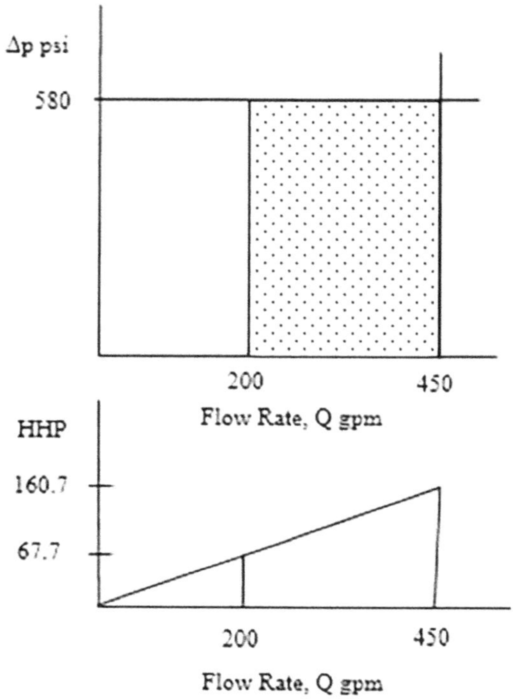

Figure 4-20. Performance diagrams for a 6¾ Moineau motor.

Table 4-5. Format of data collection

Q = constant

Torque (ft-lb)	Δp (psi)	N (rpm)

4.8 MECHANICAL FEATURES OF PDMS

It has long been recognized that only a small portion of total power (~8000 hp) at the surface is used at the drill bit for boring into a formation. For example, under normal rotary drilling operations (N = 100 rpm, T = 1500 ft-lbs), horsepower delivered to drill bits for boring into a formation is

Table 4-6. Positive displacement motor performance data.

Tool size	Pump rate gpm (min)	(max)	Bit speed rpm (min)	(max)	Max Diff. psi	Torque ft-lbs	Power range HP	HHP
3 ¾	75	85	280	700	580	385	20–51	
4 ¾	100	240	245	600	580	585	27–67	
6 ¾	200	475	205	485	580	1500	59–138	(67.7–160.7)
8	245	635	145	380	465	2090	59–152	
9 ½	395	740	195	365	695	3890	145–271	

$$P_{bit} = \frac{2\pi TN}{33{,}000} \tag{4-81}$$

$$P_{bit} = \frac{2\pi(1500)100}{33{,}000} = 28.6 \text{ hp},$$

a very small portion of total rig power. By using PDMs, bit rotary speed increases to, say 400 rpm, creating 4 times the power at the drill bit.

$$P_{bit} = \frac{2\pi(1500)400}{33{,}000} = 114.2 \text{ hp}$$

This increase in drilling power comes at a cost, which has to be weighed against rig rate ($/day) and increase in rate of penetration (ROP).

One of the issues, when PDMs were first introduced, was the need for a companion drill bit, which could operate at high rotary speeds. Roller cone drill bits operated successfully up to near 150 rpm. A companion drill bit was needed for the motor. The solution came with the introduction (1970s) of Polycrystalline Diamond Compact (PDC) drill bits, which can operate in the 400-rpm range.

Motors and turbines may be used for either straight hole or directional drilling but are typically used in directional wells. Motors are used in conjunction with deviation devices, which are integral with the motor.

Positive displacement drilling motors (PDMs) are a spin-off from the Moineau pump. This pump, sometimes called a progressive cavity pump, is still in use today for transporting: food, oil, sewage, and viscous chemicals. Fluid is trapped within cavities (characteristic of the Moineau pump), which progress axially from the intake to the outlet end. Pressure within each cavity may vary as a result of leakage between cavities.

Positive Displacement Motors come in different sizes and lobe numbers (Figures 4-21 and 4-22). In general, the higher the lobe numbers the higher the torque and the lower the speed. A 1:2 combination of rotor to stator configuration crates the highest speed. A 5:6 lobe configuration, for example, performs at a higher torque and lower speed.

Sometimes it is desirable to rotate the entire drillstring simultaneously with a PDM. Drillstring rotation disorients the tool face when PDMs with tilted bearings are used. It also reduces drillstring friction. The question concerns the amount of mechanical power delivered to the drill bit. Since mechanical power depends

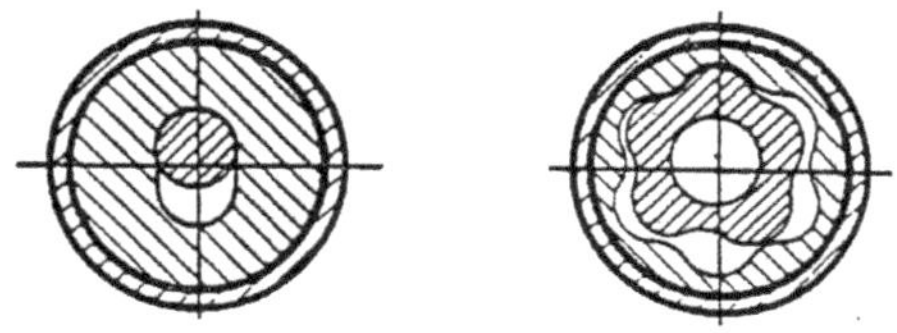

1/2 Rotor/Stator relationship   5/6 Rotor/Stator relationship

Figure 4-21. 1:2 and 5:6 lobed rotary and stator.

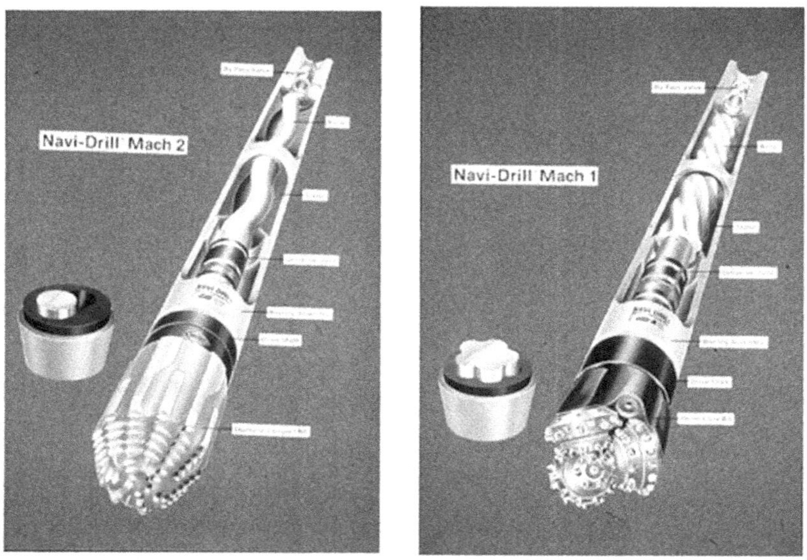

(1:2 motor)   (5:6 motor)

Figure 4-22. Down hole drilling motors (Used by Permission from Baker Hughes).

on both rotational speed and torque consider the speed first. Bit speed is the absolute speed of the motor housing plus the rotary speed of the drill bit relative to the housing. So, if the drillstring rotates at 60 rpm and the output speed of the motor (based on flow rate) is 400 rpm, then absolute bit speed is 460 rpm.

If motors operate within the recommended torque range, total bit speed is the direct sum of the two speeds. When motor torque reaches stall torque as determined from performance tests, the rotor speed relative to the housing is zero and bit speed is due only to pipe rotation; the motor is locked and forms a direct drive between bit and rotary table. It is only when bit torque goes above stall torque that the motor housing overtakes the rotor. However, this is an adverse loading condition on the motor and should be avoided.

Torque that gets to the drill bit is controlled by the motor. Bit torque can be no greater than the motors allows. Assuming zero friction, reactive motor torque gets transmitted back to the rotary table, so bit and surface torque are the same (for this idealized example). In this case, mechanical power at the drill bit is the sum of input power at the surface plus power generated by the motor.

The centerline distance between rotor and stator is important for the establishment of progressive cavities. If, for example, the stator is large, and the rotary is small, fluid would simply flow through the void space. Centerline spacing must allow continuous line contact between the rotor and stator in order to create the cavities. A universal joint is needed to transmit speed and torque from the rotor to the drill bit.

The conversion of a Moineau pump to a downhole drilling motor had two purposes: 1) a positive displacement motor (PDM) located directly above the drill bit could produce more power to the drill bit without rotating the entire drillstring at a high rate and 2) the PDM could be coupled with a bent sub, to force drill bits sideways without the use of whipstocks. Both applications provided huge cost savings. PDMs were introduced when offshore drilling was expanding rapidly during the 1960s.

Positive displacement motors have a direct relationship between output speed and through-put flow rate. This characteristic makes it possible to monitor output speed by monitoring flow rate.

$$N = CQ \tag{4-82}$$

where N is rotational speed (rpm), Q is flow rate through the motor (gpm) and C is a constant depending on geometry of rotor and stator. The proportionality constant, C, is determined from maximum rotational speed and corresponding flow rate data from the manufacture's performance information.

$$C = \frac{N}{Q} = \frac{N_{max}}{Q_{max}} \tag{4-83}$$

For the 6 ¾ motor performance data given in Table 4-6

$$C = \frac{485}{475} = 1.02$$

Rotational speed and flow rate are related by

$$N = 1.02\ Q$$

The motor's rotational speed, therefore, can be monitored by the flow rate.

Torque delivered by PDM motors is directly related to pressure differential across the motor. When the drill bit is off bottom, standpipe pressure represents total system losses plus pressure drop across the bit, say 2200 psi; depicted by the off-bottom line in Figure 4-23. This standpipe pressure also includes the effects of friction within the motor. When weight is applied to the bit, bit torque produces a corresponding increase in standpipe pressure (arrow line). The maximum allowable pressure drop across the motor (580 psi) is reached when the standpipe pressure reaches the "Max PDM Pressure" line. A pressure increase of 580 psi in the standpipe means a 1500 ft-lb bit torque has been reached. This is the recommended maximum operating pressure differential across the 6 ¾ in. motor described in Table 4-6.

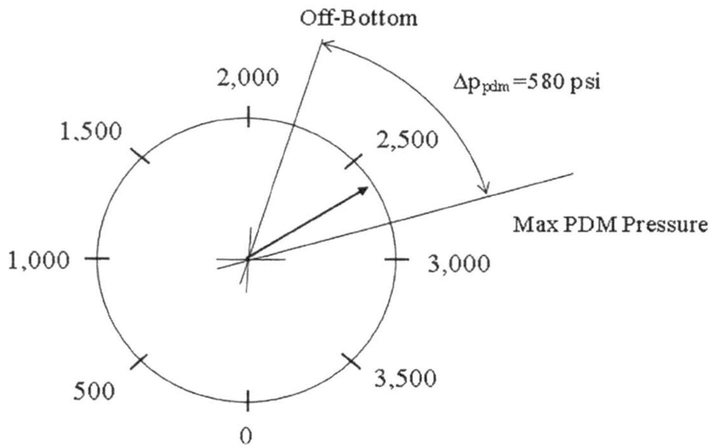

Figure 4-23. Standpipe pressure gauge showing PDM effects.

When the bit drills off some of the bit weight, differential motor pressure reduces because bit torque is reduced. If too much bit weight is applied, bit torque could reach stall torque. In this case, standpipe would be greater than the "Max PDM Pressure" line.

When diamond drill bits are used, off bottom standpipe pressure does not include bit pressure drop. About 2000 lbs is initially applied to diamond bits to establish the reference standpipe pressure for monitoring motor torque.

Motor operations are, therefore, based on torque (or standpipe pressure) rather than WOB as determined by the weight indicator.

4.8.1 Mechanics of PDMs

It is possible to configure the geometry of rotors and stators to achieve a desired C constant. The following correlates the design variables that affect the value of C.

$$C = \frac{n}{q} \tag{4-84}$$

where

n = turns of output shaft (drill bit)
q = corresponding volume thru motor

Since the rotor and stator interact as a planetary gear train, the number of turns (n) of the output shaft (equal to number of turns of the rotor) per one complete circular travel of the rotor center is

$$n = \frac{a}{b} - 1 \tag{4-85}$$

where

a = number of lobes on the stator
b = number of lobes on the rotor

For example, if a stator has two lobes and the rotor one lobe, n = 1 then the output shaft makes one turn for one circular path of the rotor center. Or, if a stator has 10 lobes and the rotor 9, the output rotation would be 1/9 rotation for each circular path of the rotor, a much smaller rotation.

Consider the volume through-put (q) for one circular movement of the rotor center. Let the two cylinders in Figure 4-24 represent the pitch cylinders of a rotor and stator. Let the length of both pitch cylinders be the length of one stage of the stator, L_s. For the sake of visualization assume:

- center lines O and O' are fixed in space
- both pitch cylinders are torsionally flexible
- top ends of both rotor and stator are fixed
- lower end of the stator makes one complete turn causing the lower end of the rotor to make a/b turns
- rotation of both rotor and stator are linear with distance from the top fixed end

Based on the principle of kinematic inversion, the volume between the rotor and stator over stator length, L_s is displaced down the motor whether point O is viewed as making a circular path about point O' or both O and O' are viewed as being fixed and the two pitch circles mesh as a gear pair. This volume is

$$q = AL_s \tag{4-86}$$

where

A = cross sectional area of the space between rotor and stator cavity
L_s = length of one stage as measured on the stator.

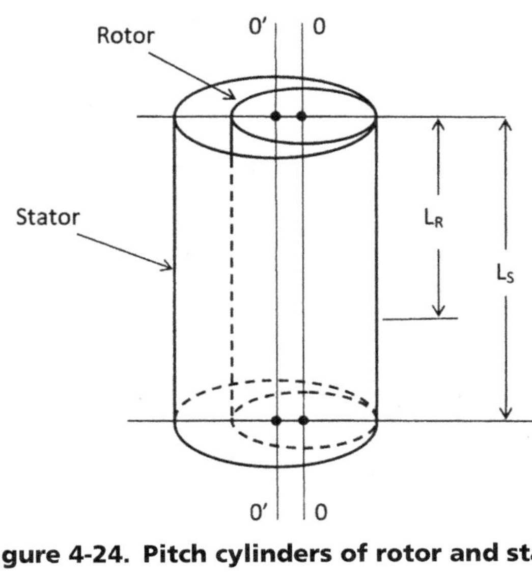

Figure 4-24. Pitch cylinders of rotor and stator.

Note, this volume is not the same as each isolated cavity entrapped between rotor and stator. It represents the total volume over one stage $\left(L_s\right)$.

Combining Eqs. (4-83) and (4-84) with Eq. (4-82) gives

$$C = \frac{n}{q} = \frac{\frac{a}{b} - 1}{AL_s} \tag{4-87}$$

Since $a = b + 1$ for PDMs

$$C = \frac{1}{bAL_s} \tag{4-88}$$

and since $L_r = \frac{b}{a} L_s$

$$C = \frac{1}{aAL_r} \tag{4-89}$$

Eq. (4-80) then becomes

$$N = \frac{1}{bAL_s} Q \tag{4-90}$$

or

$$N = \frac{1}{aAL_r} Q \tag{4-91}$$

Recall that the units of C are: rotation per volume so the units of the denominator has to be volume with units of gallons. If, for example, aAL_r are in units of in.^{2} ft then the conversion is

$$aA(in^2)L_r(ft)\left|\frac{1\ \text{ft}^2}{144\ \text{in}^2}\right|\left|\frac{7.48\ \text{gal}}{\text{ft}^3}\right| = aAL_r \frac{gal}{revolution}$$

AL_r is the total entrapped volume between rotor and stator over length L_r. In consideration of Eqs. (4-90) and (4-91), the smaller the denominator, the larger the output rotational speed for a given flow rate, Q. Also, the larger the number of lobes, the lower the output speed. That's why multi-lobed motors operate at low speeds and high torque. Note that N and Q are proportional, which is characteristic of positive displacement motors and pumps.

The dimensional units of the motor constant, C, is output revolution per gallon pumped through the motor. If, for example, 450 gpm flow rate through a one to two lobed motor produces an output rotational speed of 450 rpm,

Q = 450 gpm
b = 1

the volume of fluid between rotor and stator over one stage of the stator is

$$AL_S = 1 \text{ gal.}$$

This 1-gal volume can be achieved with either a large area and small stage length or a small area and large stage length. Almost any flow rate/rotational speed relationship can be established by various combinations of b, A, and L_S.

Consider a motor design where AL_S = 1 gal and the number of lobes on the rotor is b = 3. The motor constant in this case is

$$C = \frac{1}{bAL_S} = \frac{1}{3}$$

and

$$N = \frac{1}{3}Q$$

A flow rate of 450 gpm produces an output rotational speed of 150 rpm.

A high-speed motor could be designed by making the product bAL_S small. The smallest that "b" can be is one. A one to two lobed motor with either a small stage length or a small void area or both would be a high-speed motor. A low speed motor could be designed by making the product bAL_S large. This is why low speed/high torque motors are multi-lobed.

If a motor is to be designed to deliver a certain output speed (RPM) at a given flow rate (gpm), the motor constant C becomes a design specification. For example

$$C = \frac{N}{Q} = \frac{1}{bAL_s}$$

$$C = \frac{144(12)}{bAL_s(7.48)}L$$

or

$$bAL_s = \frac{231}{C}$$

where

b = number of lobes on rotor
A = flow area through motor, in.^{2}
C = length of stator, in.

Motor dimensions can be selected once the motor constant, C, has been specified.

As an example, assume that a motor is specified to have a rotational speed of 300 rpm at a flow rate of 400 gpm. In this case

$$C = \frac{N}{Q} = \frac{300}{400} = \frac{3}{4}$$

The motor has to be configured so that

$$bAL_s = \frac{4}{3}231 = 308$$

The selection of A and L_s depends in tool size and length. The final dimensions are made through a series of design trade-offs.

Note that by measuring A and L_s of a motor one can determine the motor constant, C, using Eq. (4-88) and estimate the relationship between output speed and flow rate through the motor by assuming no slipping.

The conversion of fluid pressure to mechanical torque is the result of the fluid wedge at the fluid entry or first stage. Fluid density and flow rate have no effect on torque generated by the motor. PDMs are typically made up of several stages and each stage contains an entrapped volume of fluid, which travels through the motor. Even though these motors have multiple stages no work is done by the constant volume cavities and only the top stage converts hydraulic pressure to torque.

Multiple stages, usually about five, perform as dynamic seals so reasonable pressure differentials across these motors can be developed. Performance tests show that even these backup stages allow a certain amount of fluid leakage. The extra stages do, however, create friction and affect the overall efficiency.

Since the hydraulic horsepower consumed by the motor is equal to the output mechanical output power (assuming no losses)

$$\Delta p Q = TN \tag{4-92}$$

So, output torque is

$$T = \frac{Q}{N}\Delta p \tag{4-93}$$

$$T = \left(bAL_s\right)\Delta p \tag{4-94}$$

This equation shows that output torque increases with number of lobes on the rotor (and stator). Torque also increases with stage length, which increases the wedging effect of fluid entering the motor.

It has long been recognized that out of the total mechanical power (~8000 hp) required to operate a drilling rig, only a small fraction is used at drill bits. For example, under normal rotary drilling operations, bit horsepower is

$$P_{bit} = \frac{2\pi TN}{33{,}000} \tag{4-95}$$

$$P_{bit} = \frac{2\pi(1500)80}{33{,}000} = 22.8 \text{ hp}$$

The output speed of a 1:2 PDM is in the range of 400 rpm. Corresponding output power is

$$P_{bit} = \frac{2\pi(1500)400}{33{,}000} = 114 \text{ hp}$$

Bit power is increased by a factor of 4. This increase in bit power comes at a cost because motors cost (C_{pdm}) is added into footage cost calculations. However, PDM cost is offset by increased rate of penetration (ROP).

Note that mud density is not a factor in torque or power generation in PDMs. This is not the case in power generation of turbines.

4.8.2 Effect of Housing Rotation on PDM Output

Remember that total torque transmitted to the drill bit across the PDM motor can be no greater than what is allowed by the motor. The maximum allowable torque depends on the maximum allowable pressure across the motor. However, the total rotary speed of the drill bit is the sum of both housing rotating and rotor rotation relative to the housing. This relative rotor rotation depends only on flow rate through the motor. Total power delivered to the drill bit is

$$P_{mechanical} = \frac{\left(N_{pdm} + N_{rotary}\right)T_{max}}{5252} \tag{4-96}$$

From the conservation of energy, the total power is the sum of input powers from both housing rotation and fluid flow.

4.8.3 Hypocycloid Motion of Rotors

The kinematic movement of a rotor within Moineau mechanisms produces a hypocycloid path relative to the housing. The hypocycloid path, interior to the stator is developed by tracking point P on the pitch circle of the rotor. As the rotor pitch circle rolls inside the stator pitch circle, point P moves in xy space according to

The vector position, $\bar{\rho}$, of point P (Figure 4-25) is

$$\bar{\rho} = \left(a\hat{i}\cos\phi + a\hat{j}\sin\phi\right) + \left(r\hat{i}\cos\theta - r\hat{j}\sin\theta\right)$$

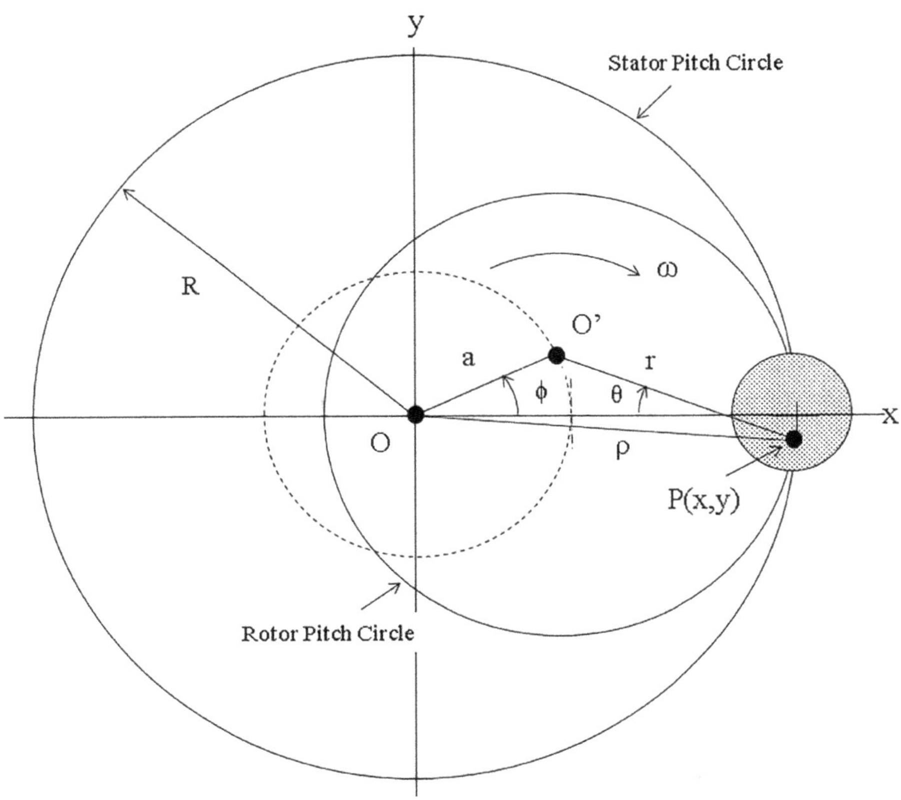

Figure 4-25. Coordinates of hypocycloid.

Gathering the x and y components gives

$$\bar{p} = \left(r \cos\theta + a \cos\phi\right)\hat{i} + \left(-r \sin\theta + a \sin\phi\right)j$$

From which

$$x = r \cos\theta + a \cos\phi$$
$$y = -r \sin\theta + a \sin\phi$$

If for example the rotor-stator ratio is 1:2 and R = 6 in., then

r = 3 in.
a = 3 in.
$\phi = \theta$

By substitution, y = 0 for all θ and x = 2r cosθ showing that point P moves horizontally.

4.9 MECHANICS OF DRILLING TURBINES

Downhole turbines were introduced into drilling programs before PDMs [8]. They generate power at a much higher rotary speed. Turbine speed is approximately double that of PDMs. Consider, again the power Eq. (4-95). Assuming an output speed of 800 rpm, then

$$P_{bit} = \frac{2\pi(1500)800}{33{,}000} = 228.4 \text{ hp}$$

a factor of 8 increase over normal rotary speeds of 100 rpm. However, maximum power of drilling turbines is developed at one particular rotational speed and this optimum speed is sometimes difficult to monitor and control.

Downhole drilling turbines develop torque fundamentally differently from positive displacement motors. Torque is developed by fluid momentum while torque in a PDM is developed by direct application of fluid pressure. Output speed of a turbine depends directly on output torque while output speeds of PDMs depend on flow rate and is independent of torque. Maximum power is developed at a particular rotor speed and herein lies a monitor/control issue. Monitoring and operating turbines at their optimum power speed is necessary for best use of drilling turbines. Since turbine torque is the result of fluid momentum, torque output depends on both drilling mud density and flow rate.

Each turbine stage has a fixed set of blades (stator) and a rotating set of blades (rotor). Stator blades direct drilling fluid onto rotor blades where output torque is generated. Output torque, output power and overall pressure drop of a turbine is a direct multiple of the number of stages in the turbine.

When the rotor blade is stationary, fluid momentum changes the most and the torque is the greatest. When the rotor blade is moving, fluid momentum is not changed as much so torque drops off. The faster the rotor turns, the lower the torque generated by the rotor blades. At a particular high speed of rotor rotation, fluid leaving the stator blades cannot catch the rotor blade, so its momentum does not change. This particular speed is the runaway speed and at this speed no torque is developed on the rotor blades.

In practice, rotational speed is not set in turbines; output speed responds to applied torque or bit weight. Torque is applied, and rotational speed automatically adjusts to the torque. Both stall torque and runaway speed change with mud weight and flow rate. Turbines are rated at maximum power. Whether or not maximum power is developed by downhole turbines depends on the ability to monitor and control bit torque.

4.9.1 Mechanics of Turbines

The mechanics of turbine power is explained by considering momentum changes across the control volume, abcd (Figure 4-26). The rate of change of fluid momentum across cd and ab requires a force vector, $\overline{R}$.

$$\overline{R} = Q_m(\overline{V}_2 - \overline{V}_1) \tag{4-97}$$

where

$\overline{V}_1$ = absolute velocity of fluid leaving stator blade boundary ab
$\overline{V}_2$ = absolute velocity of fluid leaving rotor blade boundary cd
Q_m = mass flow rate through control volume
$\overline{R}$ = resultant force on fluid in control volume
U = tangent velocity of rotor blade (center of blade)

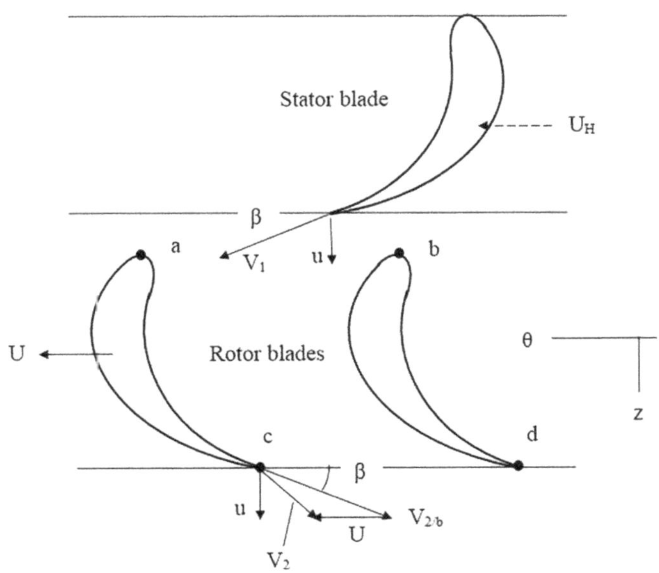

Figure 4-26. Rotor and stator turbine blades.

Letting "V" represent fluid velocity relative to each blade, the component of this vector in the tangent direction is

$$R_\theta = \left[\left(-V \cos\beta + U \right) - \left(V \cos\beta \right) \right] Q_m$$

$$R_\theta = Q_m (U - 2V \cos\beta) \tag{4-98}$$

Since $u = V \sin\beta$

$$R_\theta = Q_m \left(U - \frac{2u}{\tan\beta} \right) \tag{4-99}$$

Since this is the force acting on the fluid in the control volume, the fluid force applied to the blade is

$$F_\theta = Q_m \left(\frac{2u}{\tan\beta} - U \right) \tag{4-100}$$

Output torque is produced by this force. Total torque generated by one turbine stage is

$$T = rQ_m \left(\frac{2u}{\tan\beta} - U \right) \tag{4-101}$$

where Q_m is now the total mass flow rate through the turbine and U is the tangent speed of the rotor and r is the average radial distance. Also

$$Q = \left(\frac{\lambda}{1 - \varepsilon} \right) uA$$

$$u = \frac{1 - \varepsilon}{\lambda} \frac{Q}{A} \tag{4-102}$$

where

u = axial velocity of fluid entering (and leaving) control volume
ε = coefficient of fluid loss around outside of rotor
λ = contraction coefficient due to flow area occupied by blade thickness
A = cross sectional area of space occupied by rotor blade

In consideration of these factors, the torque generated by one turbine stage is

$$T = r\rho Q \left[\frac{2(1 - \varepsilon)Q}{A\lambda \tan\beta} - U \right] \tag{4-103}$$

where Q_m has been replaced by ρQ. U relates to rotational speed by

$$U = r\omega = r\,\frac{2\pi N}{60}$$

The actual rotational speed, N, depends on applied torque according to Eq. (4-100). In practice, rotational speed of the drill bit depends on the torque developed by the drill bit (or WOB).

From Eq. (4-100), maximum torque occurs when U = 0 or the rotor blade is stalled.

$$T_{\max} = r\frac{\gamma}{g}\frac{2(1-\varepsilon)Q^2}{A\lambda\tan\beta} \text{ (stall torque, } T_{st}) \tag{4-104}$$

On the other hand, output torque is zero when

$$U = \frac{2(1-\varepsilon)Q}{A\lambda\tan\beta} \tag{4-105}$$

This velocity is the runaway speed of the turbine. In terms of angular velocity of the turbine

$$N_{rs} = \frac{60}{2\pi r}\frac{2(1-\varepsilon)Q}{A\lambda\tan\beta} \quad \text{(runaway speed)} \tag{4-106}$$

When the blade tip angle (β) decreases from 30 deg. to 20 deg., maximum torque and runaway speed increase by a factor of 1.59.

Output mechanical horsepower is formulated in terms of rotor/stator design parameters, drilling fluid density and flow rate.

Starting with

$$T = r\frac{\gamma}{g}Q\left[\frac{2(1-\varepsilon)Q}{A\lambda\tan\beta} - U\right] \tag{4-107}$$

and

$$P_{out} = T\omega = T\frac{U}{r} \tag{4-108}$$

Power output is defined by

$$P_{out} = \frac{\gamma}{g}Q\left[\frac{2(1-\varepsilon)Q}{A\lambda\tan\beta} - U\right]U \tag{4-109}$$

Maximum power occurs at

$$\frac{d(P_{out})}{dU} = \frac{2(1-\varepsilon)Q}{A\lambda\tan\beta} - 2U = 0 \tag{4-110}$$

or

$$U = \frac{(1-\varepsilon)Q}{A\lambda\tan\beta} \quad \text{(½ runaway speed)} \tag{4-111}$$

which gives

$$P_{\max} = \frac{\gamma}{g}\left[\frac{(1-\varepsilon)}{A\lambda\tan\beta}\right]^2 Q^3 \quad \text{(mechanical output)} \tag{4-112}$$

Note that maximum power is achieved at ½ runaway speed.

Pressure drop across one turbine stage is determined by applying Bernoulli's equation to the control volume shown in Figure 4-27.

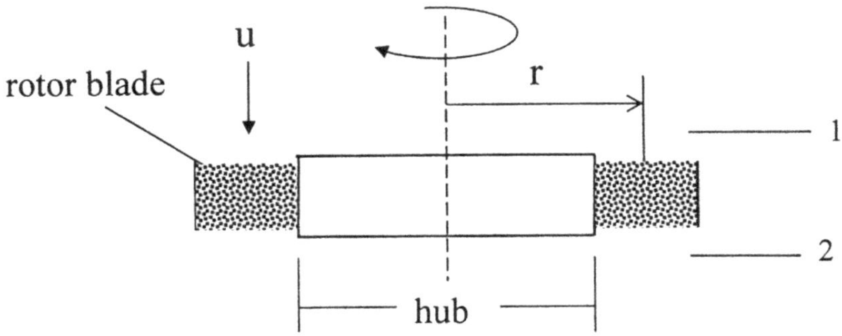

Figure 4-27. Control volume of flow across a turbine rotor.

$$\frac{p_1}{\gamma} = E_{out} + \frac{p_2}{\gamma} \qquad \text{(Bernoulli, } \frac{ft-lb}{lb}\text{)}$$

$$P = \frac{\Delta p}{\gamma}\gamma Q \quad \text{(conversion to mechanical power)}$$

At maximum mechanical power output

$$P_{\max} = \Delta p Q \tag{4-113}$$

and

$$\Delta p = \frac{\gamma}{g}\left[\frac{(1-\varepsilon)}{A\lambda \tan\beta}\right]^2 Q^2 \tag{4-114}$$

Total pressure drop across multi-rotor turbines creates a downward force that must be supported by thrust bearings.

The above equations predict the performance for drilling turbines. Actual performance is established experimentally as explained earlier (see Figure 4-18). Mechanical output is typically less than hydraulic power input to turbines. Efficiency of the power transformation is defined by

$$\eta = \frac{HP_{out}}{HHP_{in}} \tag{4-115}$$

4.9.2 Affect of Housing Rotation on Output Torque and Power

The effect of turbine housing rotation on power output is somewhat more complicated to explain than for the PDM. It is explained by considering the momentum change of the drilling fluid across the rotor blades (refer to Figure 4-28). The control volume represents fluid typical rotor and stator blades.

The effect of housing rotation on turbine output power is explained by considering momentum changes across the control volume, abcd. The rate of change of fluid momentum across cd and ab requires a force vector, $\overline{R}$.

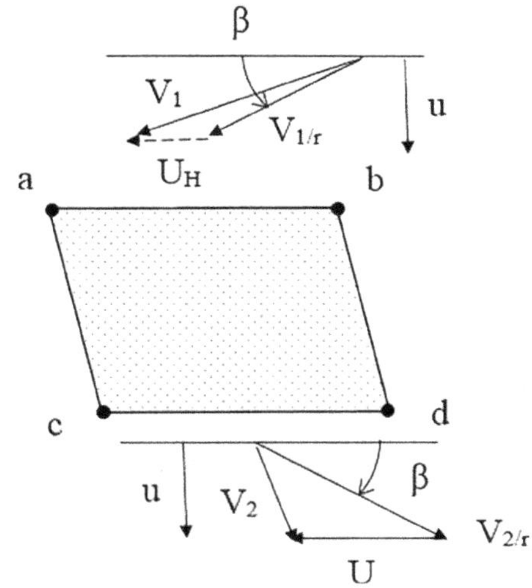

Figure 4-28. Control volume between rotor blades.

$$\overline{R} = Q_m(\overline{V}_2 - \overline{V}_1) \tag{4-116}$$

as discussed earlier. In this case, we must add the effect of housing rotation.

The component of the R vector in the tangent direction is

$$R_\theta = \left[\left(-V\cos\beta + U\right) - \left(V\cos\beta + U_H\right)\right]Q_m$$

where "V" is fluid velocity relative each blade. Housing rotation increases input velocity and a greater moment change, thus increasing output torque. Simplifying

$$R_\theta = Q_m\left[\left(U - U_H\right) - 2V\cos\beta\right] \tag{4-117}$$

Since $u = V\sin\beta$

$$R_\theta = Q_m\left(\left(U - U_H\right) - \frac{2u}{\tan\beta}\right) \tag{4-118}$$

Since this is the force acting on the fluid in the control volume, the fluid force applied to the blade is

$$F_\theta = Q_m\left(\frac{2u}{\tan\beta} - \left(U - U_H\right)\right) \tag{4-119}$$

Output torque is produced by this force. Total torque generated by one turbine stage is

$$T = rQ_m\left(\frac{2u}{\tan\beta} - \left(U - U_H\right)\right) \tag{4-120}$$

where Q_m is now the total mass flow rate through the turbine and U is the tangent speed of the rotor and r is the average radial distance. Also

$$Q = \left(\frac{\lambda}{1-\varepsilon}\right) uA$$

$$u = \frac{1-\varepsilon}{\lambda}\frac{Q}{A} \tag{4-121}$$

where

u = axial velocity of fluid entering (and leaving) control volume
ε = coefficient of fluid loss around outside of rotor
λ = contraction coefficient due to flow area occupied by blade thickness
A = cross sectional area of space occupied by rotor blade

In consideration of these factors, the torque generated by one turbine stage is

$$T = r\rho Q\left[\frac{2(1-\varepsilon)Q}{A\lambda \tan\beta} - (U - U_H)\right] \tag{4-122}$$

$$P_{out} = T\omega = T\frac{U}{r} \tag{4-123}$$

Power output is defined by

$$P_{out} = \rho Q\left[\frac{2(1-\varepsilon)Q}{A\lambda \tan\beta} - (U - U_H)\right]U \tag{4-124}$$

These equations were developed assuming turbine rotary speed is greater that housing rotation speed, i.e., $U \geq U_H$.

The output torque equation was programmed using the following input data.

d_i = 3.5 in.
d_o = 6.0 in.
λ = 0.1
ε = 0.95
γ = 12 ppg
Q = 400 gpm
N_H = 200 rpm

The output torque and power transmitted to the drill bit are shown in Figures 4-29 and 4-30. The solid lines correspond to zero housing rotation while the dashed lines correspond to the housing rotation of 200 rpm.

The housing rotation shifts the solid line to the right by N_H = 200 rpm. For a given bit torque, bit speed (output shaft speed) increases by the 200 rpm. Rotor speeds are additive.

Calculations show that maximum power out with zero housing rotation was

1.942 hp at 800 rpm (torque output is 12.75 ft-lb)

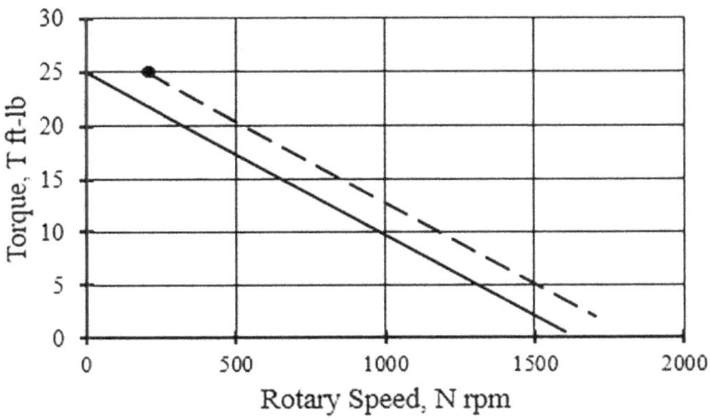

Figure 4-29. Torque with and without housing rotation.

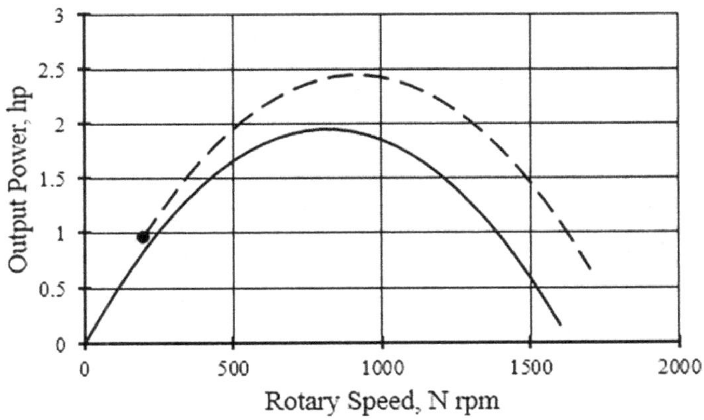

Figure 4-30. Output power with and without housing rotation.

With housing rotation of 200 rpm

2.447 hp at 900 rpm (torque output is 14.28 ft-lb)

$$HP = \frac{TN_H}{5,252} + \frac{TN_{r/s}}{5,252} \tag{4-125}$$

Mechanical power put into the turbine by rotating the housing is

$$HP = \frac{TN_H}{5,252} = \frac{14.28(200)}{5,252} = 0.544 \text{ hp}$$

Since total output power is 2.447 hp, the portion of power produced by turbine is 2.447 – 0.544 = 1.9. This is the power level at rotary speed of 900 rpm as indicated by the lower curve in Figure 4-30.

4.9.3 Performance Relationships

If performance information is known for one set of operating parameters, the performance at other operating parameters can be calculated using

$$N_2 = \frac{Q_2}{Q_1} N_1 \tag{4-126}$$

$$T_2 = \left(\frac{Q_2}{Q_1}\right)^2 \left(\frac{\gamma_2}{\gamma_1}\right) T_1 \tag{4-127}$$

$$\Delta p_2 = \left(\frac{Q_2}{Q_1}\right)^2 \left(\frac{\gamma_2}{\gamma_1}\right) \Delta p_1 \tag{4-128}$$

$$HP_2 = \left(\frac{Q_2}{Q_1}\right)^3 \left(\frac{\gamma_2}{\gamma_1}\right) HP_1 \tag{4-129}$$

These formulas are based on the preceding derivation.

In many cases performance data is given for one mud weight or a given flow rate. The above equations are useful to predict performance at different mud weights or flow rates. For example, consider the data in Table 4-7. Assume that speed (N_1 = 1010 rpm), torque (T_1 = 1430 ft-lb), pressure drop (Δp_1 = 1450 psi) and output power (HP_1 = 275 hp) are known at a flow rate of 475 gpm. Furthermore, it is desired to determine these parameters at a flow rate of 400 gpm with density staying the same at 12 ppg. The conversion is as follows:

$$N_2 = \frac{400}{475} 958 = 807 \text{ rpm}$$

$$T_2 = \left(\frac{400}{475}\right)^2 2390 = 1695 \text{ ft-lb}$$

$$\Delta p_2 = \left(\frac{400}{475}\right)^2 2241 = 1589 \text{ psi}$$

$$HP_2 = \left(\frac{400}{475}\right)^3 436 = 260 \text{ hp}$$

These results compare exactly with the numbers in Table 4-7. Converting to a different mud weight is accomplished using the above equations.

4.9.4 Maximum Output of Drilling Turbines

The torque and rotary speed are related graphically as a straight line (Figure 4-31). Both stall torque and runaway speed change with mud weight and flow rate. Mechanical power delivered by turbines can be derived from the torque-speed line as follows:

$$T = T_{st} - \frac{T_{st}}{N_{rs}} N \tag{4-130}$$

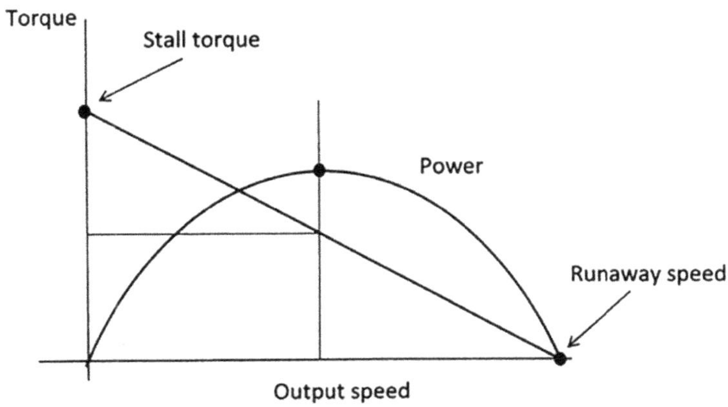

Figure 4-31. Performance curves for drilling turbines.

where

T_{st} = stall torque
N_{rs} = runaway speed

The mechanical output power is

$$P = \frac{TN}{5252}\text{ hp} \tag{4-131}$$

When Eq. (4-130) is substituted in to this equation, the result is a parabola that reaches a maximum value at ½ stall torque and ½ runaway speed (Figure 4-31).

$$P = \frac{T_{st}}{5252}N\left(1-\frac{N}{N_{rs}}\right) \tag{4-132}$$

This is a parabolic function that reaches a maximum value at one-half stall torque and one-half runaway speed. Because turbine speed depends on applied torque, mechanical power delivered by turbines depends on applied torque. When applied torque changes, output speed changes and power is delivered at particular combinations of torque and speed. A means of monitoring turbine speed is necessary to make maximum use of turbine power under drilling operations.

Consider the performance data given in Table 4-7 for a 6 ¾ in turbine operating in 12 ppg mud. Assuming a flow rate of 400 gpm;

- Bit speed is 807 rpm
- Output torque is 1695 ft-lb
- Maximum mechanical power is 260 hp
- Pressure drop across the turbine is 1589 psi

Table 4-7. Turbine performance date (12 ppg).

Pump rate (gpm)	Bit speed (rpm)	Power output (hp)	Torque (ft-lb)	Pressure drop (psi)	Thrust force (1000 lb)
250	504	64	662	621	10
300	605	110	953	894	14
350	706	175	1297	1217	19
400	807	260	1695	1589	25
450	906	371	2145	2011	31
500	1009	509	2648	2483	39

According to the equations given above;

- Runaway speed is 1614 rpm
- Stall torque is 3390 ft-lb

The amount of hydraulic horsepower consumed under the above conditions is

$$HHP = \frac{\Delta p Q}{1714} = \frac{1589(400)}{1714} = 371 \text{ hp}$$

Maximum efficiency of the turbine then is

$$\eta = \frac{HP}{HHP} = \frac{260}{371} = 0.7$$

or 70%. The efficiency curve is also parabolic because HHP is constant and HP is parabolic. Efficiency is maximum when mechanical power is maximum.

The successful application of downhole turbines depends on the amount of hydraulic horsepower available to drive a turbine. It is, therefore, important to establish the amount of available hydraulic horsepower at the bottom of the drillstring. The available hydraulic horsepower varies with flow rate. Hydraulic horsepower consumed by turbines also varies with flow rate and increases with the third power of flow rate.

For example, consider the turbine having performance data as listed in Table 4-8. The amount of hydraulic horsepower consumed by this turbine depends on both flow rate, Q, and pressure drop, Δp, according to

$$HHP = \frac{Q \Delta p}{1714} \tag{4-133}$$

Specific horsepower values are listed in Table 4-8.

Since pressure drop across turbines is fairly independent of the amount of mechanical power delivered, turbines consume the calculated hydraulic horsepower level, whether the drill bit is off or on bottom. Mechanical output power varies considerably with bit torque, however.

When this type of data is plotted, the power consumption curve for turbines can be compared with the available hydraulic horsepower to see if there is

Table 4-8. Hydraulic horsepower consumed by a turbine (12 ppg).

Flow Rate (gpm)	Pressure Drop (psi)	Hydraulic Horsepower (hp)
250	621	90.6
300	894	156.5
350	1217	248.5
400	1589	370.8
450	2011	527.9
500	2483	724.3

enough power to drive the turbine. In order to work turbines to the fullest, turbines should be run at high flow rates. This usually means that all of the available horsepower is consumed by turbines and there is little or no hydraulic horsepower left to clean the drill bit. This condition is represented by the intersection of the turbine and available power curves (Figure 4-33). In some cases, open flow diamond bits can be cleaned and cooled by high flow rates without regard to HHP when used with turbines.

4.10 HYDRAULIC DEMANDS OF DRILLING MOTORS/ TURBINES

Typical PDM power consumption is shown in Figure 4-32 along with the available HHP curve to show how much power is consumed by a PDM. The difference between to two curves represents the HHP available for bit cleaning. This difference is the basis for selecting nozzle size for best bottom-hole cleaning. This type of diagram is useful for determining best flow rate, too. Flow rate selection also has to consider annular velocity requirements.

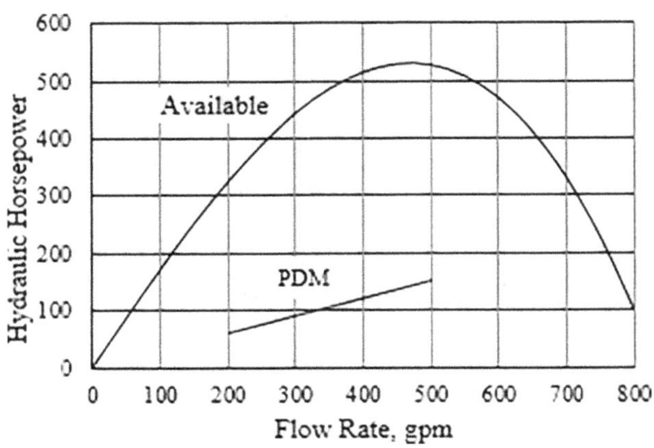

Figure 4-32. Typical PDM power consumption.

Successful application of downhole turbines depends on the amount of hydraulic horsepower available at the bottom of the drillstring. Hydraulic analysis of the total circulating system is critical. Figure 4-33 shows an available hydraulic power curve and turbine power consumption curve; assuming the turbine operates at maximum power output. Note that the power consumption curve is basically different from the PDM power curve.

Input data for Figures 4-33 and 4-34 are p_p = 3000 psi and C = 0.0166.

Turbines usually operate at high flow rates to make best use of the turbine's power capability. This usually means that all of the available hydraulic horsepower is consumed by turbines and there is little or no hydraulic horsepower left for bit cleaning. This condition is represented by the intersection of the

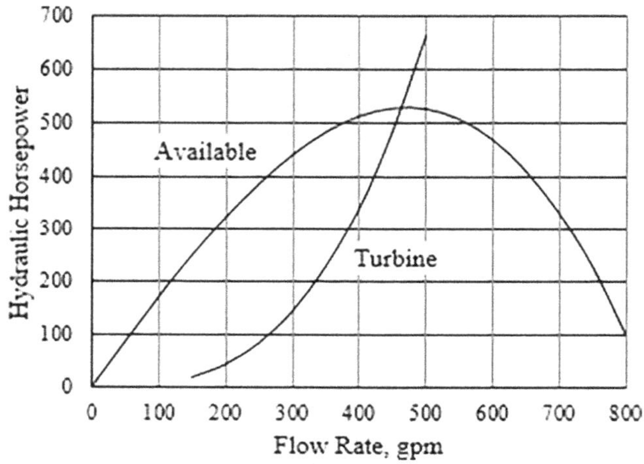

Figure 4-33. Power consumption of drilling turbine.

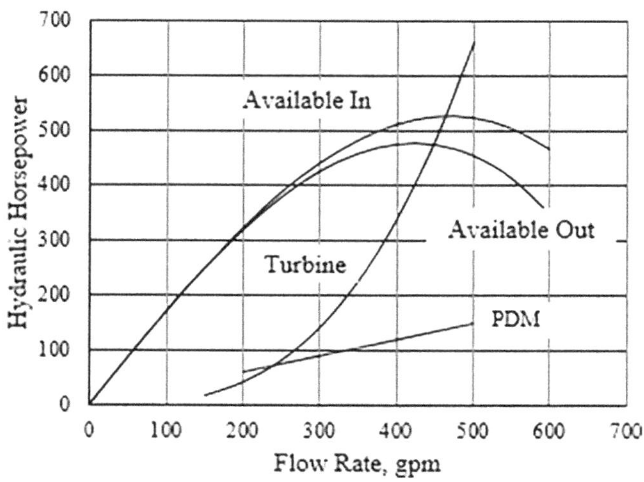

Figure 4-34. Composite hydraulic horsepower diagram.

turbine power with the available power curve. For this reason, open drill bits (no nozzles) are often used with turbines. In this case, bit cleaning and cooling is accomplished by high flow rates.

The composite hydraulics diagram (Figure 4-34) gives the available hydraulic horsepower at depth –in and depth-out and shows the amount of power consumed by both turbines and positive displacement motors. The annular velocity line is an essential part of the diagram because the final choice of flow rate, Q, is usually a compromise between bottom hole hydraulic horsepower and adequate annular velocity for hole cleaning.

4.11 SUMMARY OF EQUATIONS

$$\frac{V_1^2}{2g}+\frac{p_1}{\gamma}+y_1-E_{in}=\frac{V_2^2}{2g}+\frac{p_2}{\gamma}+y_2+E_{out}+h_f \qquad \text{(Bernoulli's equation)}$$

$$h_f=f\left(\frac{L}{D}\right)\left(\frac{V^2}{2g}\right) \text{(Darcy-Weisbach)}, \qquad f=f\left(N_R,\frac{e}{D}\right)$$

$$\frac{\Delta p}{L}=7.7(10^{-5})\frac{PV^{0.2}\gamma^{0.8}Q^{1.8}}{D^{4.8}} \qquad \text{(friction loss inside pipe)}$$

$$\frac{\Delta p}{L}=7.7(10^{-5})\frac{PV^{0.2}\gamma^{0.8}Q^{1.8}}{\left(D_h-D_p\right)^3\left(D_h+D_p\right)^{1.8}} \qquad \text{(loss in annulus)}$$

$$\Delta p=C\gamma\left(\frac{Q}{100}\right)^{1.8} \qquad \text{(loss in surface equipment)}$$

$$\Delta p=\frac{\gamma Q^2}{10{,}938A_n^{\,2}} \qquad \text{(pressure drop across bit nozzles)}$$

$$V_{an}=24.51\frac{Q}{\left(D_h^2-D_{dp}^2\right)} \qquad \text{(annular velocity, fpm)}$$

$$V_{an}=\frac{11800}{D_h\gamma} \qquad \text{(guide for needed annular velocity)}$$

$$N_2=\frac{Q_2}{Q_1}N_1 \qquad \Delta p_2=\left(\frac{Q_2}{Q_1}\right)^2\left(\frac{\gamma_2}{\gamma_1}\right)\Delta p_1$$

$$T_2=\left(\frac{Q_2}{Q_1}\right)^2\left(\frac{\gamma_2}{\gamma_1}\right)T_1 \qquad HP_2=\left(\frac{Q_2}{Q_1}\right)^3\left(\frac{\gamma_2}{\gamma_1}\right)HP_1$$

$$hhp=\frac{pQ}{1{,}714} \qquad HP=\frac{TN}{5{,}252}$$

$$N=\frac{1}{bAL_s}Q=\frac{1}{aAL_r}Q$$

4.12 PROBLEMS

1. Determine the annular velocity (fpm) of drilling mud between 17 ½ in hole and 4 ½ in drill pipe when the flow rate is 920 gpm.

2. Determine the average velocity inside drill collars (2 in bore) when flow rate is 400 gpm. Using the relation $V_1A_1 = V_2A_2$, determine mud velocity through bit nozzles having a TFA = 0.5 in.^{2}.

3. Superimpose the Blasius expression (eq. 4-32) for drilling fluid onto the Moody diagram (Figure 4-4).

4. A triplex mud pump contains a 6 ½ in liner. Pump stroke length is 12 in. Determine total output in bbl/minute at a stroke rate of 120 SPM.

5. Drilling conditions are:
   a. Maximum allowable pump pressure = 3000 psi
   b. Mud weight = 11 ppg
   c. Plastic viscosity (PV) = 15 cp
   d. Drill Pipe ID = 3.826 in. (4 ½, 16.6 lb/ft drill pipe), 9000 ft long
   e. Drill Collars = 7 in. × 2.25 in., 700 ft long
   f. Drilling depth = 9700 ft
   g. Hole size = 9 ⅞ in
   h. Surface equipment type = 3

   Determine total system losses for flow rates ranging from 100 to 500 gpm. Plot available hydraulic horsepower over this range of flow rates. Show all work. You are encouraged to write a computer program. It will be useful later.

6. If the drilled interval in Problem 5 is 1,000 ft, determine and plot the available HHP curve at the end of the drill bit run. Select best nozzle sizes (3 nozzles) for maximum bit hydraulics.

7. A triplex mud pump is rated at 1500 hp and has the following performance characteristics. Fill in the remaining columns.

Liner size (in)	Max pressure (psi)	Flow rate (gpm)	HHP (hp)	Area (in.^{2})	Force (lb)
5 ½	5555	518			
6	4670	617			
6 ½	3980	724			
7	3430	840			

8. A 17 ½ in. surface hole is drilled to 2,000 ft with the following system:
   Drill pipe = 4 ½ in. (16.6 lb/ft)
   Drill collars = 9 × 2 ½ in. (700 ft long)
   Mud pump = use data in problem 7 (assume 6 in. liner)
   Mud weigh = 9 ppg
   Surface Equip = Type 4
   Develop and plot the available hydraulic horsepower curve and annular velocity line. Determine the flow rate required for an annular velocity of 80 fpm. How much HHP is available for bit cleaning at this flow rate?

9. The diameter of a rotor used in a 1:2 lobed motor is 2.5 in. Measurements of the cross section of the stator cavity are given below. The rotor has 6 stages along its 18.75 ft length. Determine the relation between output rotational speed and flow rate through the motor.

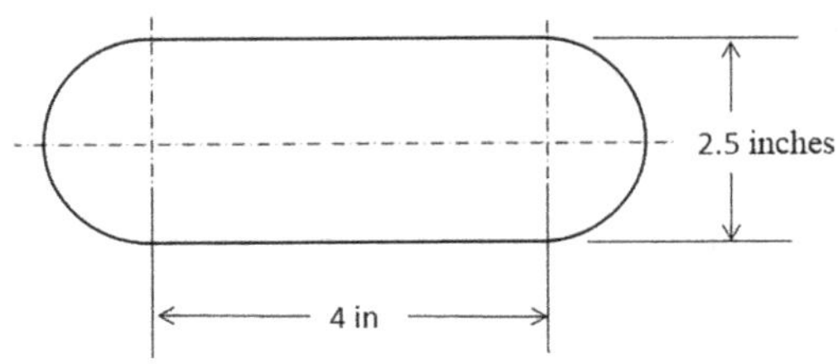

10. The cross-sectional area between rotor and stator of a 2:3 lobed motor is 8 in.². Determine the stage length so that 400 gpm flow will give an output speed of 400 rpm. If the motor is 95% efficient, and input HHP is 120 hp, determine torque output of the motor.

11. Maximum allowable pump pressure is set at p_p = 3000 psi. System losses are defined by

$$p_s = CQ^{1.8} \text{ where } C = 0.021$$

    Develop and plot the available hydraulic horsepower curve for this case. In, addition, show the power consumed by a PDM having the following performance characteristics
    Maximum Δp_{pdm} = 580 psi
    Maximum flow rate, Q_{max} = 475 gpm
    Minimum flow rate, Q_{min} = 200 gpm

12. Performance data for a given turbine is
    Drilling fluid density = 10 ppg
    Flow rate = 475 gpm
    Pressure drop = 1867 psi
    Bit Speed = 958 rpm
    Torque = 1991 ft-lb

Use this data to fill in the table below

Q (gpm)	Δp (psi)	HHP (hp)	N (rpm)	T (ft-lb)	HP (hp)
250					
300					
350					
400					
450					
500					

13. Draw the T-N and P-N diagrams for the turbine in problem 12 but for a mud weight of 12 ppg. Develop these diagrams for a flow rate of 425 gpm. Identify
    - Stall torque
    - Max power output
    - Runaway speed
    - Max torque

    The above data applies to a rotational speed of N = 1010 rpm and output torque of T = 1430 ft-lb with 10 ppg drilling fluid.

14. A given set of drilling conditions are:
    It has been determined that system losses are formulated by

    $$p_s = CQ^{1.8} \quad \text{where} \quad C = 0.021$$

    A given set of drilling conditions are:

    Mud weight = 10 ppg
    Maximum pump pressure = 3,500 psi

    Calculate and plot the available HHP curve. Show on the same diagram, the HHP consumed by: a 6 ¾ in. turbine, with performance as given in problem 12.

4.13 LIST OF REFERENCES

1. Lewis F. Moody, "Friction Factors for Pipe Flow", Trans. ASME, Vol. 66 (1944), pp. 671–684.
2. Dareing, Don W. and Dayton, Ronald D., "Non-Newtonian Behavior of Powder Lubricants Mixed with Ethylene Glycol", *STLE Tribology Transactions*, Vol. 35 (1992), 1, pp. 114–120.
3. Frank Schuh, *Drilling Equations*, Petroleum Engineer Publishing Co. 1977.

4. Moore, Preston L., *Drilling Practice Manual*, PennWell Books, Tulsa, Oklahoma, 1974.
5. Gatlin, Carl, *Petroleum Engineering Drilling and Completion*, Prentice-Hall Inc, Englewood Cliffs, N. J. (1960).
6. Security Drill String Systems — Hydraulics Manual; Dresser Industries, Inc. (Houston).
7. Kendall, H.A. and Goins, W.C. "Design and Operation of Jet-bit Programs for Maximum Horsepower, Maximum Impact Force and Maximum Jet Velocity", Trans., AIME, 1960.
8. Zamora, Mario, "Discussion Drill-Cutting Transport in Full-Scale Vertical Annuli", JPT (Nov 1974) p. 130.

Chapter 5

DRILLSTRING DESIGN AND OPERATION

Drillstrings transmit mechanical rotary power from the surface to drill bits, serve as a conduit for drilling fluid, apply force to drill bits and affect hole direction. A major portion of drillstrings is made up of drill pipe while the bottom portion (roughly 700 ft) is a heavier pipe called drill collars. The bottom portion (~150 ft) of the drill collar section contains tools (drill bit, PDM, Turbine, MWD and stabilizers) called the bottom hole assembly (BHA). Even though the total length of a drillstring may be several thousand feet long, the BHA affects everything: rate of penetration, footage cost and well bore direction.

Drillstrings are subjected to many operational loads such as direct pull, torsion, and bending. Each has a static and a dynamic component. They are designed from static load considerations following the state-of-the-art as given in API RP 7G [1]. Since 1960 much has been learned about the dynamic behavior of drill strings through downhole measurements and analytical studies. Vibrations are known to affect the performance of downhole equipment. Well bore curvature and local dog-legs also affect fatigue life.

5.1 HISTORICAL BACKGROUND [2,3]

5.1.1 Drill Pipe

Lap welded pipe made for general industrial maritime and industrial applications was first used as drill pipe. Pipe material was wrought iron having yield strength of about 25,000 psi and an ultimate strength of 40,000 psi. Joints of drill pipe were connected by threaded couplings. The threads had a slight taper (¾ in. per foot). Failures typically occurred in the last engaged thread under the coupling. Sealing across the coupling was a problem and washouts were common. Casing was also used as drill pipe and left in place when the hole was finished.

The first seamless pipe was manufactured in the early 1920s. A steel billet is rolled into a long bar while simultaneously being pierced by a die to produce the pipe bore. Seamless pipe became an API standard for drill pipe in 1927. At this time, drill pipe was made of mild steel. Grade C pipe has yield strength of 45,000 psi and was introduced around 1926. Grade D pipe, which has yield strength of 55,000 psi, was introduced in 1928. Both C and D grades are seldom used today. Currently, the industry uses four grades of steel (Grades E75, X95,

G105 and S135). Grade E75 is considered a low strength steel. The other three grades are high strength steels.

5.1.2 Rotary Shouldered Connections

The first tool joint was patented in 1910. It contained coarse tapered threads and mating shoulders to provide a seal across the pipe connection. By making tool joints separate from the tubular body, threads could be heat treated and this alleviated the galling problem. Tapered threads also allowed for quick makeup and breakout of the connections. Early tool joints were threaded onto the pipe.

Rotary shouldered connections are special threaded connections used to attach individual joints of drill pipe and drill collars. They also connect drilling tools within drillstrings. They are designed to withstand structural loads while providing a mechanical seal, separating internal pressure (~2000 psi) from annulus pressure. Threads are tapered to a 1:12 pitch to allow quick make-up and break-out. Tool joints are made separately and then attached to the drill pipe body by induction or friction welding. Threads are machined directly into the collar collars, which are made of high strength steel. Because of this, drill collar connections cannot be as strong as the drill collar body. Fortunately, drill collars are run in compression. Drill pipe is run in tension and may be subjected to severe bending in dog-legs within the curvature of well bores.

5.1.3 Drill Collars

During the early 1920s, the lower portion of drill pipe was put in compression to apply force to advance the drill bit. Short threaded collars were used to connect drill bits to drill pipe. These collars were lengthened to about 5 ft to 6 ft to facilitate the recovery of collars and drill bits in the event of a twist-off and to strengthen this area against failure. They were eventually lengthened to help maintain hole direction. After time, these collars became known as drill collars. The use of drill collars to control well bore direction was not a consideration at that time even though it was believed they helped in drilling straighter well bores.

5.2 BUOYANCY

In practice Weight-On-Bit (WOB) is established by slacking off the desirable bit force from the hook load at the surface. This puts drill collars in compression while drill pipe is in tension. A basic question is; how long should the drill collar section be to prevent drill pipe buckling? This is important because of the large difference in structural stiffness between drill pipe and drill collars. From a structural point of view, stiffness attracts bending moment and resulting stresses can be damaging, especially under stress reversals caused by drillstring rotation.

The location and definition of the neutral point was not generally understood until the 1950s [4]. The confusion was due to a misunderstanding of buoyancy and how it is applied to submerged pipe. According to Archimedes'

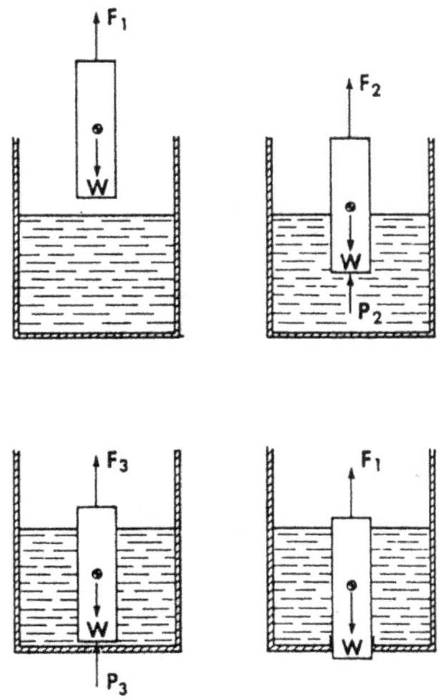

Figure 5-1. Body force (gravity) and surface force (buoyancy).

principle, the buoyant force acting on a submerged body is equal to the weight of fluid displaced by the body. However, it is important to remember that buoyancy is a surface force and not a body force. On the other hand, pipe weight is a body force.

Consider a long bar, which weighs W lbs in air (Figure 5-1). The hook force required to support the bar (position 1) is $F_1 = W$. When the bar (position 2) is partially lowered into a fluid, the hook force reduces by P_2 as a result of fluid pressure at the lower end. When the bar is lowered to the bottom of the fluid container (position 3), the hook force is reduced still further because $P_3 > P_2$. Furthermore, image that a hole is cut in the bottom of the container to allow the bar to extend through the container and at the same time seal in the fluid (position 4). In this case, the bar is supported only by the hook force, F_1. The bar displaces fluid but receives no support from the surrounding fluid. In this case, fluid is displaced but there is no buoyant force.

5.2.1 Hook Load

Applying Archimedes' principle to a string or pipe hanging freely from the draw works predicts a hook load of

$$H = W - B \tag{5-1}$$

where H is hook load, W is air weight of total string and B is the weight of the drill mud displaced by the string as stated by Archimedes' principle. In this equation, W is a body force and B is a surface force. Equation (5-1) can also be written

$$H = W\left(1-\frac{B}{W}\right) = W\left(1-\frac{\gamma_m}{\gamma_{stl}}\right) \tag{5-2}$$

$$H = WBF \tag{5-3}$$

where W is air with of pipe and BF is the buoyancy factor.

Assuming a 12 ppg mud, the buoyancy factor is 0.817 as calculated below.

$$\gamma_m = 12\,\text{lb/gal}\left|\frac{7.48\text{ gal}}{1\text{ ft}^3}\right| = 89.76\frac{\text{lb}}{\text{ft}^3} \tag{5-4}$$

so

$$BF = \left(1-\frac{89.76}{490}\right) = 0.817$$

The buoyed weight of a drillstring (including tools, etc.) is total air weight multiplied by the buoyancy factor.

The magnitude of buoyancy forces can be quite high. For example, consider 5½ in. (19.2 lb/ft) drill pipe having a cross section area of 4.9624 in.^{2}. If 10,000 ft of 5½ in. drill pipe hangs freely in γ_m = 12 ppg mud

$$\begin{aligned} p &= L\gamma_m = 0.052\,(10{,}000)\,(12) = 6240\text{ psi} \\ F &= pA \\ F &= 6240\text{ lb/in.}^2\,(4.9624\text{ in.}^2) \\ F &= 30{,}965\text{ lbs} \end{aligned} \tag{5-5}$$

It would appear that this force is great enough to buckle the drill pipe several times. However, this does not happen.

5.2.2 Definition of Neutral Point

Assume the bottom end of an open-ended pipe (Figure 5-2a) is given a virtual sidewise displacement. The static equivalence of the hydrostatic pressure forces on the pipe below section a-a is represented by the vectors B and pA. The force B is equal to the weight of fluid displaced by the pipe below section a-a. The vector pA is the product of the local fluid pressure and pipe area. The solid vector, W, represents the air weight of the drill pipe below cross section a-a. W is a body force while B is a surface force.

Moments about point "0" show there is a restoring moment that always moves the pipe back toward the vertical, assuming the density of the pipe is greater than the density of the fluid. In other words, an open-ended drill pipe will not buckle from hydrostatic pressure alone regardless of well bore depth.

Now assume the drill pipe is attached to the top of a drill collar section (Figure 5-2b) and the compression at the top of the drill collars is equal to the local hydrostatic pressure. According to the discussion above, the drill pipe will not buckle under this condition.

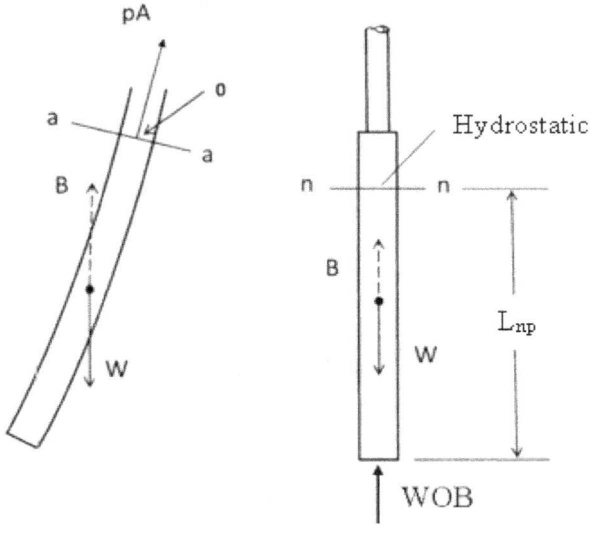

Figure 5-2. Stability of vertical pipe under hydrostatic loading.

A good definition for the neutral point is the point in the drillstring where compressive stress is equal to the local hydrostatic pressure. If the hydrostatic pressure point is located within the drill collars, the drill pipe will not buckle. Drill pipe will buckle when the internal force at the lower end reaches a critical level, which is somewhat higher than the hydrostatic force level.

The location of the neutral point depends on bit force (WOB). If bit force is zero, the neutral is located at the drill bit. As bit force is increased, the neutral point moves up the drill collars.

In practice, the neutral point is kept within the drill collars as a safety measure to avoid buckling the drill pipe. The relation between WOB and distance to neutral point is determined by

$$WOB = wL_{np} - w_f L_{np} \tag{5-6}$$

or

$$WOB = wL_{np}\left(1 - \frac{w_m}{w}\right) \tag{5-7}$$

$$WOB = wL_{np}\left(1 - \frac{\gamma_m}{\gamma}\right) \tag{5-8}$$

where

γ = steel density (490 lb/ft^3 or 65.5 ppg)
γ_m = drilling mud density (ppg)

The distance to the neutral point (or point of hydrostatic compression) from the drill bit is determined by

$$L_{np} = \frac{WOB}{wBF} \tag{5-9}$$

Note the location of the neutral point does not depend on collar length; only WOB.

Drill collar length is generally computed by

$$L_{np} = 0.85L_c \tag{5-10}$$

which means that the neutral point is 85% of drill collar length. The extra length represents a safety factor against drill pipe b1uckling.

$$L_c = \frac{WOB}{0.85wBF} \tag{5-11}$$

5.2.3 Basic Drillstring; Drill Pipe and Drill Collars

Consider a simple drillstring made up of drill pipe and drill collars. If the drillstring is hanging freely from the derrick the neutral point is located at the bottom of the drill collars. As WOB is added, the neutral point moves up the string and reaches the top of the drill collar section when

$$WOB = w_1L_1 - A_1L_1\gamma$$

Since $A_1L_1\gamma = w_fL_1$ (where w_f is weight per length of fluid displaced by the collars),

$$WOB = \left(w_1 - w_f\right)L_1 = w_1L_1BF$$

Additional bit forces move the neutral point up into the drill pipe. Letting (x) represent distance from top of drill collars to the neutral point, then

$$WOB = w_1L_1 + w_2x - w_{f1}L_1 - w_{f2}x$$

$$WOB = w_1L_1BF + w_2xBF$$

$$WOB = \left(w_1L_1 + w_2x\right)BF$$

The neutral point location from the bottom of the drill collars is $L_{np} = L_1 + x$.

5.2.4 Composite Drillstrings

Bottom hole assemblies are generally made up of many components, including motors, MWD, and stabilizers. In such cases the neutral point is located by applying

$$WOB = \sum_{i=1}\left(W_i + xw_{i+1}\right)BF \tag{5-12}$$

where W_i is the air weight of each component up to the top of the ith component and x is the distance from the top of the ith component to the neutral point. Equation (5-12) is based on the freebody diagram shown in Figure 5-3a.

An application of Eq. (5-12) with the BHA is given in Table 5-1. The numbers are based on a mud weight of 12 ppg (BF = 0.817). Assuming a WOB of 12,000 lbs

$$12{,}000 = 9909 + 150\,(0.817)\,x$$
$$x = 17.1 \text{ ft}$$

The neutral point is located 17.1 ft from the top of the MWD tool and into the nonmagnetic drill collars.

The above discussion can also include material changes in the drillstring. For example, assume the drill pipe is aluminum and the drill collar section is steel. In this case (refer to Figure 5-3b)

$$H = (Vol)_{collars}\left(\gamma_{stl} - \gamma_m\right) + (Vol)_{pipe}\left(\gamma_{alum} - \gamma_m\right) \tag{5-13}$$

$$H = \left(Vol\right)_{collars}\gamma_{stl}BF_{stl} + \left(Vol\right)_{alum}BF_{alum} \tag{5-14}$$

$$H = \left(L_c w_c\right)BF_{stl} + \left(L_p w_p\right)BF_{alum} \tag{5-15}$$

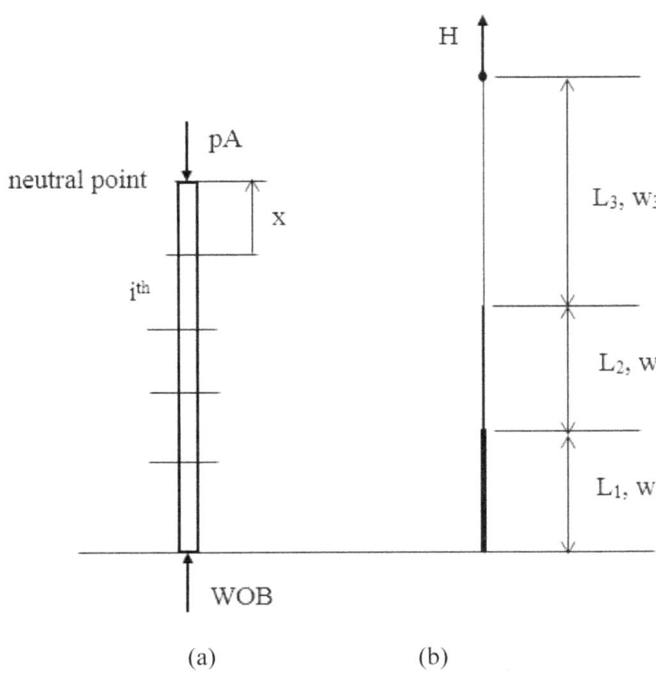

Figure 5-3. Composite drillstring.

Table 5-1. Composite bottom hole assembly.

BHA Component	Length (ft)	Unit weight (lb/ft)	Air weight (lb)	Buoyed weight, (lb)	Accumulated weight, (lb)
12¼ in. Drill Bit	1.5		200	163	163
9½ in. Motor	26.3	155	4,077	3,331	3,494
8 in. Crossover Sub	2	150	300	245	3,739
12 in. Stabilizer	5	150	750	613	4,352
8 in. N.M. Collar	11	150	1,650	1,348	5,700
12 in. N.M. Stabilizer	5	150	750	613	6,313
8 in. MWD	31	142	4,400	3,595	9,909 x
8 in. N.M. Collars	30	150	4,500	3,677	13,586
8 in. Drill Collars	90	150	13,500	11,030	24,616
8 in. Crossover Sub	2	150	300	245	24,861
5 in. Hevi-Wate D. P.	360	49.3	17,748	14,500	39,361

In general

$$H = \sum_{1}^{n} L_i w_i BF_i \tag{5-16}$$

where

w_i = weight per unit length of pipe in consideration of material density
L_i = Length of pipe section
BF_i = buoyancy factor, $BF_i = 1 - \frac{\gamma_m}{\gamma_i}$
γ_m = density of drilling mud
γ_i = density of pipe material

5.3 DRILLSTRING COMPONENTS

Drillstrings transmit torque and load to drill bits and to provide a conduit for delivering drill fluids to drill bits for the purpose of cleaning the bottom of the hole. Basic components of drillstrings are:

- Drill Pipe
- Drill Collars, including stabilizers
- Drill Bit

The drill collar section may contain special equipment, such as MWD and drilling motors in the bottom 150 ft. This assembly of special tools is called bottom hole assembly (BHA). When BHAs have to be changed out for maintenance, wear, or failure the entire drillstring must be removed and eventually replaced other bottom hole equipment.

There are two methods of transmitting power (torque, rotation) into the top of drillstring:

1. Rotary table
2. Top drive

The rotary table method is the traditional method. It is labor intensive and requires that each 30 ft joint of pipe is separated and joined by hand using special tongs. The top drive pulls and racks 90 ft stands of pipe. It has the distinct advantage of pipe rotation while pull and inserting pipe within the well bore. This is useful for reducing friction.

In either case, the mechanics of drillstring behavior within the well bore is the same. This is also true whether drilling is inland or off shore.

5.3.1 Kelly and Rotary Table Drive

The rotary table is a means of transmitting torque to the drillstring (Figure 5-4). It is used to support the freely hanging drillstring during tripping out or going into the well bore. During drilling, a special bushing, called the Kelly bushing, is used to transmit torque to the drillstring. The Kelly is a special pipe having a square or hexagonal external shape for the purpose of transmitting torque. The Kelly slides through the rotary table as well bore deepens. Drilling fluid flows through the center of the Kelly.

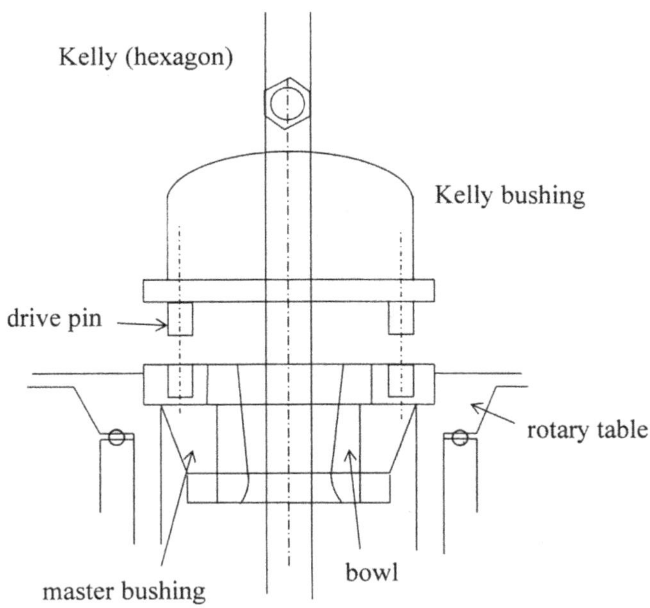

Figure 5-4. Rotary table drive.

A special sub, called a Kelly saver sub, is attached to the lower end of the Kelly to protect the lower pin-end from multiple connections with every section of drill pipe.

Another joint of drill pipe is stationed in a near-by hole, called the mouse hole. When it is time to add another pipe joint, the Kelly is broken out and made up to the separate pipe joint in the mouse hole. The new joint is then made up to the existing drillstring. When it is time to trip-out the entire drillstring, the Kelly (and Kelly Bushing) is disconnected is placed in the Rat Hole.

5.3.2 Top Drive

Modern drill rigs are now using top drives instead of rotary tables. In this case, power is supplied at the top end of the drillstring by use of an electric motor. The Kelly and rotary table are eliminated, however, the drillstring is still suspended in slips as before during pipe removal and insertion. The swivel and motor are mounted so as to move along guide rails directing pipe in and out of the well bore (Figure 5-5). Single joints of drill pipe are added as before, but pipe connections are made by robotic mechanisms.

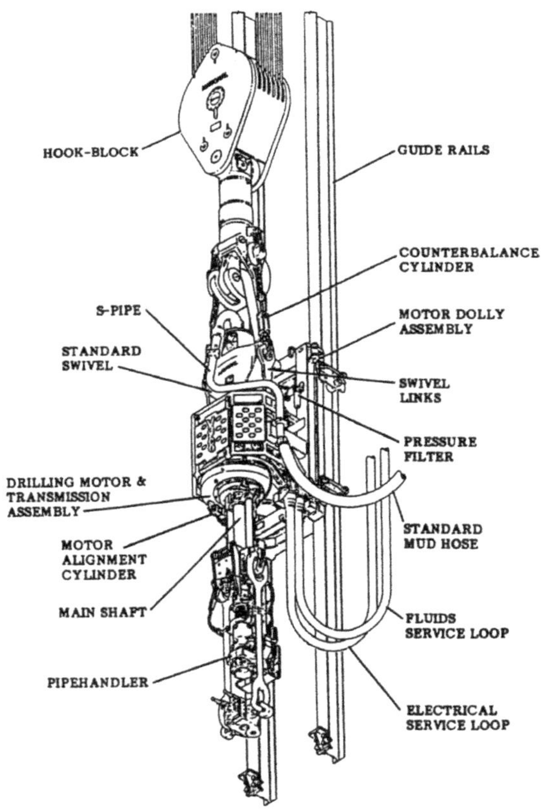

Figure 5-5. Top drive (Courtesy of NOV).

Pull-out or slack-off friction forces can sometimes be excessively high to the point of getting the pipe stuck. The standard practice to minimize pipe sticking especially while tripping is to use top drives to keep drillstrings rotating during tipping. Experience shows that pipe rotation, i.e., shearing contact friction forces in the tangent direction, reduces contact friction forces in the longitudinal direction.

Rotary power is typically transferred to drillstrings through the Kelly by means of the rotary table. Power is delivered to rotary tables by electric motors in the form of torque and rotary speed. Following conventional methods of rotary drilling, a single 30 ft joint of drill pipe is added to a drillstring by:

- setting the drillstring in the slips
- breaking out the Kelly saver sub
- attaching the Kelly onto a single 30 ft joint, which has been place in the mouse hole
- raising the Kelly assembly and attaching the new single onto the drillstring, which is hanging in the slips
- removing the slips
- circulating back to bottom and continue drilling

The top drive replaces the Kelly, Kelly bushing and rotary table. Power is transmitted to drillstrings by an electric motor, which travels with the top drive assembly. Triples or 90 ft sections of drill pipe, which have been previously racked in the derrick, can be drilled down without interrupting drilling. The ability to handle 90 ft sections of drill pipe while tripping in and out has distinct advantages over the conventional rotary table approach. Major advantages are:

- reduced connection time with both drilling and tripping
- ability to circulate and rotate out of the hole to reduce drag on the drillstring, reducing the chances of getting stuck
- ability to drill through bridges and tight spots without picking up the Kelly (adding pipe)
- helps control tool face orientation in directional drilling by capturing the trapped torque in drillstrings over a 90 ft interval as opposed to a 30 ft interval ---- reduces downhole motor orientation activity
- reduces the number of connections required by the rig crew and thus improves rig safety
- Motor drive can be calibrated against tong torque gauges as a means to quantify rotary torque

Top drives greatly reduce friction in high angle directional drilling by allowing pipe rotation while pulling pipe out of the well bore. They are essential in the drilling of extended reach wells.

Five basic subsytems are

1. Drilling motor and swivel assembly
2. Guide dolly assembly
3. Pipe handler assembly
4. Counterbalance system
5. Top drive control system

A drilling motor and swivel assembly contain and electric motor, gear drive and a main shaft, which connects directly onto a standard swivel. This assembly while supported by the swivel and traveling block is mounted on a guide dolly. The guide dolly is constrained to move up and down on vertical guide rails that as are rigidly attached to the rig. This assembly also contains an air brake capable of developing 35,000 ft-lb of static braking torque at the output shaft. One motor assembly is capable of developing 30,000 ft-lb of continuous torque at speed up to 175 rpm and can generate intermittent torque as high as 41,500 ft-lb. This assembly is capable of generating torque high enough to make up tool joints.

A pipe handler connects and disconnects a stand of pipe from the drilling motor assembly. Mechanical pipe handlers are designed to break-out pipe in the derrick at any height with torque capability up to 60,000 ft-lb.

The counterbalance system provides a 6-in. cushioned stroke to prevent damage to tool joint threads while making or breaking connections. Each of these operations can be controlled from a console located at the rig floor.

5.3.3 Physical Properties of Drill Pipe

Dimensions of three sizes of standard drill pipe are listed in Table 5-2. Each of the listed pipe sizes are made in different weights as dictated by the ID dimensions. Wall thickness for each can easily be determined as well as the weights (lb/ft) using the density of steel (490 lb/ft) and cross-sectional area. However, pipe weight is expressed in terms of nominal weight, which includes the weight of the connectors. For example, 5-in. (19.50 lb/ft) drill pipe has an ID of 4.276 in. and a plain tubular weight of 17.93 lb/ft. Drill pipe is identified by size, weight, grade and class.

Table 5-2. Dimensions of standard drill pipe.

OD (in)	ID (in)	Plain Wt (lb/ft)	Nominal Wt (lb/ft)
4½	3.958	12.24	13.75
	3.825	14.98	16.6
	3.640	18.69	20.00
	3,50	21.36	22.82
5	4.408	14.87	16.25
	4.276	17.93	19.50
	4.000	24.03	25.60
5½	4.892	16.87	19.20
	4.778	19.81	21.90
	4.670	22.54	24.70

- Size — nominal outside diameter of pipe
- Weight — weight per foot of drill length, including weight of connections
- Grade — yield strength of pipe material (E75, X95, G105, S135, numbers refer to yield strength)
- Class — New, Premium, Used

For example, 5 in. (19.50 lb/ft), Grade X95, New.

Table 5-3 lists the load and torque capacity of these standard pipe sizes corresponding to a yield strength of E75. Load and torque capacity of the other three grades can be calculated as shown below. Considering 5 (19.50) and grade E75 drill pipe, the pull capacity is 395,595 lbs.

Pull capacity is determined as follows:

$$\sigma_{yld} = \frac{F_{yld}}{Area}$$
$$Area = \frac{\pi}{4}\left(5^2 - 4.276^2\right) \tag{5-17}$$
$$F_{yld} = (5.2760)75{,}000$$
$$F_{yld} = 395{,}595 \text{ lbs}$$

Shear strength, τ_{yld}, is a determined from

$$\tau_{yld} = 0.577\sigma_{yld} \tag{5-18}$$

which is based on the von Mises energy of distortion criteria of failure. σ_{yld} is an experimental value while τ_{yld} is a calculated value.

Table 5-3. Tension and torque capacity of standard drill pipe (Grade E75).

OD (in)	Nominal Wt (lb/ft)	Tension (lbs)	Torque (ft-lb)
4½	13.75	270,034	25,907
	16.6	330,556	30,807
	20.00	412,358	36,901
	22.82	471,239	40,912
5	16.25	328,073	35,044
	19.50	395,595	41,167
	25.60	530,144	52,257
5½	19.20	372,181	44,074
	21.90	437,116	50,710
	24.70	497,222	56,574

The maximum allowable torque is determined as follows:

$$\tau_{yld} = \frac{T_{yld}}{S}$$

$$S = \frac{J}{c} \qquad \text{(torsion section modulus)}$$

$$T_{yld} = (11.415)75{,}000(0.577) \qquad (5\text{-}19)$$

$$T_{yld} = 493{,}984 \text{ in.-lbs}$$

$$T_{yld} = 41{,}165 \text{ ft-lbs} \quad \text{(checks with Table 5-3)}$$

Tension and torque limits for the other three grades (X95, G105, S135) can be determined by the above equations as well.

The collapse and burst strength of drill pipe are also given in [1,5]. For example, limiting pressures for collapse and burst for 5 in. (19.50) Grade E drill pipe, are listed as 9962 psi and 9503 psi, respectively. Drilling mud pressure at the top of drill pipe is near its maximum except for losses in the surface equipment (standpipe, hose, swivel, Kelly). Fluid pressure may be near 3000 psi. Hook load is also maximum at the top as well as the torque. Burst pressure limit should be checked from the tables and if there is a concern, a tri-axial stress analysis should be conducted. Burst pressure limits should be checked before conducting drill stem tests.

So far, the discussion has dealt with physical properties of drill pipe tubulars. Joints of drill pipe contain end connectors (tool joints) that are made separately and then attached to the tubulars. The yield strength of tool joint material is 120,000 psi, which is higher than Grades E75, X95 and G105.

A pin connector is attached to one end, while a box connector is attached to the other end. The ends of pipe sections are made thicker by forging upsets for

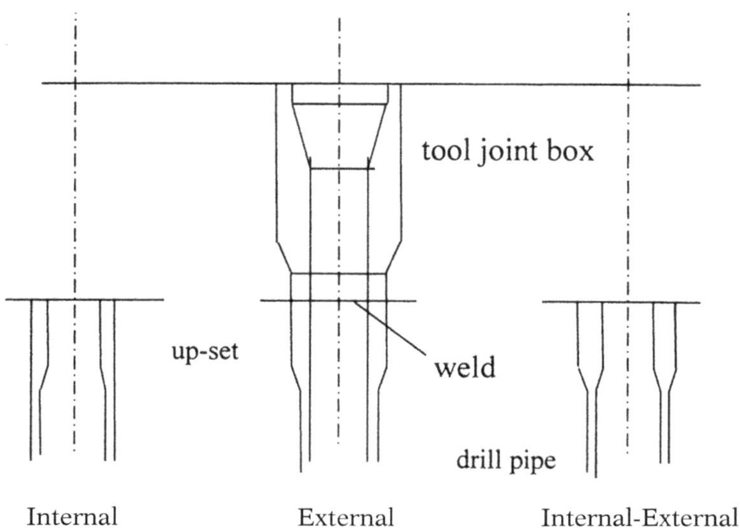

Figure 5-6. Three types of drill pipe up-sets.

the purpose of attaching the connectors. The upsets provide added thickness for welding the tool joint (pin and box) to the pipe body for a stronger connection. There are three types of up-sets

- IU — Internal Upset
- EU — External Upset
- IEU — Internal/External upset

The choice of upset must be compatible with connection type, i.e., IF, XH, FH. Connection type dictates the type of upset. For example, an NC50 (IF) requires an external upset. Each of these up-set types is illustrated in Figure 5-6.

5.3.4 Selecting Drill Pipe Size and Grade

Drill pipe is subjected to various types of loads during any drilling operation. Common loads are: direct pull, torque, bending and internal pressure. Drill pipe selection (size, grade) is typically based on direct pull type loading with a factor of safety to account for uncertainty. Other types of loading, such as dog-leg bending, should be checked for stress reversal type fatigue, especially if the dog legs are severe.

Normally drillstrings are designed to an allowable load of 90% of yield strength of drill pipe, so the allowable load is

$$H_a = 0.9H_{yld} \tag{5-20}$$

where

H_{yld} = direct pull causing material yielding
H_a = maximum allowable pull force, i.e., buoyed weight of drillstring plus margin of overpull

The corresponding factor of factor (FS) is

$$FS = \frac{H_{yld}}{H_a} = 1.11 \tag{5-21}$$

The level of the safety factor (FS) is subject to modification depending on application. High FS means higher weight and cost, while low FS could lead to unexpected failure. Factors of safety are used to cover uncertainty in the design, such as: material properties, applied loads, accuracy of stress models, etc.

The allowable pull force relates to hook load and MOP by

$$H_a = H + MOP \tag{5-22}$$

where

H = working hook load (based on buoyed weight of drillstring at maximum depth)
MOP = margin of overpull, in case of stuck pipe or excessive friction

The allowable load (H_a) is the basis for selecting drill pipe size and grade for a given Class of pipe.

Bringing these equations together gives

$$H_a = 0.9H_{yld} = \left(w_p L_p + w_c L_c\right) BF + MOP$$

This equation will be applied to the following two examples.

5.3.4.1 *Select Pipe Grade for a Given Pipe Size*

As an example, consider the following:

Total depth = 14,000 ft
Mud weight at TD = 15 ppg
Margin of Overpull = 100,000 lbs
Drill Pipe – 5 (19.50); Premium class
Drill Collars = 6¾ in. OD × 2 13/16 in. ID, 750 ft

The problem is to select the lowest grade of drill pipe that will reach TD and accommodate the specified MOP.

The static hook load (working hook load) at TD is

$$H = \left(w_p L_p + w_c L_c\right) BF$$
$$H = \left[(19.5)13,250 + (100)750\right](0.771)$$
$$H = 257,032 \text{ lbs}$$

The selected grade must have the strength to support

$$H_a = H + MOP$$
$$H_a = 257,032 + 100,000$$
$$H_a = 357,032 \text{ lbs}$$

To account for the FS pipe grade selection must satisfy

$$H_{yld} \geq 1.1(357,032)$$
$$H_{yld} \geq 396,702 \text{ lbs}$$

Grade X95 (Premium Class) shas a yield load limit of 394,612 lbs. Grade G105 has a yield load limit of 436,150 lbs. Grade G105 would be the proper choice for drill pipe grade because 436,150 is greater than the required 396,702 lb yield strength. In this case, the FS increases from FS = 1.1 to 436,150/357,032 = 1.22.

5.3.4.2 *Determining Maximum Depth for Given Pipe Size and Grade*

Now consider a situation requiring maximum depth capability of a given pipe size, class and grade. Other specifications are:

Drill Pipe size = 4½- in. (16.6 lb/ft), Grade E75, Premium Class
MOP = 57,000 lbs

Drill Collar = 6½ in. OD × 2½ in. ID, 700 ft long
Mud weight = 12 ppg

Recall

$$H_a = 0.9H_{yld}$$
$$H_a = H + MOP$$
$$H = 0.9H_{yld} - MOP$$

The maximum allowable depth is based on the allowable hook load, H. From the API RP 7G Standard [1], the pull strength for the specified pipe Grade and Class is

$$H_{yld} = 260{,}165 \text{ lbs}$$

The working hook load, H, is

$$H = 0.9(260{,}165) - 75{,}000 = 159{,}149 \text{ lbs}$$

Maximum hole depth is determined from

$$H = [w_{dp}L_{dp} + w_c L_c]BF$$

where

H = 159,149 lbs
BF = 0.817
w_{dp} = 16.6 lb/ft
w_c = 96 lb/ft (see Table 5-6)
L_c = 700 ft

Substituting these numbers into the above equation gives L_{dp} = 7687 ft. The maximum depth capability is

$$TD = L_{dp} + L_c = 8387 \text{ ft}$$

The slip limit load for 4½ (16.6) Grade E from Table 5-4 is

$$H_{SL} = 262{,}000 \text{ lbs}$$

Since $H < H_{SL}$, the slip crushing load at TD is safe.

5.3.5 Support of Drillstring in the Master Bushing

When drill pipe is added, or removed, the entire drillstring is supported in the master bushing by means of "slips" (Figure 5-7). The mechanical wedging of the slips creates crushing pressure around the pipe near the box end as shown while support the drillstrintg. The slips have small teeth, which create high friction or griping force.

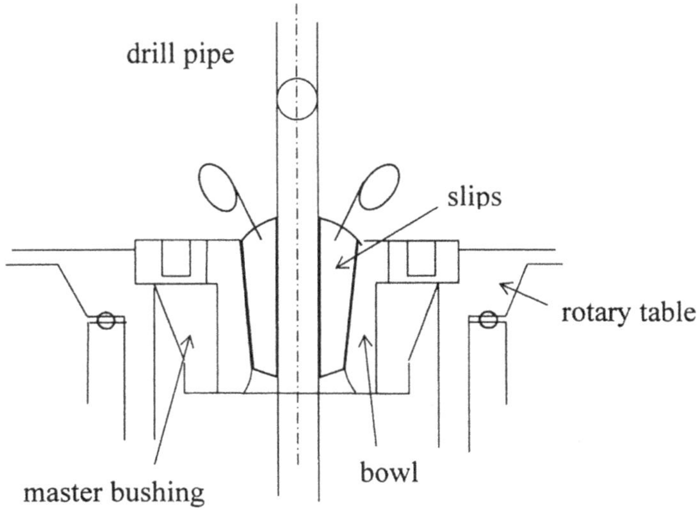

Figure 5-7. Slips holding drill pipe in the master bushing.

At total depth drill pipe must not only withstand the hook load, H, but also the slip crushing load. Slip crushing load will be due to the buoyed weight of the drillstring suspended from the slips.

The stress distribution in the drill pipe due to the combined loadings of suspended buoyed weight and the slip crushing load is complex. The freebody diagram (Figure 5-8) of the slips gives two equations of equilibrium.

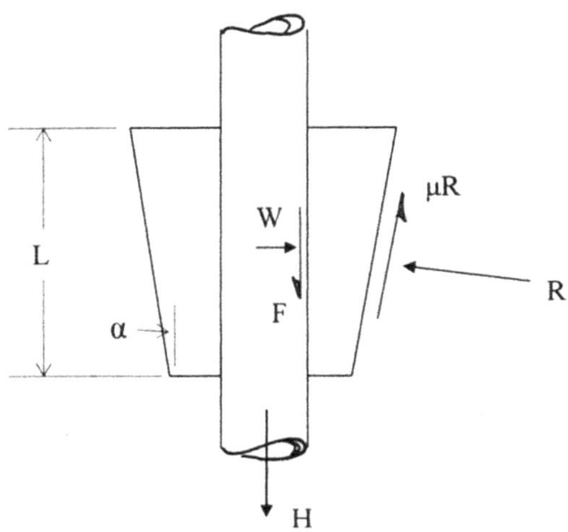

Figure 5-8. Freebody diagram of drill pipe slips.

$$\sum F_x = 0$$
$$W = R(\cos\alpha - \mu \sin\alpha) \tag{5-23}$$

and

$$\sum F_y = 0$$
$$F = R(\sin\alpha + \mu \cos\alpha) \tag{5-24}$$

where

$W = \int_0^L w(z)dz$

$F = \int_0^L \mu w(z)dz$

w = radial force per unit length

μ = coefficient of friction

The hook force, H, is supported by the grippers on the inside surface of the slips. The gripping force, F, is equal to the hook force, H.

Combining these equations gives

$$\frac{W}{F} = \frac{1 - \mu \tan\alpha}{\tan\alpha + \mu} \tag{5-25}$$

The ratio of W to F is the transverse load factor (K). The magnitude of K depends on the slip angle and the coefficient of friction between slips and master bushing. If the coefficient of friction, μ, is replaced by tan ψ, where ψ is the friction angle

$$K = \frac{1}{\tan(\alpha + \psi)} \tag{5-26}$$

For a standard API slip taper of

$$\tan\alpha = \frac{1}{6} \tag{5-27}$$

K = 4.37 when μ = 0.06 (clean, greased surfaces) and K = 1.3 when μ = 0.5 (dirty or dry surfaces). Neither case is representative of typical drilling. Industry has generally accepted a transverse load factor of K = 3 as typical for most drilling situations. This means that the total transverse load (W) on the drill pipe is about 3 times the axial pipe load (H) set in the slips.

Assuming this transverse load is distributed uniformly around the circumference of the pipe and along the slip length so that the average slip pressure on the pipe is

$$p_{ave} = \frac{W}{LC} = \frac{KH}{LC} \tag{5-28}$$

Table 5-4. Maximum safe hook load for new drill pipe (condensed from ref [6]).

Drill pipe OD, in	Weight lb/ft	Recommended safe hook load, x 1,000 lbs Grade X95	Grade G105
		K = 3	K = 3
4½	13.75	270	299
	16.6	331	366
	20	413	457
5	16.25	320	353
	19.5	386	426
	25.6		
5½	19.2	353	390
	21.9	414	458
	24.7	471	521

where

p_{ave} = average crushing pressure against drill pipe
H = buoyed weight of drillstring hanging from the slips
L = slip length
C = circumference of pipe (actually the slips are made up of three sections)

According to Spiri and Reinhold [6], drill pipe at the lower end of the slips is subjected to triaxial state of stress caused by combined loadings of slip pressure and axial load set in the slips. The von Mises criteria of failure indicates that the critical element is on the inside surface of the pipe. Critical slip hanging loads for different drill pipe sizes are given in Table 5-4.

Note that in the previous example, Grade G would be required to reach a total depth of 14,000 ft and still have a 100,000 lb overpull capability. The maximum static load to be set in the slips is 257,032 lbs. According to Table 5-4, 5 (19.5) Grade G105 drill pipe can withstand 426,000 lb of pipe load in the slip before yielding. The drill pipe in this case would not fail from slip crushing loads.

5.3.6 Rotary Shouldered Drill Pipe Connections

Rotary shouldered connections are designed for quick make-up and break-out. The basis of rotary shouldered connections are tapered threads, which force mating shoulders together to make the connection structurally sound and establish a mechanical seal.

Tool joints are, in general, stronger than the drill pipe tubular. Failure of tool joints is usually associated with improper make-up torque. If make-up torque is too low, the shoulders may not seal the inside fluid pressure.

Tool joints are designed with a clearance between the crest and root of each thread to provide a channel for thread compound and impurities when the tool joints are made up. API standards require a tight metal seal between the rating shoulders of the box and pin to seal pump pressure across tool joints. Originally, tool joints were attached to drill pipe tubulars by threads and flash welding. Today, tool joints are joined to drill pipe tubulars by friction welding. The pipe body is fixed while the tool joint is rotated and pressed into the pipe up-set. The heat generated by friction at the contact surfaces is high enough to bond the tool joint to the pipe up-set. The excess metal is machined off after the welding is complete.

5.3.6.1 Pitch Surface

The pitch surface of tool joints is tapered for fast made-up and break out. The taper is 2 in. per foot on the diameter or 1:12 taper on the radius. Pitch surfaces of all tool joint connections lie on the same conical surface. The size of a connection depends on the diameter of the pitch surface at the gage point; the gage point is located ½ in. from the pin shoulder (Figure 5-9).

API established the Number Connection nomenclature in order to alleviate confusion between the various types of tool joints such as internal flush (IF), full hole (FH), and extra hole (XH). The number used with the NC nomenclature is "the pitch diameter of the pin thread at the gage point rounded to units and tenths of inches". For example, NC 38 has a pitch diameter of 3.808 in.

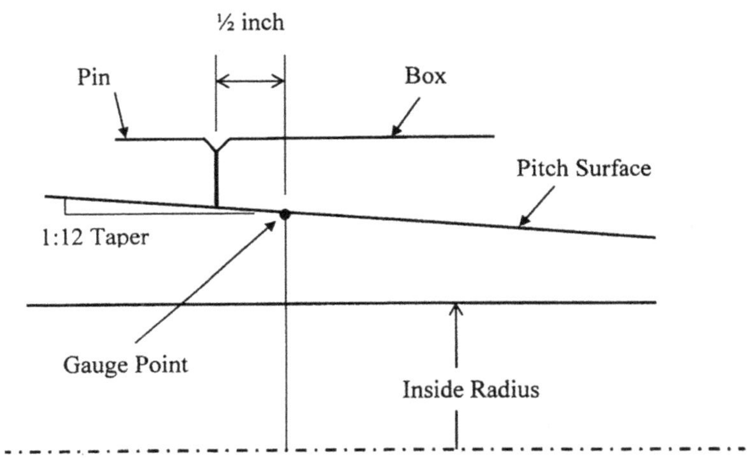

Figure 5-9. Pitch surface of rotary shouldered connection.

5.3.6.2 Thread Form

Drill pipe tool joints and drill collar connections have the same thread form; however, the pitch diameter of each connection may be different and they have to be identified in each case. Typical features of the thread form are:

- 4 threads per inch (TPI)
- 1:6 taper on the diameter or 1:12 taper on the radius
- Root radius is 0.038 in.
- Total internal thread angle is 60 deg.

The root of the threads in both pin and box has a radius of 0.038 in., which is larger than other tool joint thread forms. This larger radius reduces stress concentration at the root of the threads and makes it easier to cold work the threads in this critical area. Many rotary shouldered connections are interchangeable; see ref [1] for a list of connections that are interchangeable.

Because there are many pipe sizes and "upset" possibilities, there are many possible sizes of connections and not all are interchangeable. This can be a bit confusing. There are, however, a few features that are common to all rotary shouldered connections.

5.3.6.3 Makeup Torque

Drill pipe connections are made while the drillstring is hanging from the slips as shown in Figure 5-10. A joint of pipe to be added is stationed in the mouse hole located near the rotary table. The Kelly is disconnected from the drillstring and moved over to the stationed pipe joint. They are made up while still in the mouse hole and then lifted above the box end of the exposed drill pipe, directly above the slips. The new joint of pipe is made-up according to specified torque requirements. After the drillstring is lifted out of the bowl and the slips are removed, the Kelly and Kelly bushing are set in place then drilling continues.

Make up torque is based on the screw jack formula

$$T = \frac{F}{12}\left[\frac{p}{2\pi} + \frac{R_t f}{\cos\theta} + R_s f\right] \tag{5-29}$$

where

T = torque, ft-lbs
p = lead of threads, in. (p = ¼ in. for 4 thds/in.)
R_t = average mean radius of threads, in.
f = coefficient of friction (~0.08)
R_s = mean radius of shoulder, in.
θ = ½ of included angle of thread
F = contact force between mating shoulders, lbs

The *screw jack formula* shows that total make-up torque is distributed among three areas. One component of the torque drives the mating shoulders together

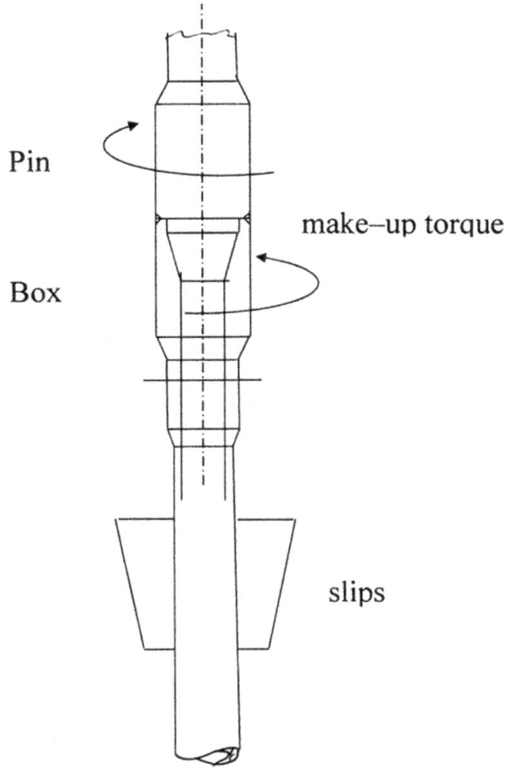

Figure 5-10. Tool joint makeup torque.

creating the contact force. If there were no friction in the connection all of the applied torque would create this force, which, in this special case, is determined by

$$2\pi T(12) = Fp \quad \text{(energy balance)} \tag{5-30}$$

The second component of torque is the torque required to overcome friction in the threads. The third torque component overcomes friction in the shoulder.

If tool joints are made up too tight, the pin could be over stretched. If tool joints are made up too low, threads can be exposed to washouts or bending fatigue. In practice, tool joints are generally made up too low. According to one drilling engineer, "if tool joints are made up too low, they will not make up downhole. On the other hand, drill collar connections will make up downhole. If drill collar connections are difficult to break out, it usually means they were made up too low initially."

The makeup force or internal contact force between the two mating shoulders directly affects the structural integrity and pressure sealing capacity of the tool joint. Consider, for example, the torque components in a new NC50 (6⅜ in. × 3¾ in.) tool joint using the following set of numbers:

p = 0.25 in.
f = 0.08
θ = 30 deg.
R_t = 2.385 in.
R_s= 2.922 in.

Substituting these numbers into Eq. (5-29) gives

$$T = \frac{F}{12}\left[\frac{0.25}{2\pi} + \frac{2.385(0.08)}{0.866} + \left(2.922\right)0.08\right]$$

$$T = \frac{F}{12}\left(0.0398 + 0.2203 + 0.233\right)$$

$$T = \frac{F}{12}\left(0.4931\right) \text{ ft-lbs}$$

This calculation shows that only 8% of makeup torque drives the shoulders together. The remaining 92% is used to overcome friction. The coefficient of friction of thread dope is, therefore, critical. Make-up torque recommendations are based on a coefficient of friction of f = 0.08.

If the actual coefficient of friction is less than 0.08, contact force between the mating shoulders will be higher than expected when the recommended API makeup torque is developed. This condition could over stretch the pin or damage the shoulder area. If the actual coefficient of friction is greater than 0.08, contact force between the mating shoulders will be lower than expected when the recommended makeup torque is developed. This condition allows bending stresses to reach the pen causing fatigue damage and creates an inadequate pressure, leading to a washout through the threads.

5.3.7 Strength of Rotary Shouldered Connections

The tensile strength of tool joints is based on a 120,000-psi stress level at the cross sectional area at the root of the thread located ⅝ in. from the shoulder in either pin or box, whichever is weakest. In the case of a new NC50 (6⅜ in. × 3¾ in.) tool joint connection, the makeup force required to yield the pin at the specified cross section is 939,095 lbs; this number can also be viewed as the force between the pin/box shoulders. This force level is generated by a makeup torque of 37,676 ft-lbs. This tabulated torque can be verified by the last equation in the previous section.

$$T = \frac{F}{12}(0.4931) = \frac{939{,}095}{12}(0.4931) = 38{,}589 \text{ ft-lb}$$

The slight difference being the input values of R_t and R_s.

Following the API RP 7G standard, allowable makeup torque for new pipe is limited to 50% of yield torque or 60, 000 psi critical stress in either pin or box. Shoulder contact force in the above example one half of 944,000 lbs or 472,000 lbs. The reason for restricting makeup torque to 50% of yield torque is to allow for additional tension stress caused by hook forces across tool joints. The initial internal forces change when hook loads are applied across tool joints as will be explained in the next section.

Recommended makeup torque for new tool joints is 50% of yield torque. The recommended makeup torque for a new NC 50 (6⅜ × 3¾) tool joint is therefore, 18,900 ft-lbs. The recommended makeup torque for worn tool joints can be established by taking 60% of the lowest yield torque.

5.3.7.1 Preload and Shoulder Separation

The API recommended preload in the shoulders of both box and pin for new pipe/tool joints is 50% of the force required to yield either box or pin. This preload is generated by makeup torque. For used drill pipe tool joints it is 60%. This level of preload allows for addition pin loading due to direct pull, such as drillstring weight and over-pull. The following explains how pin force and box shoulder force change when external pull forces are applied across tool joints. The problem is statically indeterminate, so deflection equations are required along with the static forces.

For the sake of discussion let the pin and box be represented by a cylindrical bar and tube as shown in Figure 5-11a. The tube is slightly longer than the bar. If the two heights are brought together and fixed, the preload (tension) in the pin is equal in magnitude to the preload (compression) in the box (Figure 5-11b). The pin is extended by

$$\delta_p = \frac{FL_p}{EA_p} = \frac{F}{k_p} \tag{5-31}$$

The box is compressed by

$$\delta_b = \frac{FL_b}{EA_b} = \frac{F}{k_b} \tag{5-32}$$

where

F = Preload in pin and box after make-up, lbs
A_p, A_b = Cross sectional area of pin and box at the pitch point, in.^{2}
L_p, L_b = Active length of pin and box (~0.75 in.)
k_p, k_b = Effective spring constants in pin and box, lbs/in.
E = Modulus of Elasticity, psi

Now assume that a pull force, P, is applied across the tool joint (Figure 5-11c). The pin is stretched by $\Delta\delta_p$ while the box is relaxed by $\Delta\delta_b$: note that $\Delta\delta_p = \Delta\delta_b = \Delta\delta$. Therefore, the corresponding increase in the pin force is related to the corresponding reduction in the box force by

$$\frac{\Delta F_p}{k_p} = \frac{\Delta F_b}{k_b} \tag{5-33}$$

where ΔF_p and ΔF_b are changes in pin and box forces.

Using the freebody diagram shown in Figure 5-11d,

$$P = F_p - F_b \tag{5-34}$$

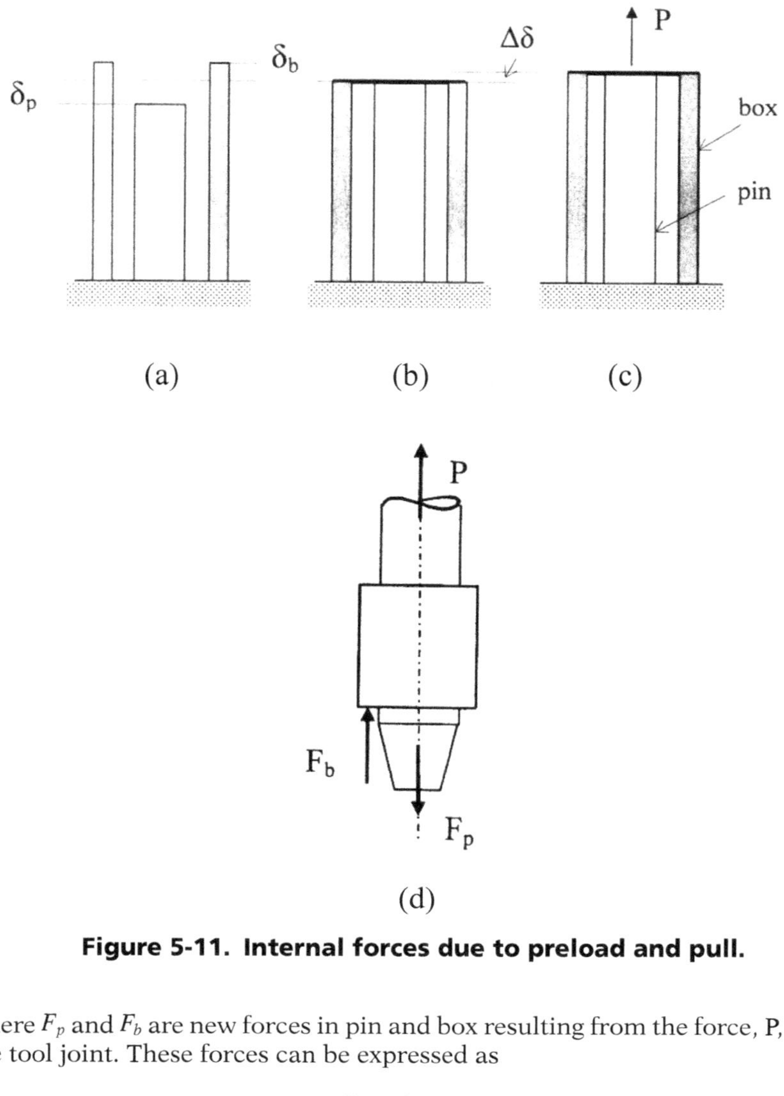

Figure 5-11. Internal forces due to preload and pull.

where F_p and F_b are new forces in pin and box resulting from the force, P, across the tool joint. These forces can be expressed as

$$F_p = F + \Delta F_p \tag{5-35}$$

$$F_b = F - \Delta F_b \tag{5-36}$$

Bringing Eqs. (5-33), (5-34), (5-35), and (5-36) together gives

$$\Delta F_p = \frac{k_p}{k_p + k_b} P \tag{5-37}$$

Therefore, the new force in the pin due to both preload and direct pull is

$$F_p = F + \frac{k_p}{k_p + k_b} P \qquad (5\text{-}38)$$

where

P = Pull force across tool joint
F = Initial preload in box/pin
F_p = New force in pin

Similarly, the new force in the box is

$$F_b = F - \frac{k_b}{k_p + k_b} P \qquad (5\text{-}39)$$

Two load conditions are of interest; 1) magnitude of P, which causes shoulder separation and 2) possible yielding in the pin caused by P.

5.3.7.2 *Shoulder Separation*

Shoulder separation occurs when $F_b = 0$. Using Eq. (5-39) with $F_b = 0$ gives

$$P = \frac{k_p + k_b}{k_b} F \qquad (5\text{-}40)$$

which is the pull force across the tool joint required for shoulder separation.

5.3.7.3 *Possible Yielding in the Pin*

Rewriting Eq. (5-38) in terms of yield force in the pin

$$\left(F_{yld}\right)_{pin} = F + \frac{k_p}{k_p + k_b} P \qquad (5\text{-}41)$$

The force across a tool joint causing yielding is

$$P = \frac{k_p + k_b}{k_p}\left(F_{yld} - F\right) \qquad (5\text{-}42)$$

This equation applies only if there is no shoulder separation.
Combining Eqs. (5-38) and (5-40) and assuming separation $\left(\Delta\delta = \delta_b\right)$, F = 0 or

$$F_P = P \qquad (5\text{-}43)$$

as expected.

As an example, consider the maximum and minimum preload conditions for a new NC 50 (IF) 6¾ × 3¾ tool joint. Note that

$$\frac{k_b}{k_p + k_b} = \frac{A_b}{A_p + A_b} \qquad (5\text{-}44)$$

and

$$\frac{k_p}{k_p + k_b} = \frac{A_p}{A_p + A_b} \qquad (5\text{-}45)$$

The dimensions for calculating these areas are:

Pitch Diameter = 5.0417 in.
Box OD = 6¾ in.
Pin ID= 3¾ in.

From these dimensions, A_p = 8.93 in.^{2} and A_b = 15.83 in.^{2} and the active pin and box lengths ($L_p = L_b \sim$ ¾ in.) are the same. Then

$$\frac{k_b}{k_p + k_b} = \frac{A_b}{A_p + A_b} = \frac{15.83}{8.93 + 15.83} = 0.639$$

$$\frac{k_p}{k_p + k_b} = \frac{A_p}{A_p + A_b} = \frac{8.93}{8.93 + 15.83} = 0.361$$

The pull force required to separate the pin/box shoulder is [using Eq. (5-40)]

$$P = \frac{1}{0.639}(472{,}000) = 738{,}700 \text{ lbs}$$

This force is felt by the pin at separation. It is less than the 944,000 lb required to yield the tool joint. This number, P, far exceeds the strength capacity of each of the four pipe grades (4½ in.), which means that the drill pipe body would fail before tool joint shoulder separation.

5.4 CONFIGURING THE DRILL COLLAR SECTION

The drill collar section is made up of pipe that is larger and much heavier than drill pipe. Its main purpose is to generate drilling force, called weight-on-bit (WOB) or simply bit weight (BW). Its length depends on expected BW, but in general its length is about 700 ft. Drill bits are attached directly to the bottom end of the drill collar section. Drill Collars may contain other tools, such as stabilizers, motors/turbines, and measurement-while-drilling (MWD) equipment, which are located directly above the drill bit. The combination of any or all of these tools is called the Bottom-Hole-Assembly (BHA). The BHA (~150 ft in length) is perhaps the most important section of the entire drillstring because it controls ROP and well bore path. BHA equipment performance, cost and reliability are key factors in footage cost.

5.4.1 Physical Properties of Drill Collars

Drill collars are produced in two grades of steel according to Table 5-5. Drill collars are made of high strength steel because connections are machined directly into the drill collar body.

Table 5-5. Physical properties of drill collars.

Drill collar OD range (in.)	Minimum yield psi	Minimum tensile strength psi
3¼ thru 6⅞	110,000	140,000
7 thru 10	100,000	135,000

Drill Collars are made in standard sizes. A list of OD-ID combinations along with weight per foot of each is given in Table 5-6. Note that weight per foot of drill collar length is important. A useful relationship, which relates pipe density to weight per unit length is

$$\gamma_{stl} A \Delta L = w \Delta L$$

$$w = A\gamma_{stl}$$

$$w = \frac{\pi}{4}\left(D_o^2 - D_i^2\right)\gamma_{stl}$$

$$w = \frac{\pi}{4}\left(D_o^2 - D_i^2\right)490\left|\frac{1\ \text{ft}^2}{144\ \text{in}^2}\right| = 2.6725\left(D_o^2 - D_i^2\right)$$

This equation was used to generate the data in Table 5-6. For example, the weight of (6 in. OD × 2 in. ID) is calculated to be

$$A = \frac{\pi}{4}\left(6^2 - 2^2\right) = 25.1327\ \text{in.}^2$$

$$w = 25.1327\ \text{in}^2\left|\frac{1\ \text{ft}^2}{144\ \text{in}^2}\right|490\frac{\text{lb}}{\text{ft}^3}$$

$$w = 85.521\ \text{lb/ft}$$

This number agrees with Table 5-6.

5.4.2 Length of Drill Collar Section

The length of drill collars depends on the expected force to be applied to the drill bit (WOB). Because of high compressive drill bit forces, the drill collar section buckles and produces bending at the connections. Makeup torque must be high enough to keep the shoulders engaged in order to protect the pin threads from cyclic bending and fluid erosion (called mud cutting).

In practice WOB is established by slacking off the desirable bit force from the hook load at the surface. A basic question is; How long should the drill collar section be to prevent drill pipe buckling for a given WOB? This is important because of the large difference in structural stiffness between drill pipe and drill collars. From a structural point of view, stiffness attracts bending moment and resulting stresses that can damage drill pipe especially under stress reversals caused by drillstring rotation.

Table 5-6. Standard drill collar sizes and weight (lb/ft).

Drill collar OD, in.	Drill collar ID, in. 2	2.25	2.5	2.75	3	3.25	3.5	3.75	4
5	56	53	50						
5.25	63	60	57						
5.5	70	67	64	61					
5.75	78	75	72	68	64	60			
6	86	83	80	76	72	68			
6.25	94	91	88	84	80	76	72		
6.5	102	99	96	93	89	85	80		
6.75	111	108	105	102	98	94	89		
7	120	117	114	111	107	103	98	93	88
7.25	130	127	124	120	116	112	108	103	98
7.5	140	137	134	130	126	122	118	113	108
7.75	150	147	144	140	136	132	128	123	118
8	160	158	154	151	147	143	138	133	128
8.25	171	168	165	162	158	154	149	144	139
8.5	182	180	176	173	169	165	160	156	150
9	206	203	200	196	192	188	184	179	174
9.5	231	228	224	221	217	213	208	204	198
10	257	254	251	247	243	239	235	230	224
11	313	310	307	303	299	295	291	286	281
12	374	371	368	365	361	357	352	347	342

The location of the neutral point depends on bit force (WOB). If bit force is zero, the neutral point is located at the drill bit. As bit force is increased, the neutral point moves up the drill collars.

In practice, the neutral point is kept within the drill collars as a safety measure to prevent drill pipe buckling. The relation between WOB and distance to neutral point is determined as follows (refer to Figure 5-12)

$$WOB = W - B \tag{5-46}$$

$$WOB = wL_{np} - w_m L_{np} \tag{5-47}$$

$$WOB = wL_{np}\left(1 - \frac{w_m}{w}\right) \tag{5-48}$$

$$WOB = wL_{np}\left(1 - \frac{\gamma_m}{\gamma}\right) = wL_{np}BF \tag{5-49}$$

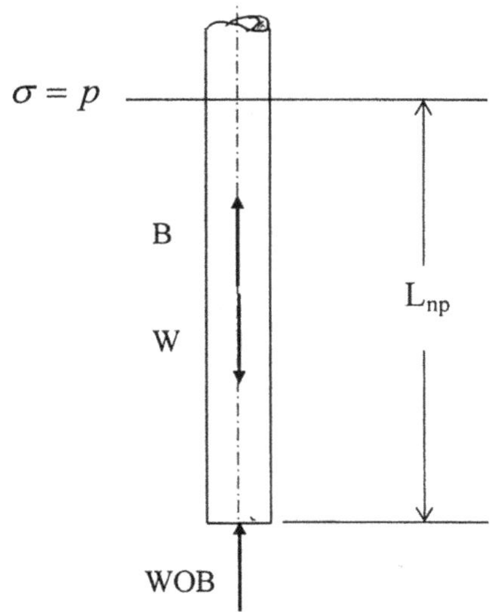

Figure 5-12. Freebody diagram of drill collar section.

where

w = weight per unit length of drill collars
w_m = weight of displaced mud by unit length of collars
γ = steel density (490 lb/ft^3 or 65.5 ppg)
γ_m = drilling mud density, ppg

The distance to the neutral point (or point of hydrostatic compression) from the drill bit is determined by

$$L_{np} = \frac{WOB}{wBF} \tag{5-50}$$

If the drill collars were separated at this location (L_{np}), neither the hook load nor WOB would change.

Drill collar length is generally computed by

$$L_{np} = 0.85L_c \tag{5-51}$$

which means that the neutral point is 85% of drill collar length. The extra length represents a safety factor against drill pipe buckling.

$$L_c = \frac{WOB}{0.85wBF} \tag{5-52}$$

As an example, assume

WOB = 50,000 lb
Collar size is [7 in. × 3 in. (107 lb/ft)]
γ_m = 12 ppg (BF = 0.817)

The distance to the neutral point is

$$L_{np} = \frac{50{,}000}{107(0.817)} = 572 \text{ ft}$$

and the length of the drill collar section is

$$L_c = \frac{50{,}000}{0.85(107)0.817} = 673 \text{ ft}$$

5.4.3 Drill Collar Buckling

Even though the neutral point is kept within drill collars, the collars themselves do buckle at some level of WOB.

During the late 1950s, Lubinski [4] began analyzing drill collar buckling by mathematically modeling drill collars as a long vertical beam-column under variable internal compression. Prior to this, little was understood about the effects of bit force, gravity, and even hydrostatic pressure on the lateral stability of drill collars. His landmark study provided understanding and insight into drill collar buckling, bit side force, side loads against the well bore and bending stresses.

Lubinski defines the condition of drill collar buckling in terms of a dimensionless unit designated as "m" where

$$m = \sqrt[3]{\frac{EI}{p}} \tag{5-53}$$

and (using Lubinski's notation)

E = modulus of elasticity, lb/ft^2
I = moment of inertia, ft^4
$p = w - w_m$ lbs/ft (also p = w BF)
m = length (feet) of one dimensionless unit

His results show that buckling of the first and second modes occurs when the neutral point is located 1.94 and 3.75 dimensionless units, respectively, above the drill bit.

$$\left(L_{np}\right)_1 = 1.94 \text{ m} \tag{5-54a}$$

$$\left(L_{np}\right)_2 = 3.75 \text{ m} \tag{5-54b}$$

Relating this formula to critical WOB

$$(WOB)_{cr} = (L_{np})_{cr} wBF \tag{5-55}$$

$$\frac{WOB_{cr}}{wBF} = 1.94\left(\frac{EI}{wBF}\right)^{\frac{1}{3}} \tag{5-56}$$

$$WOB_{cr} = 1.94\left(wBF\right)^{\frac{2}{3}}\left(EI\right)^{\frac{1}{3}} \quad \text{(first mode)} \tag{5-57}$$

$$WOB_{cr} = 3.75(wBF)^{\frac{2}{3}}(EI)^{\frac{1}{3}} \quad \text{(second mode)} \tag{5-58}$$

Figure 5-13 gives critical bit weights for two collar sizes, based on Eqs. (5-57) and (5-58). Considering the 7 × 3-in. drill collar size and a drilling mud density of 12 ppg mud

$$I = \frac{\pi}{4}\left[3.5^4 - 1.5^4\right] = 113.88 \text{ in}^4$$
$$E = 30\ 10^6 \text{ psi}$$
$$BF = 0.817$$

then by substitution into Eqs. (5-57) and (5-58) (watch units)

$$(WOB_{cr})_1 = 1.94(wBF)^{\frac{2}{3}}\left(\frac{EI}{144}\right)^{\frac{1}{3}}$$

$$(WOB_{cr})_1 = 1.94\left[107(0.817)\right]^{\frac{2}{3}}\left(\frac{EI}{144}\right)^{\frac{1}{3}}$$

$$(WOB_{cr})_1 = 10{,}980 \text{ lbs}$$

Similarly

$$(WOB_{cr})_2 = 21{,}225 \text{ lbs}$$

These points are identified in Figure 5-13.

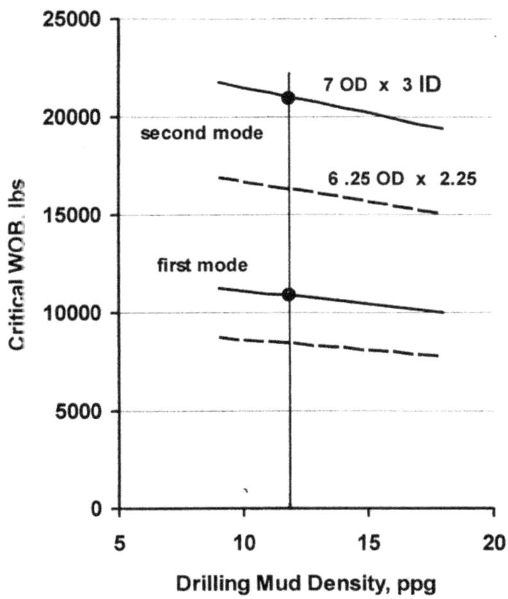

Figure 5-13. Critical WOB for two collar sizes.

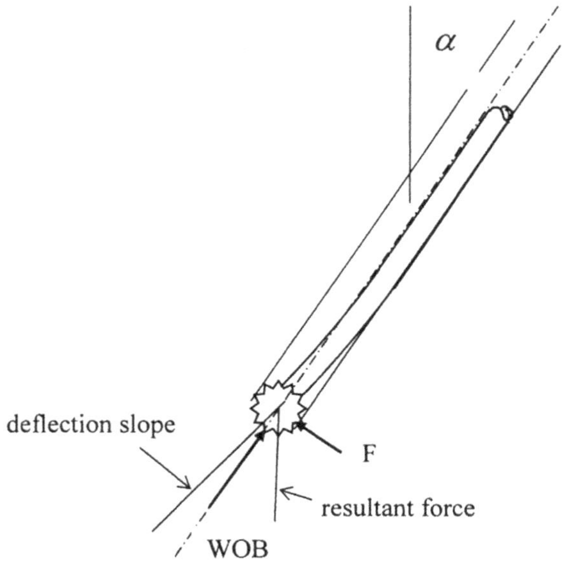

Figure 5-14. Resultant bit force.

Lubinski derived the equation of bending and showed how the solution is expressed in terms of a power series and Bessel functions. The assumption is made that the well bore is vertically straight. His analysis locates and quantifies the contact force between drill collars and well bore and predicts transverse reaction forces at the drill bit. He concludes ... "it is impossible to drill a straight vertical hole because of this side force on the drill bit". He explains well bore deviation in terms of drill bit side forces and shows that hole angle and the direction of the total bit force are, in general, different. This insightful conclusion provides the basis for current directional control technology.

A short time later, this work was extended by Lubinski and Woods [7] to show the effects of well bore inclination, α, on bit side forces (Figure 5-14). Their work also predicts the effect of formation type (isotropic and non-isotropic) on hole deviation. Graphs are given to show conditions that make bit force angle and hole inclination the same, which would be required to drill a straight hole along a given hole inclination. The data also defines drilling conditions that would either build or drop well bore angle.

5.4.4 Rotary Shouldered Drill Collar Connections

Rotary shouldered connections are also used to attach joints of drill collars. Makeup torque is equally important because the connections must be structurally strong and at the same time seal inside fluid pressures. These connections are machined directly into the drill collar body making them integral with collars. The order of magnitude of makeup torque in drill collar connections is comparable to drill pipe make up torque.

Because connections are machined directly in each drill collar, rotary shouldered drill collar connections cannot be a strong as the drill collar body. Therefore, the structural aspect of each connection and makeup torque is especially important.

The screw jack formula is also used to establish recommended makeup torque, which is based on a unit stress of 62,500 psi in the box or pin, whichever is weaker.

5.4.4.1 Bending Strength Ratio

Another important aspect of rotary shouldered drill collar connections is the relative bending stiffness of the pin and box. If, during the use of drill collars, the box is too thin the pin can force high local bending in the box and cause early failure. If the pin is to small the box becomes the dominant member. The bending strength ratio is defined as the ratio of the section modulus of the box to the section modulus of the pin. These section moduli are measured according to API standards (API RP 7G).

Experience has shown that a bending strength ratio of 2.50:1 balances the bending strength of box and pin. As drill collars wear and the outside diameter of a box reduces, the bending strength ratio (BSR) may reduce to a level that is undesirable. Therefore, BSR slightly higher than 2.50:1 may be selected, especially in abrasive formations.

5.4.5 Drill Collar Stabilization

Stabilizers are special tools that centralize drill collars at specific locations to establish a desired side load to drill bits. Stabilizer blades are roughly 2 ft long and allow fluid to flow upward in the annulus with minor flow restriction (Figure 5-15). Stabilized assemblies can build, hold, or drop hole angle, depending on stabilizer number and spacing [8,9].

The structural mechanics is very similar to the analysis of a multi spanned beams. However, in this case compression also plays a role in beam deflection and magnitude of reaction forces. The side force at the drill bit is the one parameter that affects the direction in which the drill bit will move.

Building Assemblies — Building assemblies develop an upward side force on drill bits. A near bit stabilizer acts as a fulcrum point for the collar weight to pry against. The mechanical advantage of this assembly is greatest when the first stabilizer is close to the bit. A second stabilizer is positioned some distance to establish a large free span distance to increase the strength of building assemblies. The position of the second stabilizer depends on collar size. Stiffer or larger drill collars allow a longer free span distance.

Holding Assemblies — Stabilizers are positioned within drill collars so that relatively small side forces (usually less than 200 lbs) are developed at the bit. When drill bits having long shanks or large gauge protection areas are used, it is desirable to minimize the tilt angle with stabilizer placement to alleviate bit binding.

Dropping Assemblies — Unlike building and holding assemblies, dropping assemblies have no near bit stabilizers. The drill bit stands alone and is pushed

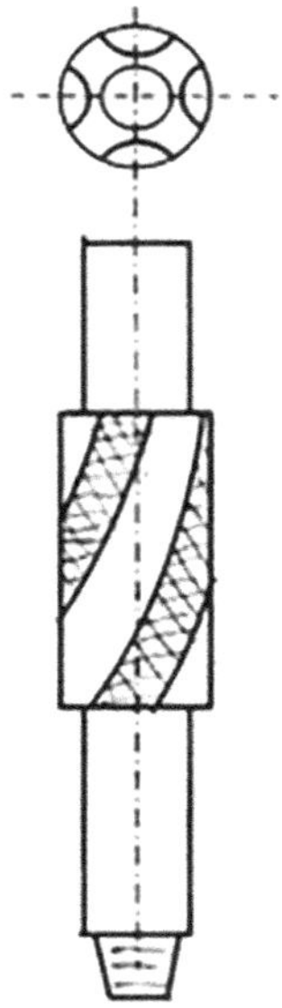

Figure 5-15. Drill collar stabilizer.

downward by the weight of drill collars suspended between the bit and the first stabilizer, which is roughly 60 ft from the bit. The first stabilizer in this case acts much like a hinge allowing the suspended collars to swing downward like a pendulum.

To make maximum use of the pendulum assembly, the freely suspended length should be as long as possible. If the first stabilizer is located too far from the bit, the suspended collars can sag too much, touch the inside of the well bore and lose the strength of the pendulum. Stiffer or larger drill collars weigh more and allow a greater suspended distance for a stronger dropping assembly.

Dropping Assemblies for Soft Formations — Special types of dropping assemblies are used in drilling vertical wells (Figure 5-16). The 60 pendulum means a single stabilizer is placed some 60 ft from the bit. The drill bit is forced downward by the upward deflection of the drill collars above the stabilizer. This causes the collars below the stabilizer to swing back to the vertical. This assembly is especially effective in soft to medium hard formations because the threshold sidewise drilling force is low in these formations. The standard pendulum assembly generates small side forces on the frill bit in near vertical holes. Bit weight is usually very low (~10,000 lbs).

Dropping Assemblies for Hard Formations — The 60 ft pendulum is slow to respond in hard formations because of low allowable bit weights. An alternative is a 30–60 pendulum assembly (Figure 5-16), which allows higher weight on bit and thus a higher rate of penetration. The distance between the first and second stabilizer is 60 ft. WOB could be near 50,000 lbs. In addition, higher bit weights tend to reduce hole angle even faster.

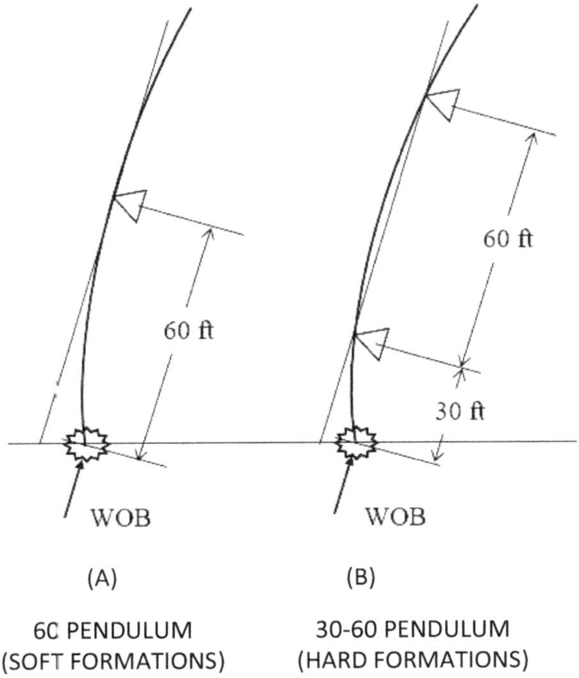

Figure 5-16. Pendulum assemblies (see Fig 12-32).

The drill collars above the second stabilizer lie on the low side of the hole due to gravity. This produces an upward bow in the drill collars between the first and second stabilizers, which forces the drill bit sideways toward the vertical. The greater the weight on bit, the greater the side force, which moves the bit toward the vertical. This assembly typically holds hole inclination within 2½ deg. from vertical. It begins the corrective action around 2-deg. inclination.

5.4.6 Differential Sticking

Differential sticking refers to pipe, normally drill collars, that is pushed against the inside of a well bore as depicted in Figure 5-17. This phenomenon occurs when a mud cake coats the walls of a well bore and well bore pressure is greater than formation pressure [10]. The latter condition is usually met so the mud cake lining becomes a key factor.

The force, F, pushes section of pipe into the mud cake lining. The magnitude of this force time a coefficient of friction defines the friction force opposing movement in either the axial or torsion directions or both. The force to break friction often exceeds the rigs pull or torque capability.

In directional drilling, drill pipe loading is usually limited by torque and not by direct pull. One practice is to allow rotary torque to reach 80% of makeup torque. Motor drive torque is calibrated against tong torque gauges as a means of quantifying rotary torque.

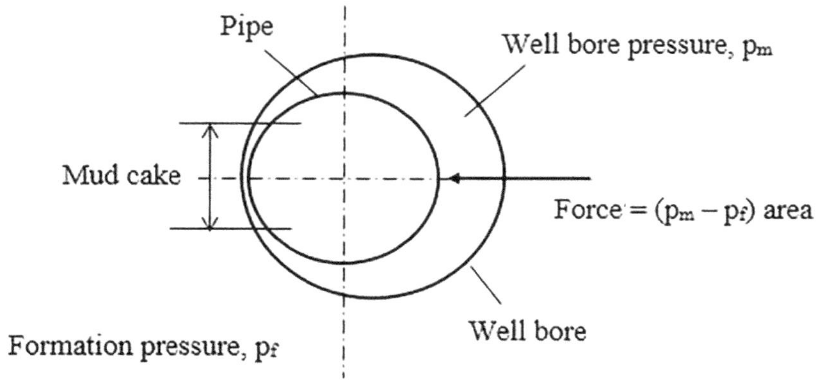

Figure 5-17. Elements of differential sticking.

5.5 OPERATIONAL CONSIDERATIONS

5.5.1 Drill Pipe Buckling

Earlier, the neutral point was identified as a location in the drillstring where internal compressive stress is equal to the local hydrostatic pressure. Furthermore, pipe above this point can be viewed as being exposed to the local hydrostatic pressure as if it were hanging freely and open ended in the fluid. Even though this compressive force can be quite high, pipe under hydrostatic end loading will not buckle. However, buckling can occur by applying an additional compressive force (F_{cr}) over and above the hydrostatic bottom end force. Drill pipe buckling under this condition is predicted by Huang and Dareing [11]. The boundary condition at the lower end was assumed to be fixed to simulate the structurally stiff interface between pipe and drill collars. The critical force for the first mode is predicted by

$$F_{cr} = \phi(EI)^{\frac{1}{3}}(w - w_m)^{\frac{2}{3}} \tag{5-59}$$

$$F_{cr} = 3.29\left(\frac{EI}{144}\right)^{\frac{1}{3}}(wBF)^{\frac{2}{3}} \quad \text{(EI having units of lb-in.}^2\text{)} \tag{5-60}$$

This buckling force is independent of hole depth (drill pipe length). The 3.29 factor applies to the case where the lower end of the drill pipe is fixed, and the upper end is pinned, which simulates the interface between drill collars and drill pipe. This explains the difference between the 3.29 constant and Lubinski's 1.94, which applies to pinned-pinned boundary conditions; which appropriately represents boundary conditions for drill collar buckling. The fixed lower end condition more closely models the connection between drill collars and drill pipe. The force predicted by Eq. (5-60) is beyond hydrostatic compression at the lower end of drill pipe (drill pipe/drill collar interface).

As an example, consider the following set of parameters, which incorporates drill collars and WOB (Figure 5-18).

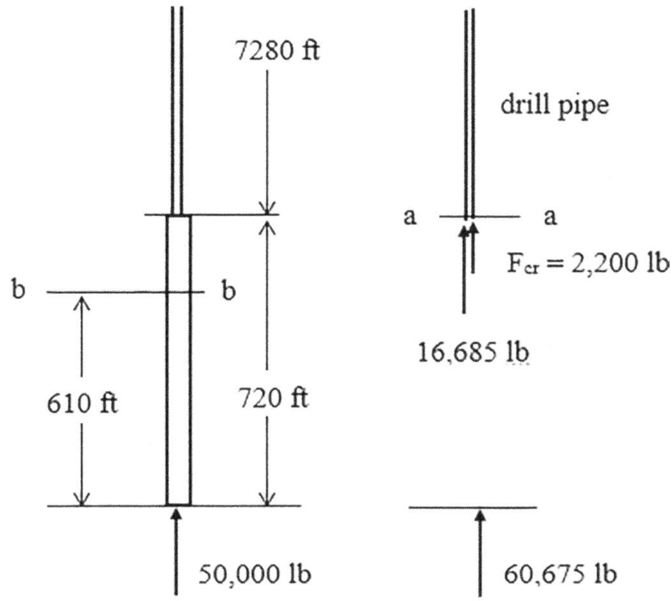

Figure 5-18. Weight on drill bit to buckle 4½ in. (16.6) drill pipe.

Drill depth = 8,000 ft
Drill Pipe = 4.5 in. OD × 3.826 in. ID
Drill Collars = 6.5 in. OD × 2.5 in. ID
Drill Collar Length = 720 ft
Mud weight = 10 lb/gal
Weight on Bit = 50,000 lb

Drill collar length is typically calculated using

$$L_c = \frac{WOB}{0.85BFw} \tag{5-61}$$

which includes a 1.15 safety factor against drill pipe buckling. The buoyancy factor, corresponding to 10 ppg mud is BF = 0.846.

In this case, collar length is 720 ft and the neutral point is 610 ft from the bit with WOB = 50,000 lb.

Using Eq. (5-60), the critical buckling force required to buckle the drill pipe into the first mode is F_{cr} = 2200 lb. This force is over and above the "hydrostatic" force associated with the neutral point at the top of the drill collars (bottom of drill pipe). The "hydrostatic" force applied to the bottom of the drill pipe is

$$F_{hydrostatic} = p_{7280}A_p = 0.052(7280)10(4.4074) = 16{,}685 \text{ lb}$$

The total internal force at the bottom of the drill pipe at the onset of buckling is 18,885 lb ($p_{7280}A_p + F_{cr}$ = 16,685 + 2200 lb). The 16,685 lb at the bottom of the drill pipe results from putting the neutral point at the top of the collars.

The applied force to the drill bit required to put the neutral point at the top of the drill collars is

$$BW = 58{,}475 \text{ lb}$$

An additional 2200 lbs would cause drill pipe buckling. Therefore, total WOB to buckle the drill pipe is 58,475 + 2200 = 60,675 lb.

5.5.1.1 Bending Stresses due to Buckling

Earlier discussion emphasized the importance of keeping the neutral point in the drill collar section to prevent buckling of drill pipe directly above the drill collar section. The above analysis was conducted to determine when drill pipe would buckle and predict expected bending stresses if buckling occurred. The results of ref [11] are used here to predict the level of bending stresses in drill pipe at the drill pipe/collar transition in the event of buckling.

The bending moments at this interface associated with the first two buckle modes are determined from [10]

$$\frac{ML^2}{EIY_0} = 200 \quad \text{(first buckle mode)}$$

and

$$\frac{ML^2}{EIY_0} = 250 \quad \text{(second buckle mode)}$$

These numbers apply for the case of $\alpha = 4000$.
Assuming 5 in. (19.5 lb/ft) drill pipe

ID = 4.276 in.
$I = 14.269 \text{ in.}^4$
$E = 30\,(10)^6$ psi
Y_0 = 2.5 in. (radial clearance between drill pipe and well bore; assumed)
γ_m = 10 ppg (BF = 0.847)

By substitution

$$\alpha = \frac{wBFL^3}{EI}\left|\frac{144 \text{ in}^2}{1 \text{ ft}^2}\right| = 4{,}000$$

$$L = 897 \text{ ft}$$

The bending moment for the first and second modes are

$$M_1 = 200\frac{EIY_0}{L^2} = 200\frac{30(10)^6 14.269(2.5)}{897^2}\left|\frac{1 \text{ ft}}{12 \text{ in}}\right|^3 = 153.9 \text{ ft} - \text{lb}$$

$$M_2 = \frac{250}{200}153.9 = 192.43 \text{ ft} - \text{lb}$$

Corresponding bending stresses are

$$\sigma_{B1} = \frac{Mc}{I} = \frac{153,9(12)2.5}{14.269} = 404.6 \text{ psi}$$

$$\sigma_{B2} = 505.7 \text{ psi}$$

These bending stress levels are quite low and support the practice of running drill pipe in compression within the vertical portion of a directional well. However, bending in this area becomes significant in high curvature situations.

5.5.2 Dog Leg Severity and Fatigue

Drill pipe is normally selected on its capability to withstand direct pull. However, bending stresses can be far more damaging, especially since drillstring rotation cycles bending stresses and can produce fatigue damage. Dog legs or wellbore curvature can be especially harmful, and this is why well plans usually limit the maximum allowable dogleg in a well. Dog legs are defined as the angle change in degrees over a 100 ft drilled interval. The variables are shown in Figure 5-19.

$$s = R\beta(rad) = R\frac{\pi}{180}\beta(\text{deg})$$
$$\frac{\beta^o}{s} = \frac{180}{\pi}\frac{1}{R}\ (\text{deg./ft}) \tag{5-62}$$

$$DLS(\text{deg./100 ft}) = \frac{18000}{\pi}\frac{1}{R} \tag{5-63}$$

This formula shows that DLS can be determined by knowing the local curvature of the well path.

Dog legs can be created accidentally or pre-planned as in directional drilling. In either case wellbore curvature is monitored closely to reduce the chances of drillstring failure and potential problems during well completion and well production. In general, the plane of curvature is not vertical but may be inclined or

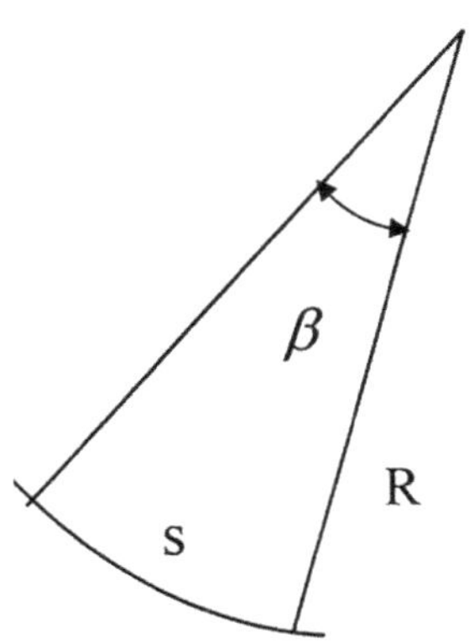

Figure 5-19. Dog leg variables.

in a space plane. Maximum permissible doglegs are specified in a well plan to avoid drilling, completion and production problems.

5.5.3 Bending Stresses in Drill Pipe Between Tool Joints

Bending caused by local dog legs can be significant. Consider the schematic (Figure 5-20) showing tool joints lying flat against the inside of a curved well bore. A drill pipe section of length, L, is assumed to be under uniform tension, T. This condition creates high localized bending at each tool joint as explained below. In this case, pipe deflection, which yields bending moments and stress are best determined by use of polar coordinates.

Consider a joint of drill pipe bent within a dog leg. The magnitude of bending stresses depends both on local hole curvature and tension in the drill pipe. When curved pipe is pulled, the middle portion tends to straighten out and shift more curvature toward tool joints. Bending stresses are greatest near tool joints. Total stress is the sum of stress caused by direct pull and bending stress.

The analysis starts with [12]

$$\frac{d^4w}{ds^4} - \lambda^2 \frac{d^2w}{ds^2} = \zeta \tag{5-64}$$

where

$$\lambda^2 = \frac{T}{EI} - \frac{1}{R^2}$$

$$\zeta = \frac{Rq + T}{REI}$$

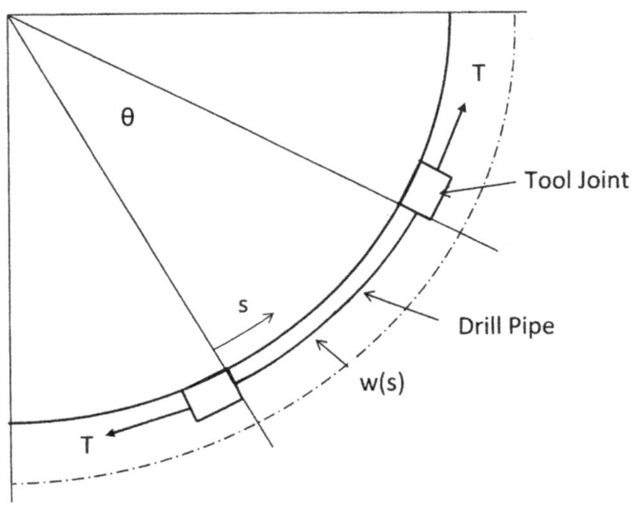

Figure 5-20. Bending of a pipe section between tool joints.

Furthermore, let

$$\chi = \frac{Rq + T}{RT - \frac{EI}{R}}$$

In applying Eq. (5-64), we assume side loading, q, is negligible. The solution to Eq. (5-64) is

$$w(s) = -\frac{1}{2}\chi s^2 + As + B + C\cosh\lambda s + D\sinh\lambda s \tag{5-64}$$

Letting

$$\chi = \frac{T}{RT - \frac{EI}{R}}$$

$$\lambda^2 = \frac{T}{EI} - \frac{1}{R^2}$$

and applying the following boundary conditions

$$w(0) = w_0 \quad w(\ell) = w_0$$

$$\left.\frac{dw}{ds}\right|_{s=0} = 0 \quad \left.\frac{dw}{ds}\right|_{s=\ell} = 0$$

establish the four constants A, B, C, and D.

$$(1) \quad B + C = 0 \quad (5\text{-}65)$$

$$(2) \quad Al + B + C\cosh\lambda l + D\sinh\lambda l = \frac{1}{2}\chi l^2 \tag{5-66}$$

$$(3) \quad A + D\lambda = 0 \quad (5\text{-}67)$$

$$(4) \quad A + C\lambda\sinh\lambda l + D\lambda\cosh\lambda l = \chi l \tag{5-68}$$

Combining (1), (2) and (3) gives

$$Al\left(1 - \frac{1}{\lambda l}\sinh\lambda l\right) - C\left(1 - \cosh\lambda l\right) = \frac{1}{2}\chi l^2 \tag{6-69}$$

Combining (3) and (4) gives

$$A\left(1 - \cosh\lambda l\right) + C\lambda\sinh\lambda l = \lambda l$$

or

$$Al\left(1 - \cosh\lambda l\right) + C\lambda l\sinh\lambda l = \lambda l^2 \tag{6-70}$$

Equations (5-69) and (5-70) are two algebraic equation with A and C as unknowns.

$$\begin{bmatrix} a_{11} & a_{12} \\ a_{21} & a_{22} \end{bmatrix} \begin{Bmatrix} A \\ C \end{Bmatrix} = \begin{Bmatrix} b_1 \\ b_2 \end{Bmatrix}$$

where

$$a_{11} = l\left(1 - \frac{1}{\lambda l}\right)\sinh \lambda l$$
$$a_{12} = \cosh \lambda l - 1$$
$$a_{21} = l\left(1 - \cosh \lambda l\right)$$
$$a_{22} = \lambda l \sinh \lambda l$$

Using Cramer's rule with l = 30 ft, λ = 0.0153 and λl = 5.502

$$A = 0.015$$
$$C = 0.9895$$
$$D = -\frac{1}{\lambda}A = -0.9815$$
$$B = -C = 0.9895$$

Displacement along the pipe.

$$w(s) = -\frac{1}{2}\chi s^2 + As + B + C\cosh \lambda s + D\sinh \lambda s \quad (\text{with } \chi = 8.334 \times 10^{-5})$$

Parameters for this example are:

$\lambda = 0.0153$
$\chi = 8.334 \times 10^{-5}$
C = 0.9895
R = 1000 ft = 12,000 in.
I = 13 in^4
E = 30 × 10^6 psi
C = 2.5 in.

By substitution and dropping W_0/R^2

$$\sigma_{max} = \frac{M_{max}c}{I}$$

$$\sigma_{max} = Ec\left(\frac{1}{R} - \chi + C\lambda^2\right)$$

$$\sigma_{max} = 30 \times 10^6(2.5)\left(\frac{1}{12{,}000} - 8.334 \times 10^{-5} + 0.9895(0.0153)^2\right)$$

$$\sigma_{max} = 17{,}372 \text{ psi}$$

From

$$M = EI\left(\frac{1}{R} + \frac{w}{R^2} + \frac{d^2w}{ds^2}\right)$$

Bending moment at s = 0 is

$$M_{\max} = EI\left(\frac{1}{R} + \frac{w_0}{R^2} - \chi\right) + C\lambda^2 EI \tag{5-71}$$

These equations were programmed to show how bending stress varies over a 30 ft pipe joint when the radius of hole curvature is R = 1000 ft. Tension is assumed to be 100,000 lbs. The highest bending occurs at the tool joints 17,336 psi (see Figure 5-21).

Figure 5-22 shows how this maximum bending stress is affected by hole radius and tension. Total stress is the algebraic sum of bending and axial stresses. In practice, an early dog-leg, where tension is highest, is not good.

These bending stresses are cyclic creating possible metal fatigue. Using the previous example, the stress condition contains an average component and a cyclic component.

$$\sigma_{dyn} = \sigma_{ave} + \sigma_{cyc} \tag{5-72}$$

where

σ_{ave} = 18,958 psi
σ_{cyc} = 17,500 psi

Fatigue tests in cases of complete reverse of stress indicate that the endurance limit of steels ranges between 35% and 60 % of the ultimate strength [13,14]. For example, the material properties for SAE 4340 (heat treated) are

σ_{yld} = 132,000 psi

σ_{ult} = 150,000 psi

σ_{end} = 76,000 psi

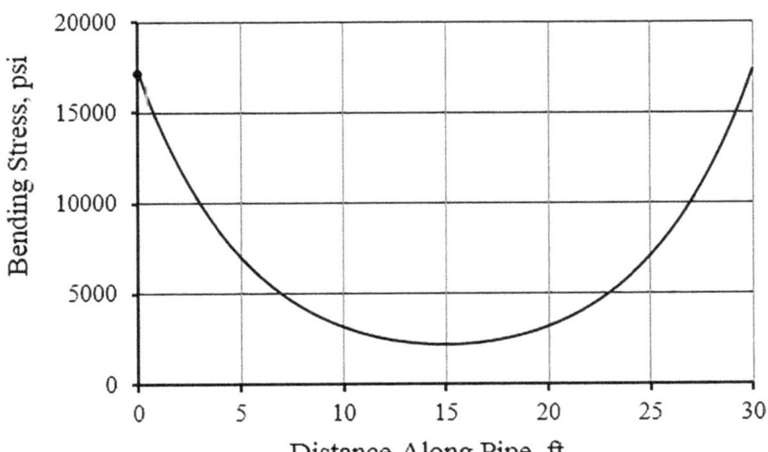

Figure 5-21. Bending stress along a section of 5 in. (16.25 lb/ft) drill pipe (R = 1000 ft, T = 100,000 lbs).

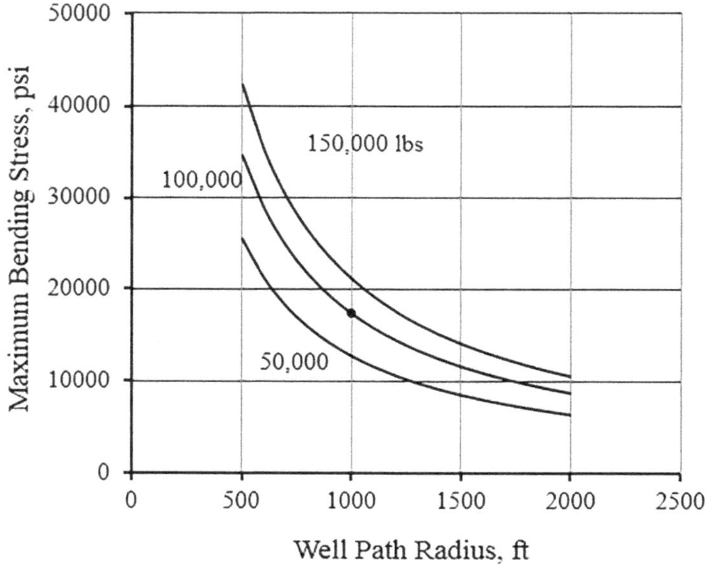

Figure 5-22. Bending stress versus tension and well path radius.

The endurance limit is 50% of the ultimate strength. Endurance limit depends on many factors such as, surface roughness, size, temperature, and stress concentration.

Fatigue tests conducted by Morgan and Roblin [15] show that endurance limits for four API grades of drill pipe are at the lower percentage level (Table 5-7). Their tests preserved the surface and size effects of these pipes.

Average stress also affects the endurance limit and has to be considered in drill pipe fatigue analysis. The modified Goodman diagram is commonly used to account for mean stress effects. Figure 5-23 shows the Modified Goodman diagram for API Grade X95. The square dot gives the location (18.9, 17.7) of the stress condition for the above example. This stress condition is below the solid line and no fatigue failure is expected.

Table 5-7. Endurance limits for drill pipe grades.

Grade	API minimum yield strength (ksi)	API yield strength maximum (ksi)	API tensile strength minimum (ksi)	Average tensile strength of Test samples (ksi)	Endurance limit: Minimum test value (ksi)	Endurance limit: Median test value (ksi)
E75	75	105	100	123	30	32
X95	95	125	105	132	32	35
G105	105	135	115	144	34	38
S135	135	165	145	167	36	40

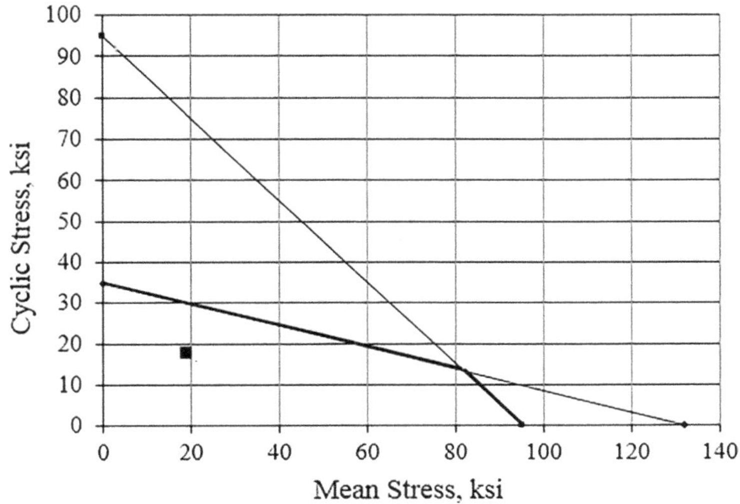

Figure 5-23. Modified goodman diagram for API grade X95.

Dog legs, especially high up where tension is high, can generate high stress reversal as indicated above.

Fatigue tests (conducted by Hughes Tool; on Grade D and Grade E drill pipe) show that the endurance limit for these grades of drill pipe for complete stress reversal is around 20,000 psi.

Lubinski used these data to develop a Goodman diagram and calculated maximum permissible dog for drill pipe.

For example, axial stress in new 4½ in. (16.6) drill pipe having a direct pull load of 100,000 lbs is 22,689 psi. According to the Goodman diagram (Figure 5-24) the endurance limit or cyclic stress level about the mean stress is 16,311 psi.

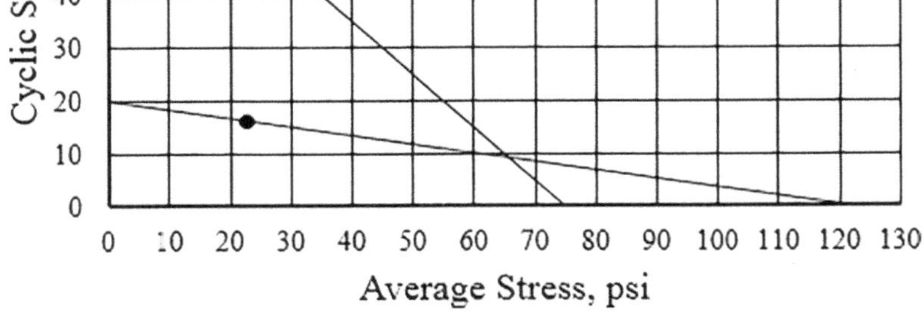

Figure 5-24. Goodman diagram based on Hughes tool test data.

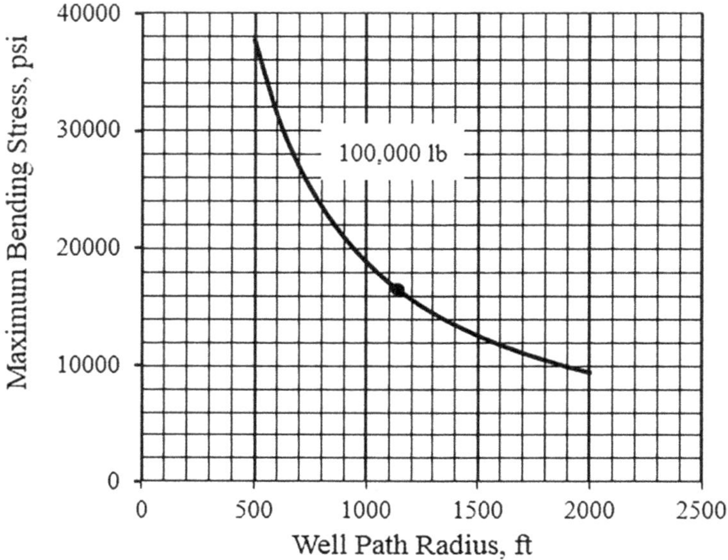

Figure 5-25. Stress data for 4½ in. (16.6 lb/ft) drill pipe.

Using the calculated bending stress shown in Figure 5-25, the radius of curvature corresponding to bending stress of 16,311 psi under a 100,000 lb a direct pull force is 1146 ft. This radius of curvature corresponds to a DLS of 5 deg./100 ft. This predicts the maximum permissible dog leg at a point in this drill pipe and is consistent with data developed by Lubinski [16].

5.6 STATIC FAILURE OF DRILL PIPE

The static analysis of drillstrings is based on uniaxial stresses caused by direction tension (compression). This is the major load to the string. However, other loads may be significant and should be considered in tri-axial stress analysis.

Drilling mud pressure at the top of the drill pipe is near maximum except for losses in the top side equipment (standpipe, hose, swivel, Kelly). Fluid pressure may be near 3000 psi. Hook load is also maximum at the top as well as the torque.

The burst limit should be checked from the tables and if there is a concern, a triaxial stress analysis should be conducted. As burst loading should be considered for drill stem testing operations.

5.6.1 Torsion Loading

Torque at the rotary table is primarily due to friction between the drillstring and wellbore. In directional drilling drill pipe loading is limited by torque and not by direct pull. One practice is to allow rotary torque to reach 80% of makeup torque. SCR motor drive torque is calibrated against tong torque gauges as a means of quantifying rotary torque.

Drilling torque is about 1500 ft-lbs. The largest amount of torque at the surface is due to friction.

5.6.2 Combined Stresses

In reality, all three of the above loads can be applied simultaneously. It is always a good idea to check the combined effects by determining principal stresses and check the triaxial state of stress against the von Mises criteria of failure. An example of the analysis of combined stresses is given below.

Figure 5-26 shows different types of stress components that can be generated in the body of drill pipe. This three-dimensional state of stress is typical directly below the rig floor. As an example, consider the following load condition directly below the rig floor.

F_x = 250,000 lbs	(hook load)
T = 20,000 ft-lbs	(surface torque)
p = 3000 psi	(standpipe pressure)

Furthermore assuming 5½ (19.2 lb/ft) drill pipe having wall thickness of 0.304 in. The $\frac{r}{t}$ ratio is 8.546, which is somewhat lower than $\frac{r}{t} \geq 10$ generally specified for thin wall pressure vessels [13]. Cross section area is 4.9624 in.^{2}. Polar section modulus, S, is 12.221 in.^{3}.

OD = 5½ in.
ID = 4.892 in.
r_0 = 2.75 in.
t = 0.304 in.
A = 4.9624 in.^{2}
S = 12.221 in.^{3}

Then stress components on the outside surface are

$$\sigma_a = \frac{F_x}{A} = 50{,}380 \text{ psi} \qquad \text{(axial stress)} \tag{5-73}$$

$$\tau = \frac{Tr_0}{J} = \frac{T}{S} = 19{,}640 \text{ psi} \qquad \text{(shear stress)} \tag{5-74}$$

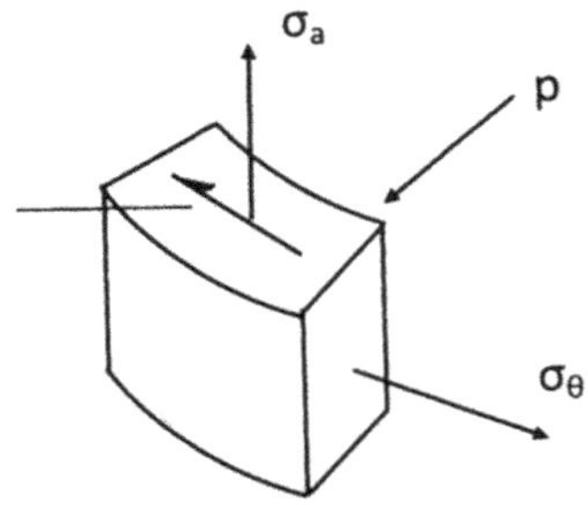

Figure 5-26. Stresses on a pipe element.

$$\sigma_\theta = \frac{r_0}{t} p = 27{,}140 \text{ psi} \qquad \text{(hoop stress)} \tag{5-75}$$

Stress components on the inside surface are the same except there is a radial component

$$\sigma_r = -3000 \text{ psi}$$

Using the above numbers, principal stresses components on the outside surface are:

$$\sigma_{p1} = 61{,}580 \text{ psi}$$
$$\sigma_{p2} = 15{,}940 \text{ psi}$$
$$\sigma_{p3} = 0$$

Principal stresses on the inside surface are

$$\sigma_{p1} = 61{,}580 \text{ psi}$$
$$\sigma_{p2} = 15{,}940 \text{ psi}$$
$$\sigma_{p3} = -3{,}000 \text{ psi}$$

5.6.3 von Mises Criteria of Failure

Assuming the von Mises criteria of failure, material yielding takes place when

$$\sigma_f \geq \sigma_{yld} \tag{5-76}$$

Substituting for the assumed loading, von Mises stress [14] on the outside surface is

$$\sigma_f^2 = \sigma_{p1}^2 - \sigma_{p1}\sigma_{p2} + \sigma_{p2}^2 \quad \text{(2D case)} \tag{5-77}$$
$$\sigma_f = 55{,}360 \text{ psi}$$

and von Mises stress on the inside surface is

$$2\sigma_f^2 = \left(\sigma_{p1} - \sigma_{p2}\right)^2 + \left(\sigma_{p2} - \sigma_{p3}\right)^2 + \left(\sigma_{p3} - \sigma_{p1}\right)^2 \quad \text{(3D case)} \tag{5-78}$$
$$\sigma_f = 57{,}500 \text{ psi}$$

Assuming a pipe grade of G105, the factor of safety on the outside surface is

$$F.S. = \frac{105{,}000}{55{,}360} = 1.9$$

The FS on the inside surface is

$$F.S. = \frac{105{,}000}{57{,}500} = 1.83$$

Bending stresses are not included here but could be important farther down the well bore.

5.7 STRETCH OF SUSPENDED PIPE

The stretch of drill pipe due to its own weight is considerable in comparison with short beams normally considered in engineering structures. This section develops the equations of drill pipe elongation while hanging in a well bore.

5.7.1 Uniform Tension

Pipe under tension (or compression) will stretch. If the tension is uniform local strain is defined by $\varepsilon = \frac{du}{dx}$. The relation between stress and strain from uniaxial tests is

$$\sigma = E\varepsilon \tag{5-79}$$

Noting that local axial stress is uniform throughout gives

$$\frac{F}{A} = E\frac{du}{dx} \tag{5-80}$$

or

$$\frac{du}{dx} = \frac{F}{EA} \tag{5-81}$$

Taking x downward from the top and u(0) = 0 gives

$$u(x) = \frac{F}{EA}x \tag{5-82}$$

The maximum displacement occurs at x = L.

$$u(L) = \frac{FL}{EA} \tag{5-83}$$

5.7.2 Drill Pipe Hanging in Air (Linear Tension)

The determination of axial displacements along vertical drill pipe hanging freely in air is determined similarly. With reference to Figure 5-27,

$$\frac{du}{dx} = \frac{\sigma}{E} = \frac{(L-x)w}{EA} \tag{5-84}$$

from which

$$u(x) = \frac{w}{2EA}\left[L^2 - \left(L-x\right)^2\right] \tag{5-85}$$

$$u(x) = \frac{wx}{EA}\left(L - \frac{x}{2}\right)$$

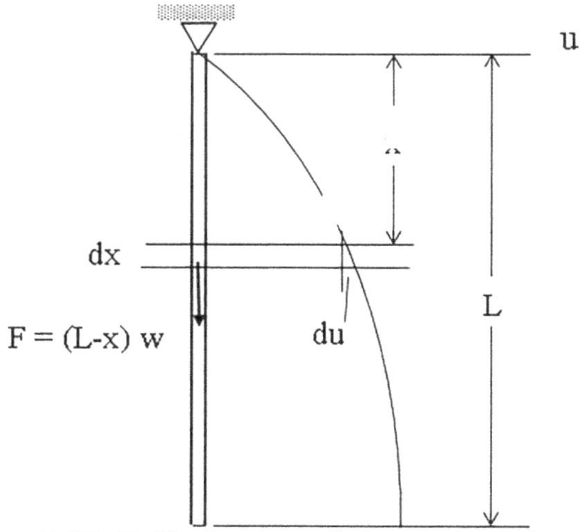

Figure 5-27. Drill pipe stretch from top to bottom.

Maximum deflection occurs at the lower end, x = L.

$$u_{\max} = \frac{L^2 w}{2EA} \tag{5-86}$$

For example, assume 4½ in. drill pipe with

w = 16.6 lb/ft
L = 5,000 ft
E = 30 $(10)^6$ psi
A = 4.4074 in.^{2}

The displacement of the lower end is

$$u(L) = \frac{5{,}000^2(16.6)}{(2)30(10)^6 4.4074} = 1.57 \text{ ft}$$

5.7.3 Drill Pipe Hanging in Drilling Mud (Linear Tension)

In consideration of the effects of hydrostatic pressure (buoyancy), a force at the bottom (P = –pA) has to be included in the analysis (see Figure 5-28). The tension (or compression) at any location, x, is

$$F = w(L - x) - pA \tag{5-87}$$

$$F = w(L - x) - \gamma_m LA \tag{5-88}$$

$$F = w(L - x) - w_m L \tag{5-89}$$

where w_m is the weight of the drilling mud that is displaced by a unit length of pipe. The reader is encouraged to verify $w_m = \gamma_m A$.

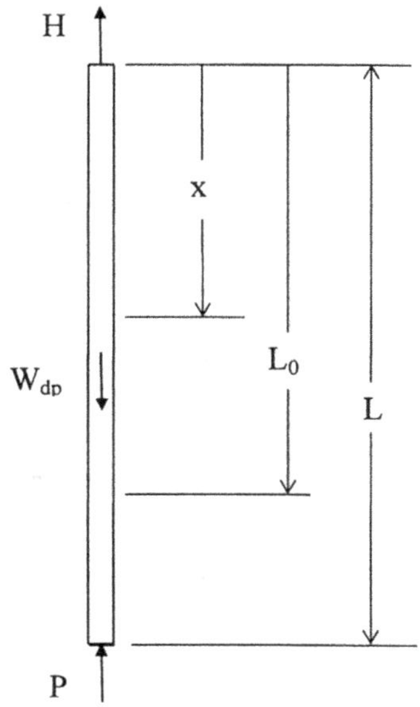

Figure 5-28. Drill pipe suspended in drilling mud.

$$F = wL(BF) - wx \tag{5-90}$$

By substitution

$$\frac{du}{dx} = \frac{\sigma}{E} = \frac{wL(BF) - wx}{EA} \tag{5-91}$$

$$\frac{du}{dx} = \frac{wL}{EA}\left(BF - \frac{x}{L}\right) \tag{5-92}$$

By integration and applying boundary conditions

$$u(x) = \frac{wL^2}{EA}\left[BF\left(\frac{x}{L}\right) - \frac{1}{2}\left(\frac{x}{L}\right)^2\right] \tag{5-93}$$

This equation is expressed in dimensionless terms as

$$U\left(\frac{x}{L}\right) = \frac{EA}{wL^2}u\left(\frac{x}{L}\right) = BF\left(\frac{x}{L}\right) - \frac{1}{2}\left(\frac{x}{L}\right)^2 \tag{5-94}$$

A plot of U versus x/L is given in Figure 5-29 for mud weights of 10, 12, and 14 ppg each having buoyancy factors of 0.847, 0.817, and 0.786, respectively.

The maximum value of U is determined as follows:

$$\frac{dU}{d(x/L)} = BF - \frac{x}{L} = 0 \tag{5-95}$$

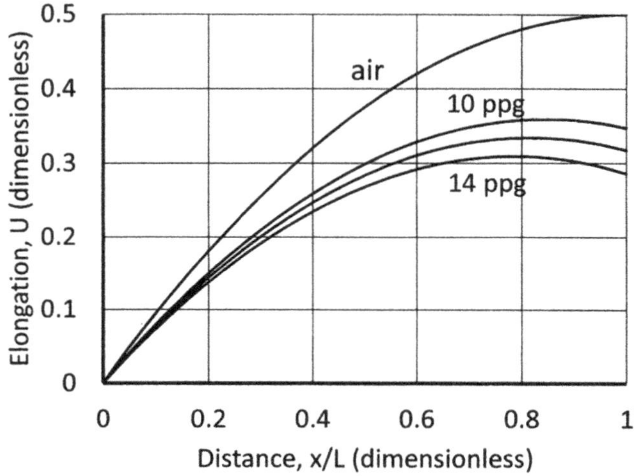

Figure 5-29. Effect of mud weight on drill pipe elongations.

The location of the maximum value of U is

$$\frac{x}{L} = BF \tag{5-96}$$

Substituting back into Eq. (5-86) gives

$$U_{\max} = \frac{1}{2} BF^2 \tag{5-97}$$

Letting the location of the maximum displacement be L_0, then

$$\frac{L_0}{L} = BF \tag{5-98}$$

This is also verified in Figure 5-29. Note, L_0 is measured from the top down. L_0 is also the location of the point of zero compression/tension.

5.8 SUMMARY OF EQUATIONS

$$L_{np} = \frac{WOB}{wBF}, \quad L_c = \frac{WOB}{0.85wBF} \quad \text{(neutral point, collar length)}$$

$$T = \frac{F}{12}\left[\frac{p}{2\pi} + \frac{R_t f}{\cos\theta} + R_s f\right] \quad \text{(makeup torque)}$$

$$H_a = 0.9H_{yld} = \left(w_p L_p + w_c L_c\right) BF + MOP \quad \text{(drillstring design)}$$

$$\tau_{yld} = 0.577\sigma_{yld}$$

Location of neutral point in composite BHA

$$WOB = \sum_{i=1} \left(W_i + xw_{i+1}\right)BF$$

$$DLS = \frac{18{,}000}{\pi R}\text{deg. per 100 ft}$$

Drill Collar Buckling

$$WOB_{cr} = 1.94\left(wBF\right)^{\frac{2}{3}}\left(EI\right)^{\frac{1}{3}} \quad \text{(first mode)}$$
$$WOB_{cr} = 3.75\left(wBF\right)^{\frac{2}{3}}\left(EI\right)^{\frac{1}{3}} \quad \text{(second mode)}$$

Drill Pipe Buckling (first mode)

$$F_{cr} = 3.29\left(\frac{EI}{144}\right)^{\frac{1}{3}}(wBF)^{\frac{2}{3}} \qquad \text{(EI having units of lb-in.}^2\text{)}$$

Beam Deflection in terms of polar coordinates

$$\frac{d^4w}{ds^4} - \lambda^2\frac{d^2w}{ds^2} = \zeta \quad \text{where} \quad \lambda^2 = \frac{T}{EI} - \frac{1}{R^2} \quad \text{and} \quad \zeta = \frac{Rq + T}{REI}$$

Elongation of Pipe in Fluid

$$u(x) = \frac{wL^2}{EA}\left[BF\left(\frac{x}{L}\right) - \frac{1}{2}\left(\frac{x}{L}\right)^2\right]$$

5.9 PROBLEMS

1. Determine the buoyancy factor for 13 ppg mud.

2. Determine the hook load for a drillstring containing 9000 ft of 4½ in. (16.6) drill pipe plus 700 ft of 8 × 2½ drill collars suspended in 12 ppg drilling mud.

3. Determine the maximum depth that can be reached with a drillstring containing 600 ft of 8 × 2½ in drill collars and 5 (19.5) drill pipe, assuming 14 ppg mud and
   a. New Drill Pipe; Grade E
   b. New Drill Pipe; Grade G
   c. Repeat a. and b. allowing for 75,000 lbs overpull.

4. Select lowest grade of drill pipe for the following case:
   a. TD = 18,000 ft
   b. Drill pipe = 4 ½ (16.6)
   c. Drill Collars = 6½ × 2¼ in.; 700 ft long
   d. Drill pipe class = New
   e. Overpull = 75,000 lbs

   Assume a mud weight of 14 ppg.

5. Determine the maximum depth that can be reached with 5½ (19.2) drill pipe, Grade X95. Assume
   a. 7 × 2½ drill collars (650 ft long section)
   b. 12 ppg mud weight
   c. Drill Pipe Class (New)
   d. Overpull = 100,000 lbs

6. Using the information from Problem 5, determine the allowable depth assuming the freely hanging drillstring is limited by slip crushing force. Use Table 5-4 to obtain the crushing force limit.

7. How much torque is required to develop 500,000 lbs force in the shoulder of an NC 46 threaded connection, assuming zero friction in the connection.

8. What percent of total make-up torque recommended for New 5 (19.50) drill pipe having NC 50 threads goes into overcoming friction in the: a) threads, b) shoulder?

9. Using the following numbers, which describe a NC50 (6¾ × 3¾) connection

   p = 0.25 in.
   θ = 30 deg.
   R_t = 2.385 in.
   R_s = 2.922 in.

   Develop a graph showing how contact force, F, varies with the coefficient of friction, f, assuming a torque value of 15,000 ft-lbs. Let f vary between 0.08 and 1.

10. Determine the location of the neutral point in 8 × 2½ inch drill collars if

    BW = 40,000 lbs
    Mud weight = 12 ppg

    Determine the length of the drill collar section if $L_{np} = 0.85L_c$.
    Fill-in the following table.

BW, lb	L_{np}, ft
10,000	
20,000	
30,000	
40,000	

11. Determine the buoyant force acting on 700 ft of 7½ × 2¼ drill collars hanging from the drawworks using hydrostatic pressure and area. Mud weight is 12.5 ppg. Calculate the buoyant weight of the collars using: 1) pressure area method and 2) Archimedes principle.

12. Determine the bit weight (WOB) required to buckle (7.5" × 3.25") drill collars into the 1st and 2nd modes. Assume 12 ppg mud.

13. Prove the working hook load of a drillstring made up of drill pipe and drill collars is

$$H = \left(w_p L_p + w_c L_c\right) BF$$

14. Determine the location of the neutral point for a bottom hole arrangement containing 500 ft of 7 × 3 in. drill collars and 600 ft of heavy weight drill pipe (5 × 3 in; 50 lb/ft). Drill pipe weight is 19.5 lb/ft. Let WOB take on four different values.
    Determine for four different bit weights

    a. WOB = 40,000 lbs
    b. WOB = 50,000 lbs
    c. WOB = 60,000 lbs
    d. WOB = 70,000 lbs

    Mud weight is 13 ppg.

15. Prove that the weight per foot of drill collars can be determined by

$$w = 2.67\left(d_0^2 - d_i^2\right) \text{ lb/ft}$$

    where the outside and inside diameters have units of inches.

16. If the neutral point is located at the bottom end of 5 (19.50) drill pipe, calculate the additional bit weight required to buckle the drill pipe. Mud weight is 12 ppg.

17. Consider the following set of operational parameters

Drill depth	10,000 ft
Drill Pipe	5.0 in. OD × 4.276 in. ID (19.5 lb/ft)
Drill Collars	6.5 in. OD × 2.5 in. ID (96 lb/ft)
Drill Collar Length	720 ft
Mud weight	10 lb/gal (BF = 0.847)
Weight on Bit	50,000 lb

    Determine the WOB required to buckle the drill pipe.

18. The pump off force under a diamond drill bit is 7500 lbs. Drilling bit weight for this bit is to be a net 12,000 lbs. Determine the length of 6½ × 2 (102 lb/ft) drill collars so that the neutral point is 0.85 of drill collar length.

19. A well bore is drilled along a curved path with a radius of 1000 ft. What is the dog leg severity? What is the curvature of the well

path? What is the nominal bending stress in 5½ in. (19.2 lb/ft) drill pipe?

20. Determine and plot the axial displacements along a 5" (19.5) drill pipe that is 10,000 ft long. Assume the pipe is hanging in 1) air and 2) 11 ppg mud.

21. Derive the elongation equation, u(x), for drill pipe stretch including the effects of the weight of drill collars (W_{dc}). Neglecting the stretch of drill collars, determine the maximum deflection of the lower end of the drillstring. Assume a mud weight of 11.5 ppg.

22. A cable is to be strapped along the length of a 10,000 ft pipe. The question is — Will the cable stretch or slack as it is lower into the well bore. Consider the cable is strapped every 50 ft. Assume the following pipe conditions:

    - 5 in OD, 4.75 in ID
    - Drilling fluid, γ_m = 11 ppg
    - Modulus of Elasticity, 30 × 10^6 psi

    Determine the change in a 50 ft strapped section located at 6,000 ft after the pipe is completely lowered into the well.

5.10 LIST OF REFERENCES

1. Recommended Practice for Drill Stem Design and Operating Limits (API RP 7G), American Petroleum Institute, 12th Ed., 1987.
2. *History of Petroleum Engineering*, American Petroleum Institute, 1961.
3. Brantly, J. E., *History of Oil Well Drilling*, Gulf Publishing Company, 1971.
4. A. Lubinski, "A study of the Buckling of Rotary Drilling Strings", *Drilling and Production Practice* (API), **17**, 1950, pp. 178–214.
5. Technical Report on Equations and Calculations for Casing, Tubing, and Line Pipe Used as Casing or Tubing; and Performance Properties Tables for Casing and Tubing; ANSI/API Technical Report 5C3; First Edition, December 2008; TN 879.53. A64.
6. Spiri, W. H. and Reinhold, W. B.: "Why Does Drill Pipe Fail in the Slip Area?", *World Oil*, October 1959.
7. A. Lubinski and H. B. Woods, "Factors Affecting the Angle of Inclination and Dog-Legging in Rotary Bore Holes", *Drilling and Production Practice* (API), **20**, 1953, pp. 222–250.
8. Bruce H. Walker; "Some Technical and Economic Aspects of Stabilizer Placement", *Journal of Petroleum Technology*, June 1973.
9. D. Eickelberg, W. Rischmuller, E. Burger, R. Jurgens; "Controlled Directional Drilling, Instruments and Tools", 9th International Symposium 82.
10. Adam T. Bourgoyne, Keith K. Millhelm, and M. E. Chenevert: *Applied Drilling Engineering*, SPE Textbook Series, Vol 2 (1986).

11. T. Huang and D. W. Dareing; "Buckling and Lateral Vibration of Drill Pipe", Trans. ASME, *J. of Engineering for Industry*, Vol. 90, Series B, No. 4, 1968, pp. 613–619.
12. Don W. Dareing; *Mechanics of Drillstrings and Marine Risers*, ASME Press, 2012.
13. H. O. Fuchs and R. I. Stephens, *Metal Fatigue in Engineering*, Wiley, NY, 1980 (69–71).
14. Shigley, Joseph Edward and Larry D. Mitchell, *Mechanical Engineering Design*, McGraw-Hill 1983 (p. 275).
15. Morgan, R. P. and Roblin, M. J., "A Method for the Investigation of Fatigue Strength in Seamless DrillPipe," ASME Conference, Tulsa, Oklahoma, September 22, 1969.
16. Lubinski, Arthur; "Maximum Permissible Dog-Legs in Rotary Boreholes", *Journal of Petroleum Technology*, Feb. 1961.

Chapter 6

MECHANICS OF TUBULARS

This chapter covers special topics related to the mechanics of tubulars. Its purpose is to go beyond standard practices and review issues that may be worth considering in the future. Loads and responses are explained in terms of short beam theory. Long beam theory, which is covered in ref [1], is more complex and goes beyond the scope of this manuscript.

6.1 REVIEW OF BEAM THEORY

It is useful to review a few basic concepts of classical beam theory, before discussing bending and buckling of tubulars. Consider a simply supported beam that is uniformly loaded as shown in Figure 6-1.

Several questions are of interest, such as, deflection, slope, shear, bending moments and associated stresses. Following Euler's theory [2] (a cross section plane after bending remains a plane)

$$\sigma = \frac{Mc}{I} \quad \text{(bending stress)} \tag{6-1}$$

$$M = \frac{EI}{\rho} = EI\frac{d^2y}{dx^2} \quad \text{(assuming small deflections)} \tag{6-2}$$

From equilibrium of the differential element, dx, shown in Figure 6-1 (unit load, shear and moment are shown in their positive directions)

$$\frac{dV}{dx} = w \tag{6-3}$$

$$\frac{dM}{dx} = V \tag{6-4}$$

The first equation yields

$$V_2 - V_1 = \int_{area} w dx \tag{6-5}$$

The second equation gives

$$M_2 - M_1 = \int_{area} V dx \tag{6-6}$$

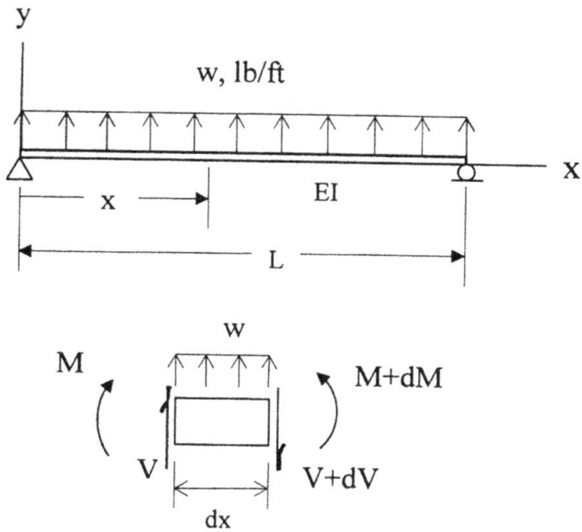

Figure 6-1. Loading of simply supported beam.

All four equations are useful in constructing shear and bending moment diagrams. For example, the shear and bending moment diagrams for uniform loading, $w(x) = -w_o$ are (Figure 6-2).

Beam deflections are determined by solving

$$EI \frac{d^2y}{dx^2} = M(x) \tag{6-7}$$

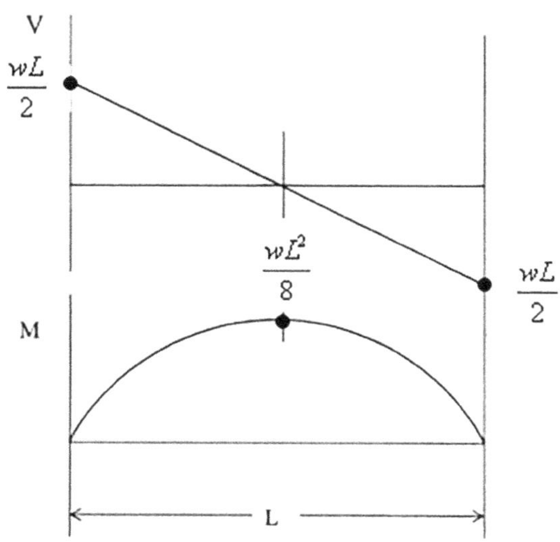

Figure 6-2. Shear and bending moment diagrams.

In terms of the distributed load, w, Eq. (6-7) becomes

$$EI\frac{d^4y}{dx^4} = w(x) \tag{6-8}$$

Boundary conditions have to be satisfied in either case. If the distributed load, w(x), is constant [$w(x) = -w_o$] and assuming pinned-pinned supports,

$$y(x) = -\frac{w_o}{24EI}\left[x^4 - 2Lx^3 + L^3x\right] \tag{6-9}$$

Everything about this beam is predictable.

6.1.1 Column Buckling Under Uniform Compression

Buckling is a condition under which a critical force, Q_{cr}, is reached causing the column to become unstable, i.e., the column takes a different configuration. Columns are usually assumed to have uniform compression throughout. End supports, or boundary conditions may vary as shown in Figure 6-3.

Consider a column simply supported as shown in Figure 6-4.

The differential equation of bending from the freebody diagram of the lower end is

$$EI\frac{d^2y}{dx^2} + Qy = 0 \tag{6-10}$$

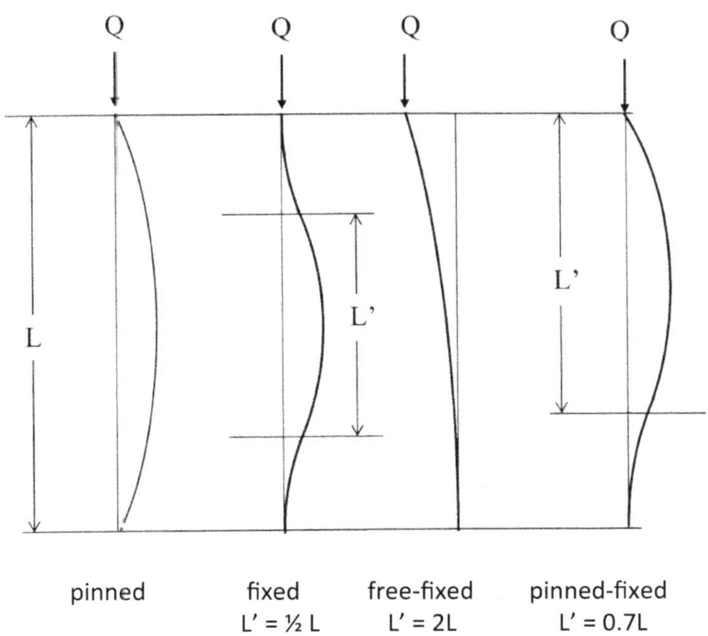

Figure 6-3. Columns with different end constraints.

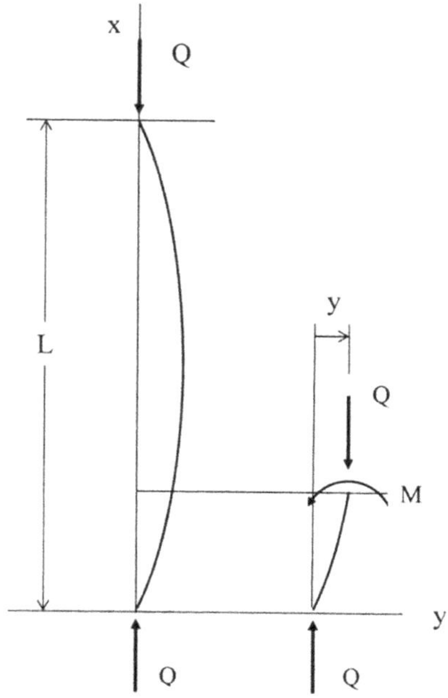

Figure 6-4. Simply supported column.

$$\frac{d^2y}{dx^2} + \beta^2 y = 0 \tag{6-11}$$

where

$$\beta^2 = \frac{Q}{EI}$$

Here, Q, is a compressive load. The solution to Eq. (6-11) is

$$y = A \sin \beta x + B \cos \beta x \tag{6-12}$$

With simply supported boundary conditions, y(0) = 0 giving B = 0, leaving

$$y = A \sin \beta x \tag{6-13}$$

Applying y(L) = 0 gives

$$0 = A \sin \beta L \tag{6-14}$$

For a non-trivial solution

$$\beta L = n\pi, \quad n = 1,2,3,\text{------} \tag{6-15}$$

For the first buckling mode, n = 1 and

$$\frac{Q}{EI} = \left(\frac{\pi}{L}\right)^2 \tag{6-16}$$

giving

$$Q_{cr} = \left(\frac{\pi}{L}\right)^2 EI \tag{6-17}$$

In terms of compressive stress, Eq. (6-17) becomes

$$\sigma_{cr} = \frac{\pi^2}{(L/r)^2} E \tag{6-18}$$

where L/r is the slenderness ratio of the column cross section and r is the radius of gyration of the smallest moment of inertia.

Figure 6-5 shows how critical buckling stress increases with reduction in slenderness ratio. However, for short compressive members, material yielding will occur prior to buckling, as indicated by the horizontal line (assuming σ_{yld} = 60 ksi; L/r = 70.2). This condition establishes a limit to Euler's buckling equation. Euler was the first to recognize the two different modes of failure.

Testing of columns show failure follows along a curved line between compression failure and Euler buckling (Figure 6-6). The transition between these two failure modes is defined by three slenderness ratio ranges:

- Compression block range
- Intermediate range
- Slender range

As a guideline, the compression block theory covers L/r < 40. Euler buckling theory applies to L/r > 140. Failure over the intermediate range is defined by empirical formulas, which depend on material type (steel, aluminum, wood). These empirical formulas are defined in column codes [3–5].

The above equations apply to simple supports. The buckling equations also apply for other boundary conditions providing L is replaced by L' (see Figure 6-3).

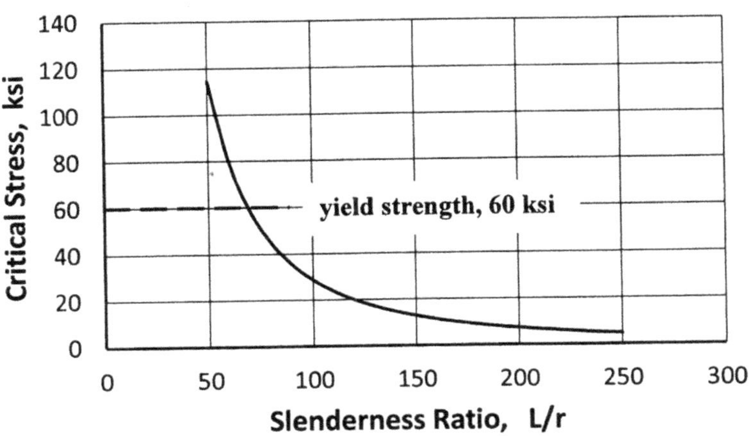

Figure 6-5. Euler buckling curve.

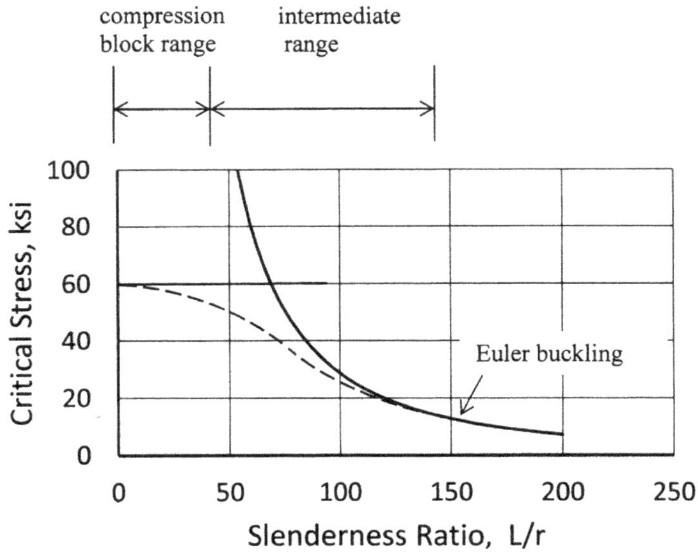

Figure 6-6. Failure theory ranges.

Consider, for example, a section of drill collars (6" × 2", 85 lb/ft) between two stabilizers separated by 60 ft. The minimum yield strength for this collar size is $\sigma_{yld} = 110$ ksi.

$$I = \frac{\pi}{4}\left(R_0^4 - R_i^4\right) = 62.83 \text{ in.}^4$$
$$A = \pi\left(R_0^2 - R_i^2\right) = 25.13 \text{ in.}^2$$

r = 1.58 in. (radius of gyration)
L/r = 455.69 (slenderness ratio)

Using Eq. (6-18)

$$\sigma_{cr} = \left(\frac{\pi}{455.69}\right)^2 30 \times 10^6 = 1426 \text{ psi}$$

This stress-load level is well within the yield limits of the collar material so Euler buckling theory would apply. The corresponding collar force is $F_{cr} = 1426$ (25.13) = 35,832 lbs. This force level is within typical WOB levels, so buckling is possible.

Next consider a larger drill collar size (8" × 2", 160 lb/ft; $\sigma_{yld} = 100$ ksi).

I = 200.27 in.^{4}
A = 47.12 in.^{2}
r = 2.06 in. (radius of gyration)
L = 60 ft
L/r = 349.5 (slenderness ratio)

Using Eq. (6-18)

$$\sigma_{cr} = \frac{\pi^2}{(349.5)^2} 30 \times 10^6 = 2424 \text{ psi}$$

This stress-load level is well within the yield limits of the collar material so Euler buckling theory would apply. The corresponding collar force is F_{cr} = 2424 (47.12) = 114,219 lbs. This force level is higher than normal WOB so the drill collars would not buckle between the two stabilizers by axial force alone.

6.2 BUCKLING ON AN INCLINED PLANE

Drill pipe is sometimes put in compression under certain conditions of directional drilling in order to generate WOB especially in horizontal well bores. While bit thrust is achieved in this manner, drill pipe fatigue and torsion friction can be significant.

Pasley and Bogy [6, 7] give a mathematical basis for predicting drill pipe buckling under conditions of hole size and hole inclination. Later, Dawson and Pasley [8] developed a formula for approximating buckling within an inclined hole. The formula relates critical buckling force to hole size, pipe size and hole angle. It is assumed that the pipe is lying on the low side of the hole.

$$F_{crit} = 2\left(\frac{EIw \sin\theta}{r}\right)^{\frac{1}{2}} \tag{6-19}$$

where

EI = flexure stiffness, lb-in.^{2}
w = weight per unit length
r = radial clearance between pipe and hole
θ = hole angle measured from vertical

This equation is used to generate stability curves of the type shown in Figure 6-7. These curves (based on pipe size) show regions of stability and instability. They show that pipe stability increases with well bore size and hole inclination. Stability curves can easily be generated for other pipe sizes. This information allows drill pipe to carry high levels of compression, which eliminate the need for additional drill collars and greatly reducing friction torque and drag.

6.3 TUBULAR BUCKLING DUE TO INTERNAL PRESSURE

Consider the buckling of pipe with internal pressure acting alone as illustrated in Figure 6-8. The surface pressure force is shown in the left drawing. The right element represents the statically equivalent force system, which is obtained by adding and subtracting pressure forces over the inside area across an imaginary plane. The net effect of the pressure within an enclosed volume is zero,

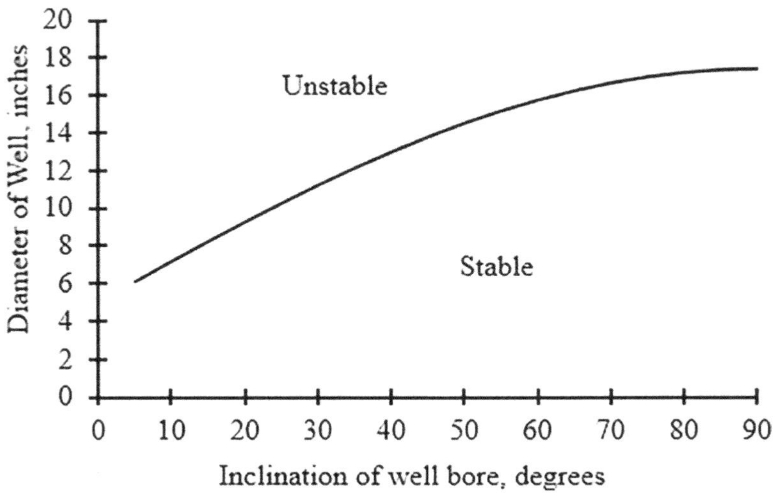

Figure 6-7. Stability data for 5-in. (19.5 lb/ft) drill pipe (Compressive Force, F_{crit}, is 20,000 lbs).

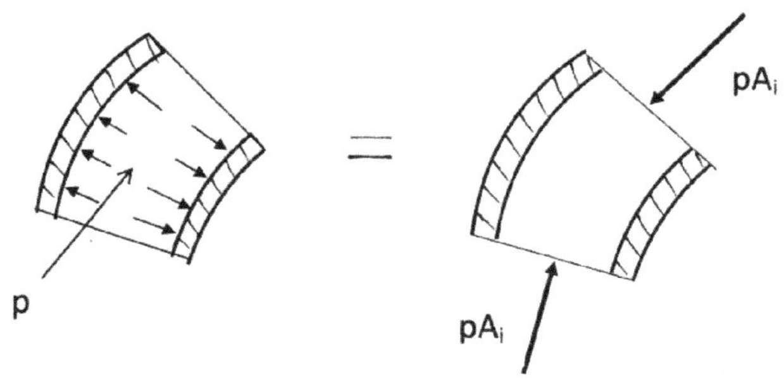

Figure 6-8. Statically equivalent constant pressure effects.

leaving the end force vectors as shown. The equivalent force system is easily incorporated into the bending equation.

This is a special case of Eq. (6-10). The solution, using Eq. (6-17) and noting that $Q = p_i A_i$ is

$$(p_i)_{cr} A_i = \left(\frac{\pi}{L}\right)^2 EI \tag{6-20}$$

where

p_i = pressure inside pipe, psi
A_i = inside area of pipe, in.^{2}

The pipe is assumed to be simply supported as shown in Figure 6-9 and not capped. The pipe in this case cannot be a compression block but burst pressure limits may have to be considered.

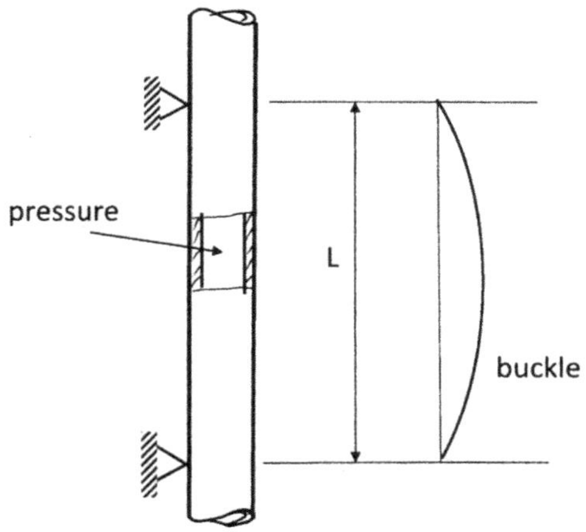

Figure 6-9. Buckled mode from internal pressure.

If the pipe is capped at the both ends and free of the supports nothing happens because the pressure forces at both ends cancel out the effects of the side pressure. However, if the tubular is supported as shown, the inside pressure can cause the pipe buckling. In this case, forces outside of the supports are not a factor in the lateral buckling of the supported section.

Consider a section pipe (4" × 3.5") 100 ft long supported as shown in Figure 6-9. Applying Eq. (6-20)

$$p_{cr} = \left(\frac{\pi}{L}\right)^2 \frac{EI}{A_i} \tag{6-21}$$

$$p_{cr} = \left(\frac{\pi}{1200}\right)^2 \frac{30 \times 10^6 (5.20)}{9.621} = 111.13 \text{ psi}$$

This is equivalent to a buckling force of

$$F_{cr} = (111.3)9.621 = 1{,}069 \text{ lbs}$$

6.3.1 Effective Tension in Pipe

Pressure applied to the outside surface acts to stabilize pipe. Figure 6-10 shows the statically equivalent force systems, which translates external pressure into an *effective* axial tension force.

The net effect of both internal and external pressures combined with an axial force is an effective tension defined by

$$T_{eff} = T - \left(p_i A_i - p_o A_o \right) \tag{6-22}$$

where plus T is the actual internal pull force and a minus T is a compressive force. This term is call *effective tension* because it occupies the space in the

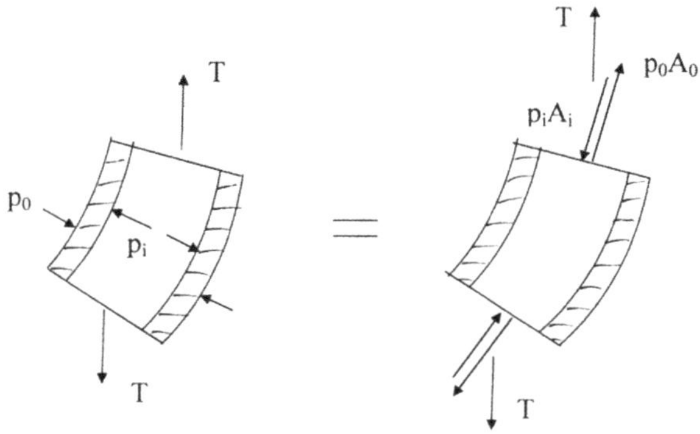

Figure 6-10. Statically equivalent internal/external pressure system.

differential equation of bending reserved for tension. It will vary with vertical location.

A common application of effective tension is in the bending vertical pipe in a hydrostatic situation. In this case, inside and outside pressures are equal, but vary with depth.

$$p_0 A_0 = z\gamma_m A_0$$

$$p_i A_i = z\gamma_m A_i$$

The pressure terms are now expressed

$$p_0 A_0 - p_i A_i = z\gamma_m A_c$$

Noting that the volume displaced by a unit length of pipe is

$$\Delta L(A_c \gamma_m) = \Delta L w_m$$

The pressure can now be written

$$p_0 A_0 - p_i A_i = z w_m$$

where w_m is the weight of drill fluid displaced by a unit length of pipe. Effective tension, in this case is

$$T_{eff} = T + w_m z$$

where z is measure from the top downward. If the pipe is hanging freely then

$$\left(T_{eff}\right)_{top} = H \quad \text{(hook load)}$$

At the lower end, with the pipe open,

$$\left(T_{eff}\right)_{bottom} = T + Lw_m = -Lw_m + Lw_m = 0$$

If the pipe is not open at the lower end but is subjected to an internal force of F_B (tension), then effective tension at location L is

$$T_{eff} = F_B + Lw_m$$

6.3.2 Combined Effects of Axial Force and Internal/External Pressures

Using the previous notation for the compressive force, the effective tension expression, when the axial force is compressive, is

$$Q_{eff} = Q + \left(p_i A_i - p_o A_o\right) \qquad (6\text{-}23)$$

where +Q is compression. By substitution

$$EI\frac{d^2y}{dx^2} + Q_{eff}y = 0 \qquad (6\text{-}24)$$

$$\frac{d^2y}{dx^2} + \beta^2 y = 0 \qquad (6\text{-}25)$$

where

$$\beta^2 = \frac{Q_{eff}}{EI}$$

and following the earlier solution, buckling occurs when

$$\left[Q + (p_i A_i - p_o A_o)\right]_{cr} \geq \left(\frac{\pi}{L}\right)^2 EI \qquad (6\text{-}26)$$

where the left term is effective compression. In this case, loads are uniform with x. Note that external pressure stabilizes the beam while internal pressure destabilizes the beam.

6.3.3 Buckling of Drill Collars Between Stabilizers

In the previous chapter, we considered the buckling of drill collars without stabilizers. The analysis considered the variable internal compression moving upward from the drill bit. We will now consider the buckling of drill collars between stabilizers or between the drill bit and first stabilizer. The analysis will be based on the application of Eq. (6-26) assuming uniform effective compression between lateral supports, which will be taken as the "average effective compression" between sections 1 and 2 (Figure 6-11). In this case, section 1 represents drill bit location.

The effective compression at any location x is [1]

$$Q_{eff} = \left(WOB + Lw_m\right) - wx + (p_i A_i - p_0 A_0) \qquad (6\text{-}27)$$

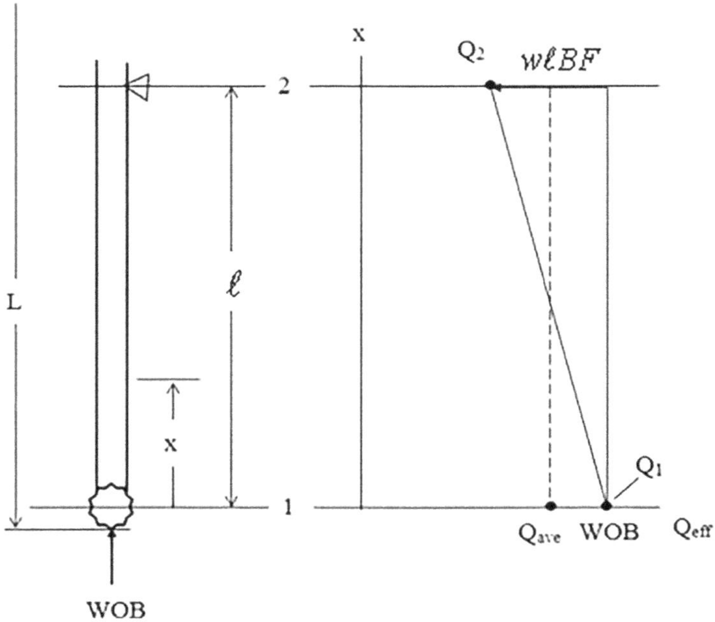

Figure 6-11. Equivalent force between drill bit and first stabilizer.

The equation can be used to determine effective compression between two stabilizers located anywhere within the drill collars.

Considering the effects of hydrostatic pressure

$$p_i A_i - p_0 A_0 = -p(x) A_c = -(L - x) w_m \tag{6-28}$$

By substitution

$$Q_{eff} = \left(WOB + L w_m\right) - wx - (L - x) w_m \tag{6-29}$$

The effective compression at location 1 (x = 0) is

$$Q_{eff,1} = WOB \tag{6-30}$$

The effective compression at location 2 (x = ℓ) is

$$Q_{eff} = WOB - (w - w_m)\ell \tag{6-31}$$

$$Q_{eff,2} = WOB - w\ell BF \tag{6-32}$$

The average of effective compression between locations 1 and 2 is

$$Q_{ave} = \frac{1}{2}\left(Q_{eff,1} + Q_{eff,2}\right) \tag{6-33}$$

$$Q_{ave} = WOB - \frac{1}{2} w \ell BF \tag{6-34}$$

Buckling occurs between two stabilizers (assuming pinned-pinned supports) and assuming a constant compression is

$$Q_{ave} \geq \left(\frac{\pi}{\ell}\right)^2 EI \tag{6-35}$$

6.4 EFFECT OF UNIFORM TENSION/COMPRESSION ON BEAM DEFLECTION

The addition of axial tension, T, to the beam loading affects the analysis significantly (Figure 6-12). In this case, the equations of equilibrium for freebody, dx, now become

$$\frac{dV}{dx} = w \tag{6-36}$$

$$\frac{dM}{dx} = V + T\frac{dy}{dx} \tag{6-37}$$

Note this moment equation is different from Eq. (6-4).

Combining these equations along with Euler's equation gives

$$EI\frac{d^4y}{dx^4} - T\frac{d^2y}{dx^2} - w = 0 \tag{6-38}$$

The solution, along with boundary conditions, establishes beam deflection distribution. Grouping the parameters gives

$$\frac{d^4y}{dx^4} - \beta^2\frac{d^2y}{dx^2} = \frac{q_0}{EI} \tag{6-39}$$

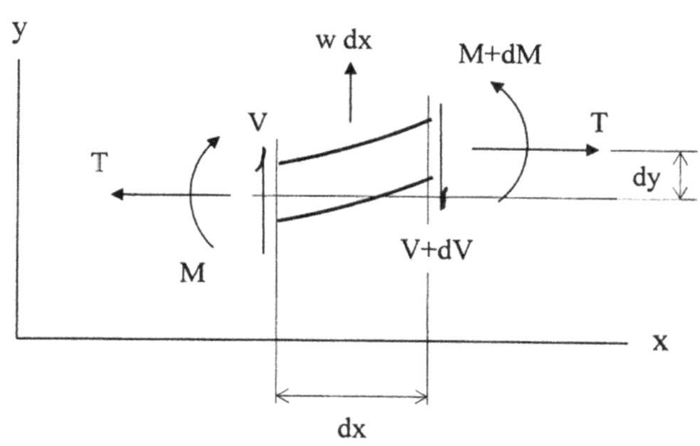

Figure 6-12. Beam under tension and side load.

where

$$\beta^2 = \frac{T}{EI}$$

6.4.1 Inside and Outside Fluid Pressure Combined With Tension/Compression

This section shows how to incorporate the effects of inside and outside fluid pressure into the equations of bending. The differential equation of bending is developed from the freebody diagram given in Figure 6-13.

Equating force components in the y direction to $ma_y dx$ gives

$$\frac{dV}{dx} = -ma_y + q + P\frac{d^2y}{dx^2} \tag{6-40}$$

where

q = force per unit length
$P = p_0 A_0 - p_i A_i$
F = axial internal force (actual) within pipe

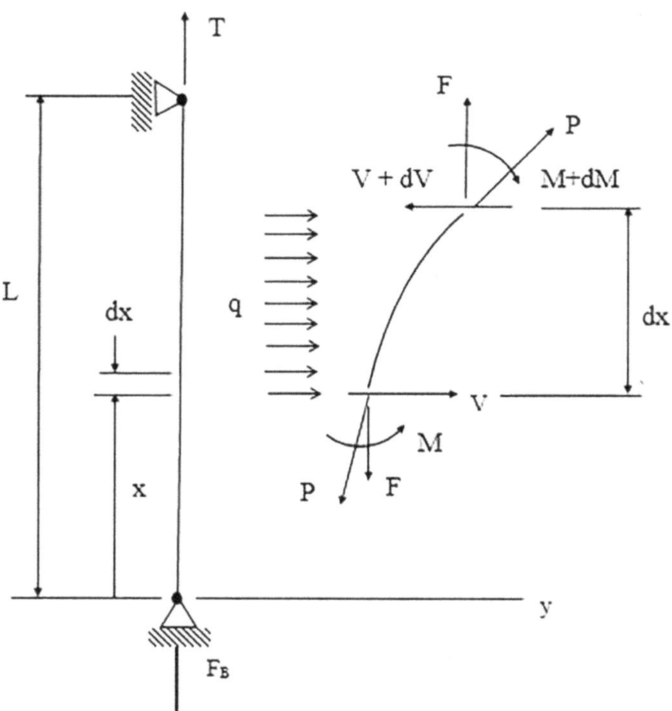

Figure 6-13. Freebody of beam element.

Summation of moments gives

$$\frac{dM}{dx} - F\frac{dy}{dx} - V = 0 \tag{6-41}$$

Combining Eqs. (6-40) and (6-41) and assuming F and P are constant with x, gives

$$\frac{d^2M}{dx^2} - (P + F)\frac{d^2y}{dx^2} + ma_y = q(x) \tag{6-42}$$

Assuming Euler bending, $EI\frac{d^2y}{dx^2} = M$, Eq. (6-42) produces

$$EI\frac{\partial^4 y}{\partial x^4} - (P + F)\frac{\partial^2 y}{\partial x^2} + c\frac{\partial y}{\partial t} + m\frac{\partial^2 y}{\partial t^2} = q(x,t) \tag{6-43}$$

For static bending, Eq. (6-43) simplifies to

$$EI\frac{d^4y}{dx^4} - (P + F)\frac{d^2y}{dx^2} = q(x) \tag{6-44}$$

The term F + P is *effective tension* where F is the actual internal force and P accounts for pressure effects.

6.4.1.1 *Uniform Side Loading*

Now consider a short beam column with uniform side loading, q_o. In this case Eq. (6-44) becomes

$$EI\frac{d^4y}{dx^4} - T_{eff}\frac{d^2y}{dx^2} = q_0 \tag{6-45}$$

where q_o is a constant and $T_{eff} = F + (p_oA_o - p_iA_i)$. The force, F, is assumed to be tension and a plus number.

$$\frac{d^4y}{dx^4} - \beta^2\frac{d^2y}{dx^2} = \frac{q_0}{EI} \tag{6-46}$$

where

$$\beta^2 = \frac{T_{eff}}{EI}$$

The assumption is that T_{eff} is a positive number. Integrating of Eq. (6-46) gives

$$\frac{d^2y}{dx^2} - \beta^2 y = \frac{1}{2}\frac{q_0}{EI}x^2 + C_1x + C_2 \tag{6-47}$$

The particular solution is determined using (D) operators, where D stands for differentiation and its inverse stands for integration.

$$\left(D^2-\beta^2\right)y=\frac{1}{2}\frac{q_0}{EI}x^2+C_1x+C_2 \tag{6-48}$$

$$y=-\left(\frac{1}{\beta^2}+\frac{D^2}{\beta^4}\right)\left[\frac{1}{2}\frac{q_0}{EI}x^2+C_1x+C_2\right] \tag{6-49}$$

The particular solution is

$$y_p=-\frac{1}{\beta^2}\left[\frac{1}{2}\frac{q_0}{EI}x^2+C_1x+C_2\right]-\frac{1}{\beta^4}\frac{q_0}{EI} \tag{6-50a}$$

The complimentary solution is

$$y_c=C_3\sinh\beta x+C_4\cosh\beta x \tag{6-50b}$$

The total solution to Eq. (6-46) is

$$y=y_c+y_p \tag{6-51}$$

Applying the pinned-pinned set of boundary conditions as illustrated in Figure 6-13.

$$y(0)=0 \tag{6-52a}$$

$$\frac{d^2y}{dx^2}(0)=0 \tag{6-52b}$$

$$y(L)=0 \tag{6-52c}$$

$$\frac{d^2y}{dx^2}(L)=0 \tag{6-52d}$$

gives

$$C_1=-\frac{q_0}{EI}\frac{L}{2} \tag{6-53a}$$

$$C_2=0 \tag{6-53b}$$

$$C_3=\frac{1}{\sinh\beta L}\frac{1}{\beta^4}\frac{q_0}{EI}\left(1-\cosh\beta L\right) \tag{6-53c}$$

$$C_4=\frac{1}{\beta^4}\frac{q_0}{EI} \tag{6-53d}$$

Consider, for example, pipe and loading as follows:

OD = 2.5 in.
ID = 2.25 in.
A_i = 3.9761 in.^{2}

$I = 0.6954 \text{ in.}^4$
$q_0 = 2$ lb/ft
p = 1000 and 2000 psi (internal)
L = 30 ft
F = 10,000 lb (tension)

In this case, $F + P = F - p_i A_i$ is positive so Eq. (6-45) applies. Figure 6-14 clearly shows the effect of internal pressure on beam deflection.

6.4.1.2 *Deflection Caused by Internal Pressure and Side Loading*

Now consider a slightly different case where F = 0 and $P = -p_i A_i$. Here F + P is negative and Eq. (6-45) becomes

$$EI\frac{d^4y}{dx^4} + p_i A_i \frac{d^2y}{dx^2} = q_0 \quad (p_i A_i \text{ can also be } Q_{eff}) \tag{6-54}$$

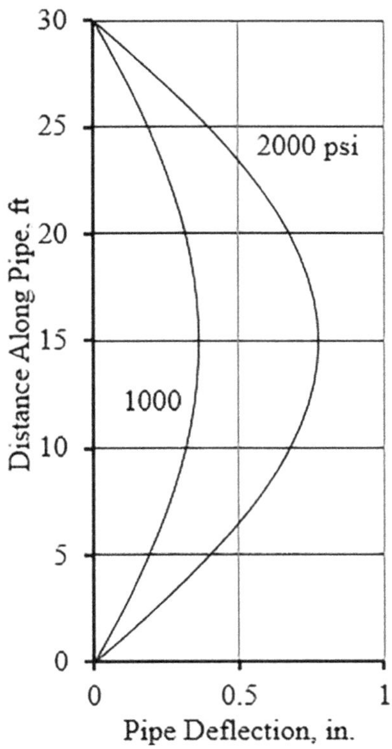

Figure 6-14. Pipe deflection caused by side load, direct pull, and internal pressure.

and

$$\frac{d^4y}{dx^4} + \beta^2 \frac{d^2y}{dx^2} = \frac{q_0}{EI} \tag{6-55}$$

where

$$\beta^2 = \frac{p_i A_i}{EI}$$

Integrating Eq. (6-55) twice gives

$$\frac{d^2y}{dx^2} + \beta^2 y = \frac{1}{2}\frac{q_0}{EI}x^2 + C_1 x + C_2 \tag{6-56}$$

Proceeding as before,

$$\left(D^2 + \beta^2\right) y = \frac{1}{2}\frac{q_0}{EI}x^2 + C_1 x + C_2 \tag{6-57}$$

$$y = \left(\frac{1}{\beta^2} - \frac{D^2}{\beta^4}\right)\left[\frac{1}{2}\frac{q_0}{EI}x^2 + C_1 x + C_2\right] \tag{6-58}$$

The particular solution becomes

$$y_p = \frac{1}{\beta^2}\left[\frac{1}{2}\frac{q_0}{EI}x^2 + C_1 x + C_2\right] - \frac{1}{\beta^4}\frac{q_0}{EI} \tag{6-59a}$$

The complimentary solution is

$$y_c = C_3 \sin \beta x + C_4 \cos \beta x \tag{6-59b}$$

The total solution is

$$y = y_c + y_p \tag{6-60}$$

Applying the boundary conditions gives

$$C_1 = -\frac{q_0}{EI}\frac{L}{2} \tag{6-61a}$$

$$C_2 = 0 \tag{6-61b}$$

$$C_3 = \frac{1}{\sin \beta L}\frac{1}{\beta^4}\frac{q_0}{EI}\left(1 - \cos \beta L\right) \tag{6-61c}$$

$$C_4 = \frac{1}{\beta^4}\frac{q_0}{EI} \tag{6-61d}$$

Figure 6-15 shows deflections for three values of internal pressure. Side loading is $q_0 = 2$ lb/ft and other input data is the same as before.

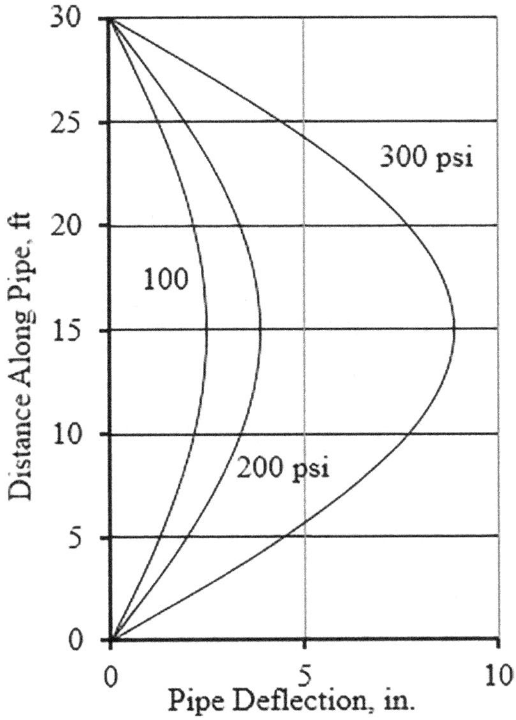

Figure 6-15. Pipe deflection caused by side load and internal pressure.

From Eq. (6-21), the critical bucking pressure is

$$p_{cr} = \frac{\pi^2 EI}{A_i L^2} \tag{6-62}$$

$$p_{cr} = \frac{\pi^2 30(10^6)0.6954}{(3.976)360^2} = 399.6 \text{ psi}$$

In this example, the pipe becomes unstable as internal pressure approaches the critical buckling pressure. Already, at a pressure of 300 psi pipe deflections start to increase at a rapid rate.

6.4.1.3 Drill Collar Deflection Between Stabilizers

In this example, we examine the lateral deflection of drill collars between two stabilizers in a well bore having an inclination, θ (Figure 6-16). Uniform compression, equal to the average effective compression between the two stabilizers is assumed. The uniform lateral load, q_o, is generated by gravity. In this case, Eq. (6-55) applies where

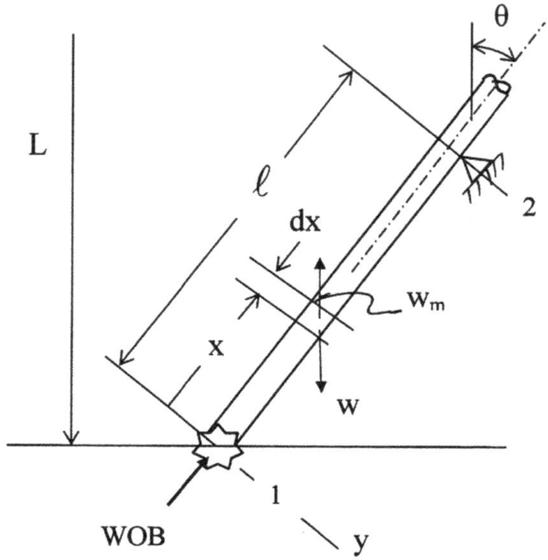

Figure 6-16. Drill collar deflections between stabilizers.

$$\beta^2 = \frac{Q_{ave}}{EI}$$

The solution given by Eqs. (6-46) thru (6-51) apply. The rate of lateral loading is defined by

$$q_0 = wBF \sin\theta \tag{6-63}$$

The average effective compression is determined as follows. In general, effective compression at any section, x, is

$$Q_{eff} = \left(WOB + Lw_m\right) - wx\cos\theta + \left(p_iA_i - p_0A_0\right) \tag{6-64}$$

$$Q_{eff} = \left(WOB + Lw_m\right) - wx\cos\theta - p(x)A_c$$

$$Q_{eff} = \left(WOB + Lw_m\right) - wx\cos\theta - \left(L - x\cos\theta\right)w_m$$

$$Q_{eff} = WOB - wxBF\cos\theta$$

$$Q_{eff,1} = WOB$$

$$Q_{eff,2} = WOB - w\ell BF\cos\theta$$

$$Q_{ave} = \frac{1}{2}\left(Q_{eff,1} + Q_{eff,2}\right) \tag{6-65}$$

$$Q_{ave} = WOB - \frac{1}{2} w \ell BF \cos\theta \tag{6-66}$$

The drilling mud pressure is assumed to be hydrostatic inside and out so effective tension is simply Q_{ave}.

Boundary conditions stated by Eqs. (6-52) also apply according to our assumed model.

6.5 SYNCHRONOUS WHIRL OF ROTATING PIPE

Synchronous whirl of rotating pipe refers to motion where the inside surface of the deflected pipe is always pointing inward toward the axis of the support bearings. The outside surface is always pointing outward away from the bearing axis.

Assuming there is no tension in the pipe and applying Eq. (6-8), the equation of motion is

$$EI \frac{d^4 y}{dx^4} - m\omega^2 y = 0 \tag{6-67}$$

$$\frac{d^4 y}{dx^4} - \alpha^4 y = 0 \tag{6-68}$$

where

$$\alpha^4 = \frac{m\omega^2}{EI}$$

The solution to Eq. (6-68) is

$$y = A \sin\alpha x + B \cos\alpha x + C \sinh\alpha x + D \cosh\alpha x \tag{6-69}$$

Boundary conditions for a simply supported beam are:

$$\mathrm{y(0)} = 0 \tag{6-70a}$$

$$\left.\frac{d^2 y}{dx^2}\right|_{x=0} = 0 \tag{6-70b}$$

$$\mathrm{y(L)} = 0 \tag{6-70c}$$

$$\left.\frac{d^2 y}{dx^2}\right|_{x=L} = 0 \tag{6-70d}$$

Constants B, C, D are zero leaving

$$y = A \sin\alpha L \tag{6-71}$$

For a non-trivial solution

$$\alpha L = n\pi, \quad n = 1, 2, 3, \text{-----}$$

$$\frac{m\omega^2}{EI} = \left(\frac{n\pi}{L}\right)^4 \tag{6-72}$$

The critical rotary speed for the first mode (n = 1) is

$$\omega_{crit}^2 = \left(\frac{\pi}{L}\right)^4 \frac{EI}{m} \tag{6-73}$$

To illustrate consider a section of drill collars (6" × 2", 85 lb/ft) between two stabilizers separated by 60 ft.

$$I = \frac{\pi}{4}\left(R_0^4 - R_i^4\right) = 62.83 \text{ in.}^4$$

$$A = \pi\left(R_0^2 - R_i^2\right) = 25.13 \text{ in.}^2$$

$$m = \frac{w}{g} = \frac{85\, lb/ft}{32.2\, ft/\sec^2}\left|\frac{1\, ft^2}{144\, in^2}\right| = 0.0183\frac{lb - \sec^2}{in} \text{ per inch}$$

Applying Eq. (6-73),

$$\omega_{cr}^2 = \left(\frac{\pi}{720}\right)^4 \frac{30 \times 10^6(62.83)}{0.0183} = 37.33 \text{ (rad/second)}^2$$

$$\omega_{cr} = 6.11 \text{ rad/second}$$

Since

$$\omega = 2\pi \frac{N}{60}$$

$$N_{cr} = \frac{60(6.11)}{2\pi} = 58.34 \text{ rpm}$$

which is within the range of normal rotary speeds.

6.5.1 Whirl of Long Pipe Considering Linear Tension and Damping

The previous discussion applies to short pipe (or shafts) where tension of assumed to be zero. This model probably gives reasonable engineering prediction for, say, drill collars between stabilizers. However, in reality it would not apply to long vertical pipe, such as the drillstring. This case was considered by Huang and Dareing [9].

The prediction of whirling motion in drillstrings is complicated by multiple drilling parameters and hole confinement. Factors that can affect pipe whirl are torque, axial force, rotary speed and damping caused by fluid viscosity. This section addresses the special case of vertical pipes rotating and transmitting torque in a viscous media. The results predict conditions of forward or backward whirl, stable or unstable whirl and synchronous whirl.

It is difficult to account for every constraint, but much insight can be obtained by considering tubular behavior in an open space. It is apparent that when portions of drillstrings contact a well bore, contact forces and resulting kinematics greatly affect whirling behavior. The analysis that follows is based on conditions of non-confinement with an initial concentric position in a vertical well bore.

The differential equations of motion for x(z) and y(z) transverse displacement components are (see Figure 6-17)

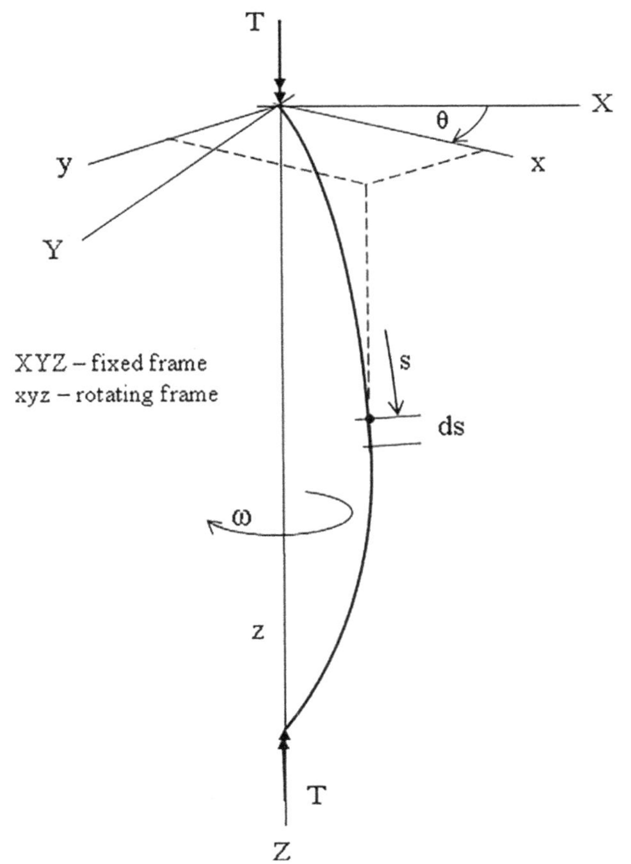

Figure 6-17. Coordinate systems for rotational stability analysis.

$$EI\frac{\partial^4 x}{\partial z^4} - T\frac{\partial^3 y}{\partial z^3} - \frac{\partial}{\partial z}\left(\left[\left(F_B + w_m L\right) + \left(w - w_m\right)(L - z)\right]\frac{\partial x}{\partial z}\right) \tag{6-74a}$$

$$+m\left(\frac{\partial^2 x}{\partial t^2} - 2\omega\frac{\partial y}{\partial t} - \omega^2 x\right) + c\left(\frac{\partial x}{\partial t} - \omega y\right) = 0$$

$$EI\frac{\partial^4 y}{\partial z^4} + T\frac{\partial^3 x}{\partial z^3} - \frac{\partial}{\partial z}\left(\left[\left(F_B + w_m L\right) + \left(w - w_m\right)(L - z)\right]\frac{\partial y}{\partial z}\right) \tag{6-74b}$$

$$+m\left(\frac{\partial^2 y}{\partial t^2} + 2\omega\frac{\partial x}{\partial t} - \omega^2 y\right) + c\left(\frac{\partial y}{\partial t} + \omega x\right) = 0$$

The x and y displacements are measured relative to a reference frame rotating about the Z axis. The solution to these equations is beyond the scope of this book, however, the details are given in ref [9].

One of the results of this paper is shown in Figure 6-18 illustrating conditions of stable and unstable whirl as predicted in the analysis.

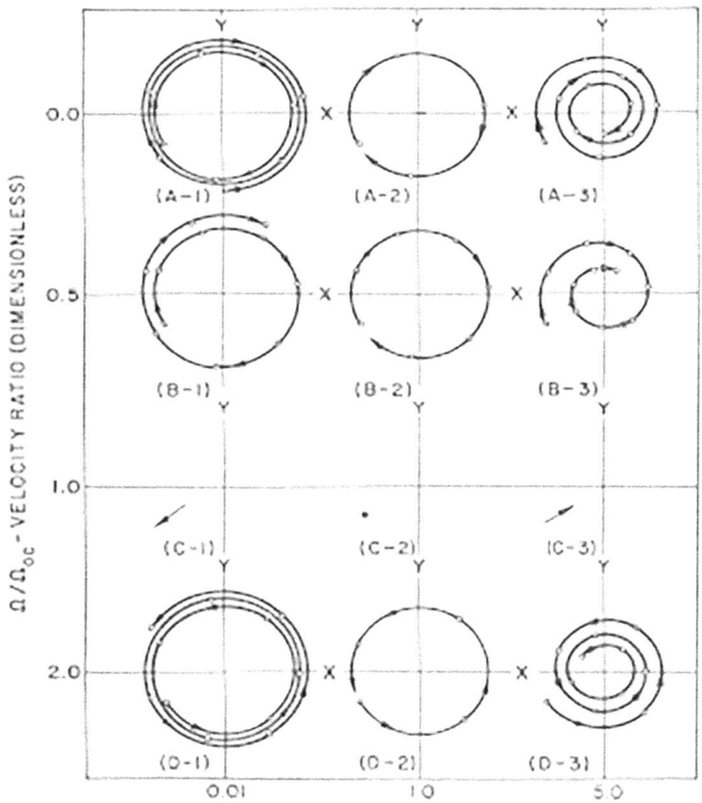

Figure 6-18. Centerline motion of pipe relative to rotating reference frame.

6.6 TORSION BUCKLING OF PIPE

Drill pipe can undergo high levels of torque during directional drilling operations. This is mainly due to wall friction along the entire length of the drillstring. Friction torque develops over curved portions of a planned well path, local dog legs and drill bits. If drill pipe buckles under applied torque, the post buckling configuration will add further wall contact and friction. It will also increase bending stress.

In 1895 Greenhill [10, 11] developed and solved the differential equation of bending defining the lateral stability of shafts under end torque. The coordinate system is shown in Figure 6-19 and the differential equations of bending are

$$M_t \frac{dy}{dx} = -EI\frac{d^2z}{dx^2} \tag{6-75a}$$

$$M_t \frac{dz}{dx} = +EI\frac{d^2y}{dx^2} \tag{6-75b}$$

Later, Webb [12] obtained a solution to these equations using a complex function. His solution starts by defining

$$u = y + iz \tag{6-76}$$

Equations (6-75a) and (6-75b) are combined by multiplying the second equation by the complex number, i, and adding it to the first giving

$$\frac{d^2u}{dx^2} + i\frac{M_t}{EI}\frac{du}{dx} = 0 \tag{6-77}$$

This is a first order equation in terms of $\frac{du}{dx}$ having a general solution of

$$\frac{du}{dx} = Ce^{-i\frac{M_t}{EI}x} \tag{6-78}$$

$$\frac{du}{dx} = C_1\cos\left(\frac{M_t}{EI}x\right) - iC_1\sin\left(\frac{M_t}{EI}x\right) \tag{6-79}$$

$$u(x) = C_1\frac{EI}{M_y}\left[\sin\left(\frac{M_t}{EI}x\right) + i\cos\left(\frac{M_t}{EI}x\right) + C_2\right] \tag{6-80}$$

Using the boundary condition at the left end, y = z = 0 at x = 0 gives $C_2 = -i$ and

$$u = C^*\left\{\sin\left(\frac{M_t}{EI}x\right) - i\left[1 - \cos\left(\frac{M_t}{EI}x\right)\right]\right\} \tag{6-81}$$

Applying the boundary condition at the right end, y = z = 0 at x = L gives

$$\sin\left(\frac{M_t}{EI}L\right) - i\left[1 - \cos\left(\frac{M_t}{EI}L\right)\right] = 0 \tag{6-82}$$

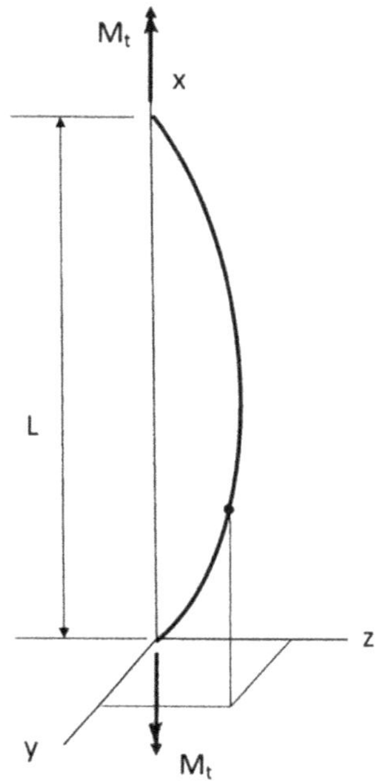

Figure 6-19. Coordinate systems for greenhill's formulation.

Both real and imaginary parts have to be zero simultaneously. This condition is satisfied when

$$\frac{M_t}{EI}L = 2\pi, 4\pi, 6\pi, etc. \tag{6-83}$$

The eigenvalues from this equation give the critical buckling torque, assuming the pipe is structurally pinned at both ends.

The lowest critical torque for the *first buckle mode* is

$$\left(M_t\right)_{crit} = \frac{2\pi}{L}EI \tag{6-84}$$

At that time Webb developed his solution, the feeling was — "only extremely long and thin shafts run in danger of buckling; the question hardly is a practical one, except possibly for long, thin wires transmitting torque."

Again, consider the above drill collar section (6" × 2") between a 60 ft stabilizer spacing,

$$\left(M_t\right)_{crit} = \frac{2\pi}{720}30 \times 10^6(62.83) = 16{,}448{,}842 \text{ in-lb}$$

$$\left(M_t\right)_{crit} = 16.45 \times 10^6 \text{ in-lb}$$

$$\left(M_t\right)_{crit} = 1.37 \times 10^6 \text{ in-lb}$$

Drill collar connections would fail long before torsion buckling.

6.6.1 Buckling of Pipe Due to Torque and Compression/ Tension

When the same problem is solved for both torque and compressive force, Greenhill's equations become

$$EI\frac{d^2z}{dx^2} = -Qz - M_t\frac{dy}{dx} \tag{6-85a}$$

$$EI\frac{d^2y}{dx^2} = -Qy + M_t\frac{dz}{dx} \tag{6-85b}$$

The solution follows Webb's approach [12] and is summarized by den Hartog [13, p295]. The result is presented in a compact form as

$$\left(\frac{M_t}{M_{t,Greenhill}}\right)^2 + \frac{Q}{Q_{Euler}} \geq 1 \quad \text{(buckling occurs)} \tag{6-86}$$

where

$$M_{t,Greenhill} = \frac{2\pi}{L}EI$$

$$Q_{Euler} = \left(\frac{\pi}{L}\right)^2 EI$$

6.6.2 Torsional Buckling of Long Vertical Pipe

Greenhill's equation for predicting stability under pure torque [Eq. (6-84)] assumes zero axial force throughout. His solution applies to slender and relative long wire as indicated above. On the other hand, drillstrings are several thousand feet long and tension varies with depth. In a vertical well, tension is greatest at the top and reduces linearly with depth. These effects were considered by Dareing [1].

The equations of bending are

$$EI\frac{d^4x}{dz^4} - T\frac{d^3y}{dz^3} - \frac{d}{dz}\left[T_{eff}\frac{dx}{dz}\right] = 0 \tag{6-87a}$$

$$EI\frac{d^4y}{dz^4} + T\frac{d^3x}{dz^3} - \frac{d}{dz}\left[T_{eff}\frac{dy}{dz}\right] = 0 \tag{6-87b}$$

The symbol, T, without the subscript, represents applied torque, which is assumed to be constant over the entire length. Effective tension, T_{eff}, however, is assumed to vary linearly along the pipe (Figure 6-20). This complicates the mathematics as T_{eff} is inside the bracket of the last term.

Force, F_B represents the actual internal force in the pipe at z = L. The effective tension is expressed in terms of the independent variable, z, is

$$T_{eff} = \left[F_B + w\left(L - z\right)\right] + zw_m \tag{6-88}$$

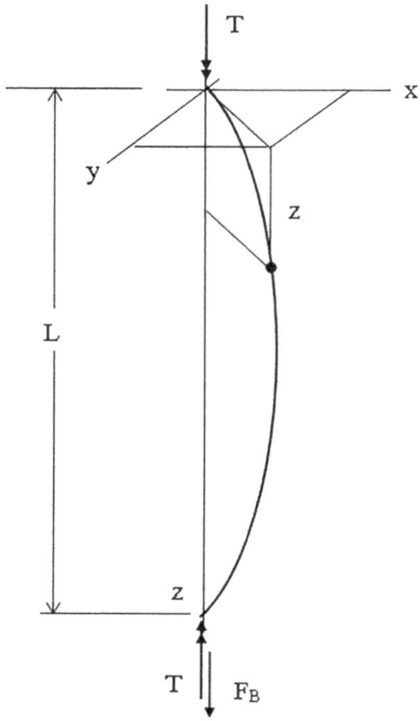

Figure 6-20. Torsion stability model.

The first term is the actual tension in the pipe at location z, while the last term represents the effects of hydrostatic fluid pressure at location z. Here we assume the fluid density is the same both inside and outside the pipe. If the pipe is open to the fluid, $F_B = -Lw_m$ and

$$T_{eff} = (L - z)(w - w_m) \tag{6-89}$$

This is the case if the pipe is open ended and only torque is applied. Going back to the more general case and rearranging Eq. (6-88) (add and subtract w_mL) gives

$$T_{eff} = (F_B + w_mL) + (w - w_m)(L - z) \tag{6-90}$$

Boundary conditions for pinned conditions at both top and bottom are

$$x(0) = y(0) = 0 \tag{6-91a}$$

$$x(L) = y(L) = 0 \tag{6-91b}$$

$$EI\frac{d^2x}{dz^2}(0) = T\frac{dy}{dz}(0) \tag{6-91c}$$

$$EI\frac{d^2y}{dz^2}(0) = -T\frac{dx}{dz}(0) \tag{6-91d}$$

$$EI\frac{d^2x}{dz^2}(L) = T\frac{dy}{dz}(L) \quad (6\text{-}91e)$$

$$EI\frac{d^2y}{dz^2}(L) = -T\frac{dx}{dz}(L) \quad (6\text{-}91f)$$

The dual equations are now combined using the complex variable

$$u = x(z) + iy(z) \quad (6\text{-}92)$$

Multiplying Eq. (6-87b) by the imaginary number i and adding the resulting equation to Eq. (6-87a) gives

$$EI\frac{d^4u}{dz^4} - iT\frac{d^3u}{dz^3} - \frac{d}{dz}\left\{\left[\left(F_B + w_mL\right) + \left(w - w_m\right)(L - z)\right]\frac{du}{dz}\right\} = 0 \quad (6\text{-}93)$$

Ref [1] explains how this equation is solved to give critical buckling torque.

The general solution is represented in Table 6-1 by columns 1 and 3 (α and Θ). The numbers in these two columns represent the general solution to Eq. (6-93) assuming pinned-pinned boundary conditions, where Θ is a function of α. They can be used to determine critical torque for any pipe size.

By assuming the pipe is open ended, $F_B = -w_mL$. Using the numbers in Table 6-1 an interesting plot is generated to show the effects of pipe length on the critical buckling torque. The data applies to 5½ in. (4.892 in. ID, 19.2 lb/ft) drill pipe in both air and fluid (Figure 6-21).

Drilling fluid density has little effect on critical buckling as shown by the two curves in Figure 6-21.

A comparison is also given in Figure 6-22 of critical torque for 4½ (16.6) and 5½ (19.2). The graph shows buckling torque fairly constant for pipe lengths greater than 400 ft. Beyond this point the critical torque for each are

5½	57 kip-ft
4½	37 kip-ft

Drill pipe lengths of, say, 400 ft and higher have about the same buckling torque indicating drill pipe length has little effect on critical buckling torque. In each case, buckling displacement is greatest near the bottom end where tension is lowest.

Yield torque for the two drill pipes are: (from API RP 7G)

	5½" (19.2 lb/ft)	4½" (16.6 lb/ft)
Grade E	44,074 ft-lb	30,807 ft-lb
Grade X	55,826 ft-lb	39,022 ft-lb
Grade G	61,703 ft-lb	43,130 ft-lb
Grade S	79,332 ft-lb	55,453 ft-lb

Table 6-1 Critical Buckling Torque for 5½ in. Drill Pipe
Open ended, $F_B = -w_m L$

α	L, ft	Θ	T, lb-ft	T, kip-ft
0		6.28318		
1	56.04602	4.99873	3.02E+05	301.8
2	70.61192	4.60423	2.21E+05	220.7
3	80.82938	4.39605	1.84E+05	184.1
4	88.96339	4.27309	1.63E+05	162.6
5	95.8322	4.19655	1.48E+05	148.2
6	101.8363	4.14753	1.38E+05	137.8
7	107.2052	4.11579	1.30E+05	129.9
8	112.0843	4.09544	1.24E+05	123.7
9	116.5719	4.08285	1.19E+05	118.5
10	120.7382	4.07574	1.14E+05	114.2
15	138.2089	4.08336	1.00E+05	100.0
20	152.1171	4.12117	9.17E+04	91.7
25	163.862	4.16903	8.61E+04	86.1
30	174.1283	4.22072	8.20E+04	82.0
35	183.3085	4.27381	7.89E+04	78.9
40	191.6511	4.32723	7.64E+04	76.4
45	199.3244	4.38048	7.44E+04	74.4
50	206.4484	4.43331	7.27E+04	72.7
100	260.1026	4.92777	6.41E+04	64.1
150	297.7392	5.36535	6.10E+04	61.0
200	327.7012	5.75751	5.95E+04	59.5
250	353.0028	6.11295	5.86E+04	58.6
300	375.1192	6.43826	5.81E+04	58.1
350	394.8959	6.73847	5.77E+04	57.7
400	412.8681	7.01756	5.75E+04	57.5
450	429.3985	7.27919	5.74E+04	57.4
500	444.7454	7.52138	5.72E+04	57.2
750	509.0995	8.57198	5.70E+04	57.0
1000	560.3311	9.4216	5.69E+04	56.9
1500	641.4104	10.77679	5.69E+04	56.9

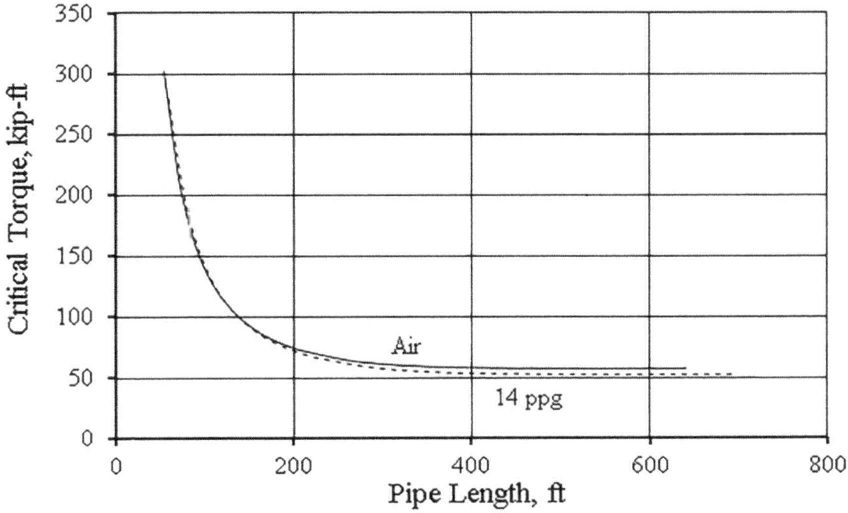

Figure 6-21. Critical buckling torque for first mode. (5½ in. Drill Pipe, 19.2 lb/ft).

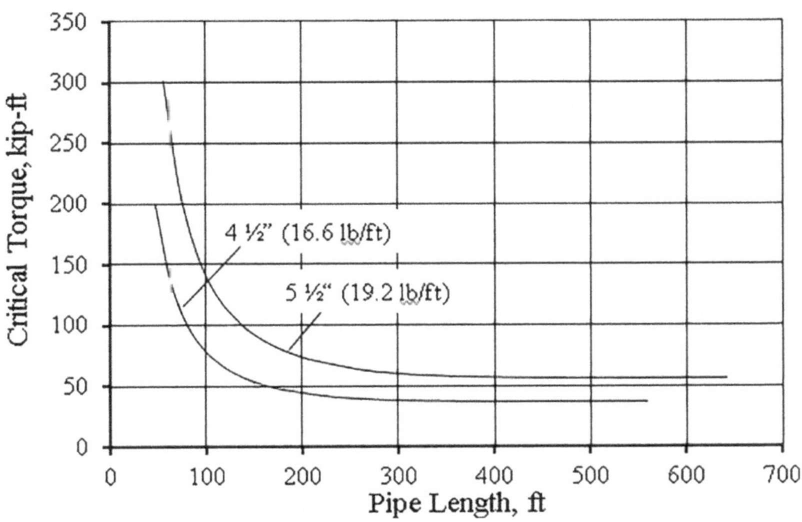

Figure 6-22. Comparison of buckling torques.

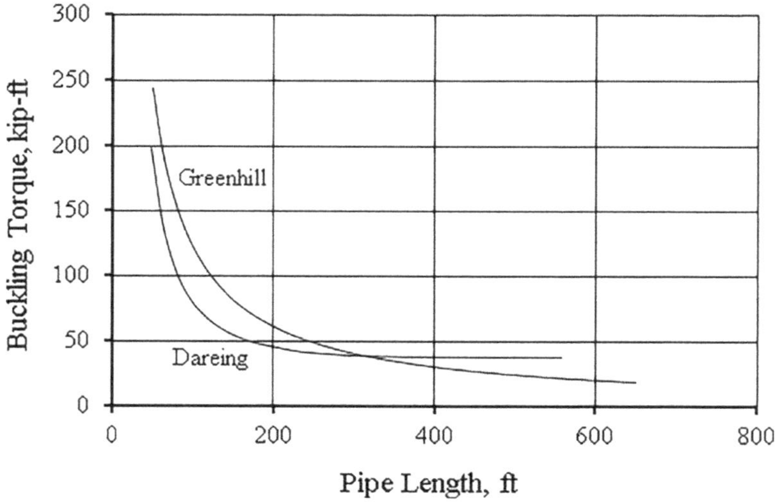

Figure 6-23. Comparison of critical buckling torques. (5½ in. Drill Pipe, 19.2 lb/ft).

Considering the 5½" pipe: Grades E and X, the pipe would yield before buckling occurs. For grades G and S, buckling could occur before yielding.

Considering the 4½" pipe: Grade E would yield before buckling. For grades X, G, and S, the pipe would buckle before yielding.

A comparison is also made between Greenhill's solution for short pipe and for long pipe explained here (Figure 6-23). There is a significant difference between the two critical torque predictions because they are based on two completely different math models. By comparison, the Greenhill mode shape [Eq. (6-82)] extends over the entire pipe while the linear tension case deflects mainly over the bottom 400 ft.

6.6.3 Combined Effects of Compression/Tension and Whirl on Pipe Stability

A similar solution is given by den Hartog [13] for a rotating beam (pipe) subjected to compressive end loading, Q. If the shaft rotates slowly and jarred sidewise, the elastic stiffness will restore the shaft to its original straight position. At some higher rotational speed, the shaft, if disturbed laterally, will not return. Under this condition, the shaft has buckled from the rotation. This scenario is different from the whirling of unbalanced rotating shafts, which experience radial displacement once the shaft begins to rotate.

The differential equation of bending for this case is similar to the previous case except the applied end force is compressive and the side loading inertia is due to centrifugal forces based on synchronous whirl.

$$\frac{d^4y}{dx^4} + \frac{Q}{EI}\frac{d^2y}{dx^2} - \frac{m\Omega^2}{EI}y = 0 \tag{6-94}$$

The end force, Q, is compressive and carries a (+) sign. It can also be viewed as $\left(Q_{eff}\right)_{ave}$. The notation for the speed of rotation is Ω rad/second.

Lateral buckling is imminent when [9]

$$\frac{Q}{2EI} + \sqrt{\left(\frac{Q}{2EI}\right)^2 + \frac{m\Omega^2}{EI}} = \frac{\pi^2}{L^2} \tag{6-95}$$

Equation (6-95) is put into the form

$$\frac{Q}{Q_{cr}} + \left(\frac{\Omega}{\Omega_{cr}}\right)^2 \geq 1 \quad \text{(buckling occurs)} \tag{6-96}$$

where

$$Q_{cr} = \frac{\pi^2 EI}{L^2}$$

$$\Omega_{cr} = \left(\frac{\pi}{L}\right)^2 \sqrt{\frac{EI}{m}}$$

If Q is tension instead of compression, Q in Eq. (6-96) changes signs and the rotary speed required to create whirling motion by the centrifugal force increases as a result. If the left side of Eq. (6-96) is greater than one (1), the rotating column (pipe) is dynamically unstable.

Consider, for example, a 60 ft span of drill collars between two stabilizers. Assuming pinned support at each stabilizer and

OD = 7 in.
ID = 2 in.
I = 117.07 in.^{4}
E = 29 $(10)^6$ psi
w = 120 lb/ft
m = 3.73 slugs/ft
N = 100 rpm
Ω = 10.47 rad/second; $\Omega = \dfrac{2\pi N}{60}$
WOB = 50,000 lbs

then

$$Q_{cr} = \frac{\pi^2 29(10^6)117.07}{\left[12(60)\right]^2} = 64,636 \text{ lbs}$$

and

$$\Omega_{cr} = \left(\frac{\pi}{60}\right)^2 \sqrt{\frac{29(10^6)117.07}{3.73}\left|\frac{1 \text{ ft}^2}{144 \text{ in}^2}\right|} = 6.78 \text{ rad/second}$$

Applying Eq. (6-96)

$$\frac{50,000}{64,636} + \left(\frac{10.47}{6.78}\right)^2 = 0.774 + 2.38 = 3.15$$

Based on these numbers the collar section between the two stabilizers is unstable and will experience synchronous whirl. The compressive force contributes to this instability but collar rotary speed is the main culprit.

6.6.4 Combined Effects of Torque, Compression, and Whirl on Pipe Stability

Southwell and Dough [14] developed the solution for synchronous whirl of shafts subjected to combinations of torque, rotation, and end thrust. Lateral displacement coordinates are x and y with s coordinate being along the length of the shaft.

Differential equations for out-of-plane bending are

$$EI\frac{d^4x}{ds^4}+T\frac{d^3y}{ds^3}+Q\frac{d^2x}{ds^2}-m\omega^2x=0 \tag{6-97a}$$

$$EI\frac{d^4y}{ds^4}-T\frac{d^3x}{ds^3}+Q\frac{d^2y}{ds^2}-m\omega^2y=0 \tag{6-97b}$$

The compressive force, Q, is taken as a plus number. Steps to a solution require a determination of four roots to an algebraic equation, which are determined numerically by trial.

Two sets of boundary conditions are considered by Southwell and Dough: pinned-pinned and fixed-fixed. Results in both cases are presented in graph form, which relate three dimensionless parameters

$$A=\frac{TL}{2EI}$$

$$B=\frac{QL^2}{4EI}$$

$$C=\frac{m\omega^2L^4}{16EI}$$

The authors show that A, B, and C can be related by a simple formula to give a reasonable approximation.

$$\left(\frac{A}{\pi}\right)^2+\frac{4B}{\pi^2}+\frac{16C}{\pi^4}=1 \tag{6-98}$$

The equation is more accurate at small values of C and exact if either A or C is zero. A similar express is given for the fixed-fixed case with each term being multiplied by 0.49, 0.25 and 0.195, respectively.

Based on Eqs. (6-86) and (6-96) it would appear that Eq. (6-98) could be casted in the following form.

$$\frac{Q}{Q_{Euler}}+\left(\frac{M_t}{M_{t,Greenhill}}\right)^2+\left(\frac{\omega}{\omega_{crit}}\right)^2\geq 1 \tag{6-99}$$

The authors state — this equation is not exactly correct but does give a decent approximation.

6.7 SUMMARY OF KEY EQUATIONS

Euler Buckling $$Q_{cr} = EI\left(\frac{\pi}{L}\right)^2$$

Effective Tension $$T_{eff} = T - \left(p_i A_i - p_o A_o\right)$$

Effective Compression $$Q_{eff} = Q + \left(p_i A_i - p_0 A_0\right)$$

For hydrostatic conditions $$\left(p_0 A_0 - p_i A_i\right) = pA_c = z\gamma_m A_c = zw_m$$

(z measured from surface)

Effective Buckling Force $$\left[Q - (p_0 A_0 - p_i A_i)\right]_{cr} = \left(\frac{\pi}{L}\right)^2 EI$$

(watch units)

$$EI\frac{d^4y}{dx^4} + \beta^2\frac{d^2y}{dx^2} = \frac{q}{EI}$$

Synchronous Whirl Speed $$\omega_{crit}^2 = \left(\frac{\pi}{L}\right)^4 \frac{EI}{m},\ m = \frac{w}{g}$$

Critical Torque $$\left(M_t\right)_{crit} = \frac{2\pi}{L}EI \text{ (Greenhill)}$$

Buckling $$\left(\frac{M_t}{M_{t,Greenhill}}\right)^2 + \frac{Q}{Q_{Euler}} \geq 1 \text{ (exact)}$$

Buckling $$\frac{Q}{Q_{cr}} + \left(\frac{\omega}{\omega_{cr}}\right)^2 \geq 1 \quad \text{(exact)}$$

Buckling $$\frac{Q}{Q_{Euler}} + \left(\frac{M_t}{M_{t,Greenhill}}\right)^2 + \left(\frac{\omega}{\omega_{crit}}\right)^2 \geq 1$$

(approximation)

Formulas assume T_{eff} (or Q_{eff}) is constant over pipe length.
Effective compressive force between drill bit and first stabilizer

$$\left(Q_{eff}\right)_{ave} = WOB - \frac{1}{2}w\ell BF$$

6.8 PROBLEMS

1. Determine the reactions at the two supports. Draw the shear and bending moment diagrams for the beam assuming:
   a. w = 127 lb/ft
   b. L = 90 ft

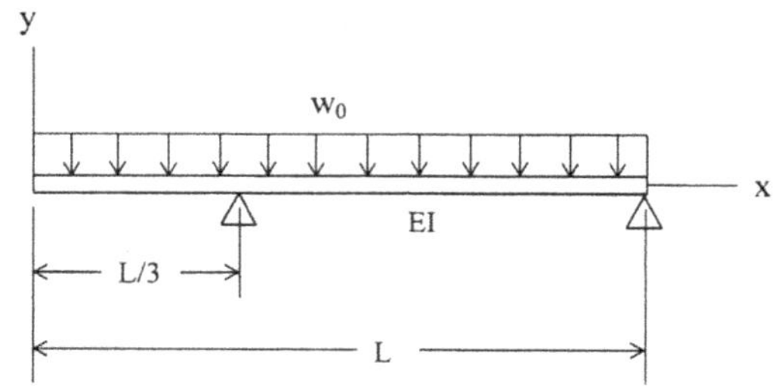

2. Determined the WOB that will buckle the following drill collar section assuming a uniform compression force equal to the average of $(Q_{eff})_{ave}$.
   a. OD = 6½ in.
   b. ID = 2 in.
   c. w = 102 lb/ft
   d. L = 90 ft

   Assume the drill collars are vertical. Solve for the two cases below:
   a. Air
   b. 11 ppg drilling mud (BF = 0.832)

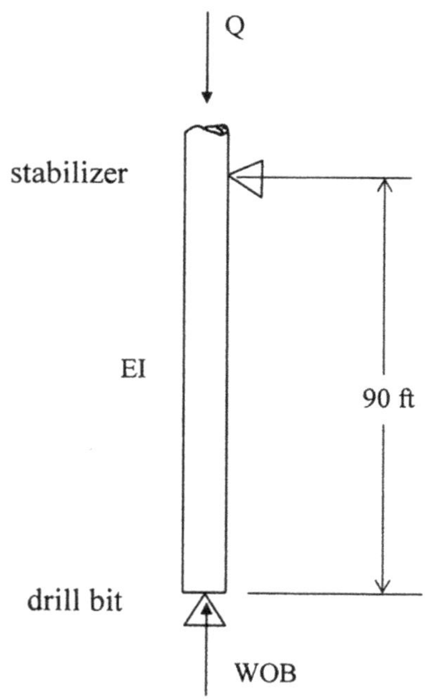

3. Compare WOB_{cr} for slick and stabilized assemblies for the drill collars in problem 2.

4. Horizontal tubing (3" × 2.5") is simply supported by two stabilizers separated by 90 ft.

Determine:
a. Internal pressure required to buckle the tubing
b. Axial uniform force required to buckle the tubing
c. Torque required to buckle the tubing
d. Rotary speed (rpm) the will cause synchronous whirl.
Assume short beam theory.

5. Determine the external force required to buckle (or stabilize) 4" × 3.75" pipe simply supported assuming internal pressure is 1000 psi and external pressure is 3000 psi. The pipe span between supports is 35 ft.

6. Determine the rotary speed that will produce synchronous whirl in 40 ft (stabilizer span) of 7" × 2" (120 lb/ft) drill collars.

7. Using Greenhill and Webb's equation, determine the critical torque for (3" × 2.5") tubing assuming: a) no axial force and b) with an axial tension force of 5000 lb. The tube is 100 ft long.

8. Determine the maximum deflection of the drill collars between the two stabilizers depicted in the attached figure.

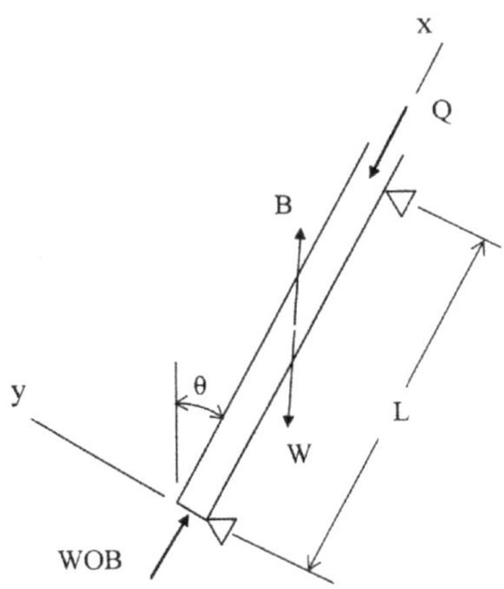

Assume the drill collar span is simply supported and

Hole inclination, θ = 30 deg.
L = 60 ft
Drill Collar size is 8" × 2½" (154 lb/ft)
Mud weight is γ_m = 12 ppg (BF = 0.817)

9. Using the solution for critical torque for long vertical pipe (Table 6-1), Determine critical torques for 2 3/8 in. (4.85 lb/ft) drill pipe in 12 ppg drilling fluid. Plot critical torque verse length ranging from 50 ft to 1000 ft in 50 ft increments.

6.9 LIST OF REFERENCES

1. Dareing, D. W.; *Mechanics of Drillstrings and Marine Risers*, ASME Press, NY, 2012.
2. William Riley, Leroy Sturges, and Don Morris; *Mechanics of Materials*, John Wiley and Sons Inc (6th) ed. 2007.
3. *Manual of Steel Construction*, 9th ed, American Institute of Steel Construction, New York, 1989.
4. *Specification for Aluminum Structures*, Aluminum Association Inc, Washington, DC, 1986.
5. *Timber Construction Manual*, American Institute of Timber Construction, John Wiley & Sons Inc, New York, 1985.
6. Bogy, D. B. and Paslay, P. R.: "Buckling of Drill Pipe in an Inclined Hole", Trans., ASME, Series B (May 1964) 214–220.
7. Paslay, P. R. and Bogy, D. B., "The Stability of a Circular Rod Laterally Constrained to be in Contact with an Inclined Circular Cylinder," *J. Appl. Mech*. (Dec 1964) **31**, 605–610.
8. Rapier Dawson and P. R. Paslay, "Drill Pipe Buckling in Inclined Holes", *Journal of Petroleum Technology*, October 1984, pp. 1734–1738.
9. Huang, T. and Dareing, D. W., "Predicting the Stability of Long Vertical Pipe Transmitting Torque on a viscous Medium, "Trans., ASME, *Journal of Engineering for Industry*, Vol. 88, Series B, No.2, May 1966, 191–200.
10. Stephen P. Timoshenko and James M. Gere; *Theory of Elastic Stability*, McGraw-Hill Book Company, 1961, p. 157.
11. A. G. Greenhill, "On the Strength of Shafting When Exposed both to Torsion and End Thrust", Proceedings of the Institution of Mechanical Engineers, London, England, Vol. 6, 1883, pp. 182–209.
12. H. A. Webb solution is taken from the English text by John Case, Strength of Materials, Edward Arnold & Co. London 1925 (from ref. 9).
13. J. P. Den Hartog, *Advanced Strength of Materials*, McGraw-Hill Book Co., 1952 (see p. 297).
14. R. V. Southwell and B. S. Dough; "On the Stability of a Rotating Shaft Subjected Simultaneously to End Thrust and Twist", *British Association for the Advancement of Science*, Vol. 8, (1921) pp. 345–355.

Chapter 7

DRILLSTRING DYNAMICS AND VIBRATION CONTROL

Vibrations in drillstrings have been a concern for many years. Equipment failures, especially those that occur in the bottom-hole assembly, can be linked directly to severe vibrations. Equipment, associated with directional drilling has become more and more complex and vulnerable to every mode of vibration. Reliability and maintainability have become important considerations in downhole tool design. This chapter considers the causes and control of mechanical vibrations along drillstrings, especially in the bottom hole assembly region.

For many years drill bit life dictated drilling time between trips. The bearing cavity in early roller bits was unsealed and the drilling mud environment led to early bearing failure. In many cases, bearings failed before the teeth wore out. Bearing failure, no doubt initiated the destruction of the bit shown in Figure 7-1. Once the cones separated from the bit body, the bit became totally dysfunctional. Vibrations may have played a role in initiating this failure.

The overall dynamic behavior of drillstrings is a combination of axial, torsional and lateral modes of vibration occurring simultaneously. Mathematically coupling of these vibrations is complicated by wellbore confinement. Much can be learned, however, through mathematical models of the various modes (axial, torsion, lateral) separately.

7.1 AXIAL VIBRATIONS IN DRILLSTRINGS

Axial vibrations in drillstrings have been seen for many years at the surface. Garrett [1] measured vertical vibrations and studied the effects of shock absorbers in hard rock areas. In one test, an accelerometer was attached to the swivel to measure Kelly motion. Measurement were taken with and without a special tool (shock absorber) inserted directly above the drill bit. His measurements showed a reduction in surface vibrations with a shock absorber as well as a frequency of vibration of 3 cycles per revolution (Figure 7-2). This frequency can be related to a three-lobed pattern commonly found in core tops (Figure 7-3).

Assuming 3 cycles per bit revolution then

$$f(cps) = \frac{N(rpm)}{60\,\mathrm{sec/min}} 3(cycle/rotation)$$

Figure 7-1. Total destruction of a drill bit.

$$f = \frac{N}{20} \text{ cps}$$

where N is bit rotational speed in rpm. From Figure 7-2, measured frequency of the vibration is about 4.5 cps. The above equation indicates a bit rotational speed of 90 rpm.

7.1.1 Natural Frequencies of Axial Modes

Natural frequencies and mode shapes are important information in design and operation of equipment. Resonance and severe vibrations can be avoided by

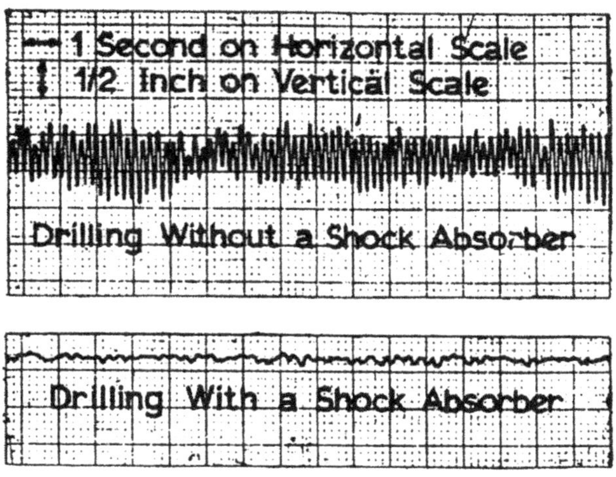

Figure 7-2. Early measurements of the effects of using shock absorbers.

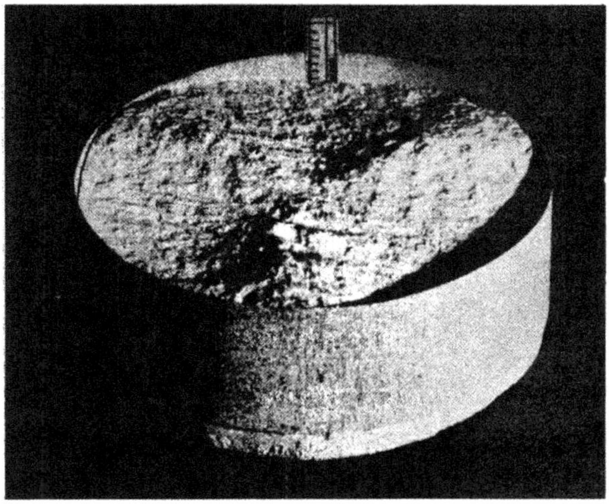

Figure 7-3. Three lobed undulations commonly found in core tops.

simply avoiding frequency tuning between the frequency of the driving force and the natural frequency of any of the vibration modes. This fundamental theory also applies to drillstrings and BHAs.

Drillstrings can be viewed as having two different sets of boundary conditions. In one case the bottom end is fixed as shown in Figures 7-4a and 7-4b. In the second case, the bottom end of the drill collar section is free ended.

With reference to Figure 7-4, consider a drillstring made up of a drill collar section, a drill pipe section and a top mass representing the swivel and traveling block, which typically weigh about 50,000 lbs. The top spring accounts for the flexibility of the drawworks cable, which is typically around k = 50,000 lb/in. The spring constant for any drawworks can be determined by changing the hook load (picking up more drillstring weight) and measuring vertical travel of the Kelly. The added stretch of the drawworks, due to the added hook load, is the difference between measured travel and calculated travel assuming no stretch in the draw lines.

Natural frequencies are determined from undamped free vibrations analysis. The equation of motion for the axial mode becomes

$$AE\frac{\partial^2 u}{\partial x^2} = m\frac{\partial^2 u}{\partial t^2} \tag{7-1}$$

$$\frac{\partial^2 u}{\partial x^2} = \frac{1}{c^2}\frac{\partial^2 u}{\partial t^2}$$

where

u(x,t) = axial displacement of any point within the pipe
A = cross section area of pipe
E = modulus of elasticity
m = mass per unit length of pipe
c = acoustic velocity of compression (tension) wave (16,850 ft/second in steel pipe)

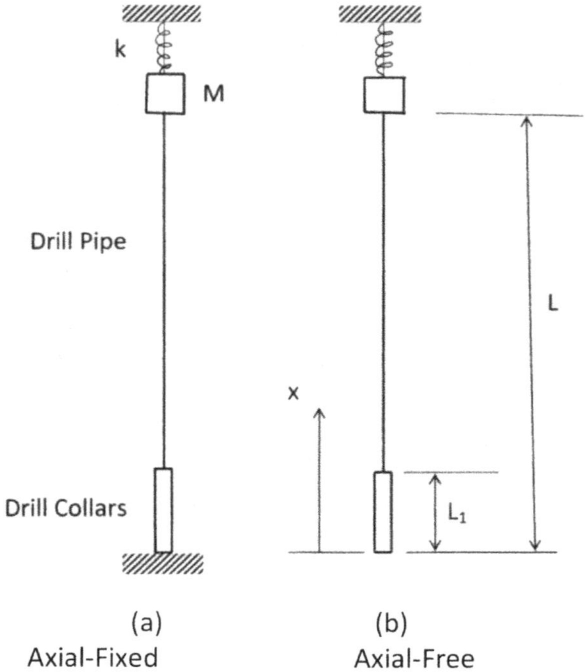

Figure 7-4. Axial vibration models of drillstrings.

This equation applies over both drill pipe and drill collar sections.

$u_1(x,t)$ = drill collar displacements

$u_2(x,t)$ = drill pipe displacements

The determination of mode shapes and corresponding natural frequencies proceeds as follows.

Boundary conditions are

$$u_1(0,t) = 0 \quad \text{(drill bit)} \quad (7\text{-}2a)$$

$$u_1(L_1,t) = u_2(L_1,t) \quad \text{(interface; pipe/collars)} \quad (7\text{-}2b)$$

$$A_1E\frac{du_1}{dx}(L_1,t) = A_2E\frac{du_2}{dx}(L_1,t) \quad \text{(interface; pipe/collars)} \quad (7\text{-}2c)$$

$$M\frac{\partial^2 u_2}{\partial t^2} + A_2E\frac{\partial u_2}{\partial t} + ku_2 = 0 \quad \text{(drawworks)} \quad (7\text{-}2d)$$

The fourth boundary condition is derived by direct application of the second law to the top mass.

Natural frequencies are determined, following the method of separation of variables as applied in ref [2] starting with

$$u(x,t) = X(x) \sin \omega t \tag{7-3}$$

Figure 7-5 shows how natural frequencies change with hole depth for the following set of drillstring parameters. Multiple roots to the characteristic equation give the frequencies of each of the modes of vibration.

Drill pipe size (5 in. OD by 4.276 in. ID)
Drill collar size (7 in. OD by 2.937 in. ID)
Length (L_1) = 550 ft
Acoustic velocity = 16,850 ft/second
Spring Constant of Drawwork Cables (k) = 640,000 lb/ft (53,333 lb/in.)
Mass of Swivel and Traveling Block (M) = 466 slugs (15,000 lbs)
Drillstring Length (L) is a variable.

Considering a well depth of 5000 ft, the natural frequency of the fourth mode is 6.13 cps and a critical rotary speed of 122.6 rpm (Table 7-1). The fourth mode would resonant with a roller cone bit at this speed. Note also that the frequency band width of the response is quite large so any of the axial modes can easily be excited.

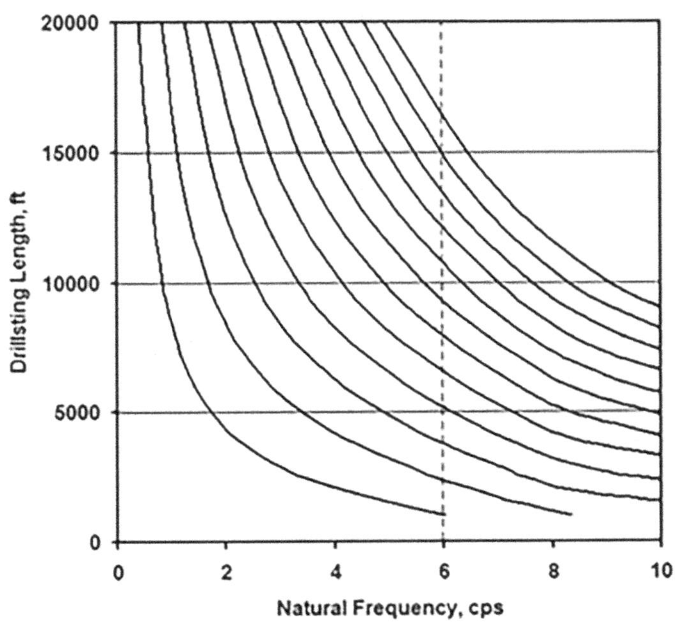

Figure 7-5. Natural frequencies of axial modes (rotary speed equals 20 times frequency).

Table 7-1. Natural frequencies of axial modes [2].

	Length — 5000 ft		Length — 20,000 ft	
Mode	Natural Frequency (cps)	Critical Rotary Speed* (rpm)	Natural Frequency (cps)	Critical Rotary Speed* (rpm)
1	1.72	34.4	0.414	8.3
2	3.39	67.8	0.84	16.8
3	4.87	97.4	1.26	25.2
4	6.13	122.6	1.67	33.4
5	7.26	145.2	2.09	41.8
6	8.30	166.0	2.50	50.0
7	9.77	195.4	2.92	58.4

*Based on 3 cycles per bit rotation.

On the other hand, a rotary speed of 80 rpm, would fall between critical speeds of the 2nd and 3rd axial modes so a roller cone bit should operate relatively smooth at this speed.

According to Figure 7-5, the frequency band width of any mode is of the order of $\Delta f \sim 0.1$ cps ($\Delta N \sim 2$ rpm). Since the exciting frequency is related directly to bit rotation, rotary speed has to be maintained within a $\pm$ 1.0 rpm variance in order for energy to be fed into the mode. It takes time to put energy into a mode because of drillstring length. On the other hand, the frequency band width of drill collar response is much larger, plus the drill collars are much shorter making it easier to excite drill collar modes as demonstrated in shallow test wells. The frequency band width at 20,000 ft is much smaller and it is more difficult for bit speeds to latch onto these modes.

Figure 7-5 also shows how natural frequency changes with depth. For example, at a rotary speed of 120 rpm (6 cps), mode number changes about every 1500 ft.

7.1.2 Damping of Axial Modes

The transient free vibration taken on a drill site (Figure 7-6) provides some insight into the effects of mechanical friction on the damping coefficient, γ. Damping is best determined experimentally [3]. The following discussion is based on the damped free vibration equation.

$$AE\frac{\partial^2 u}{\partial x^2} = \gamma\frac{\partial u}{\partial t} + m\frac{\partial^2 u}{\partial t^2} \tag{7-4}$$

The frequency of this response is about 4 cps. Assuming this is the response of a particular mode in the drillstring, then the damping factor, ζ_n, for this mode can be determined from the log decrement

$$\delta = \frac{1}{n}\ln\frac{X_i}{X_{i+1}} \tag{7-5}$$

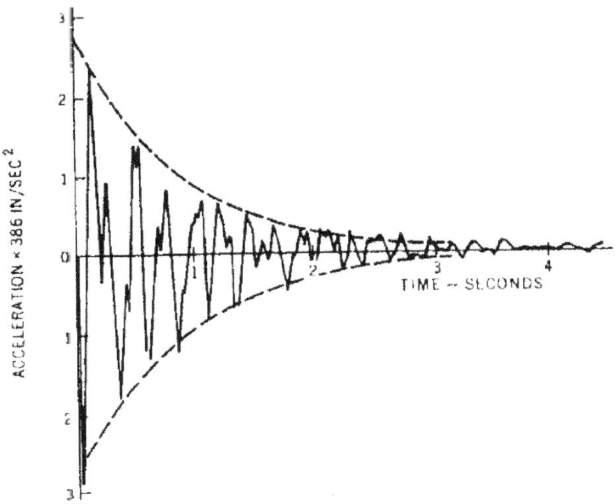

Figure 7-6. Damped free vibration of drillstring taken at the surface [3].

From the graph

$$\delta = \frac{1}{4}\ln\frac{1.5}{0.5} = 0.2747$$

Using

$$\delta = \frac{2\pi\zeta}{\sqrt{1-\zeta^2}} \cong 2\pi\zeta \tag{7-6}$$

$$\zeta = 0.0437$$

Recall that the modal damping factor is

$$\zeta = \frac{C}{C_{cr}} \tag{7-7}$$

from which

$$C_{cr} = 2M\omega \tag{7-8}$$

$$C = \zeta(2M\omega) \tag{7-9}$$

where, C and M are modal damping and modal mass, which can be related to local damping, γ, and local mass, ρ as follows.

Using modal analysis techniques, the solution to Eq. (7-4) can be represented by

$$u(x,t) = \sum_{i=1}^{\infty} X_i(x)\eta_i(t) \tag{7-10}$$

The substitution of this function and its derivatives into Eq. (7-4) and taking advantage of the orthogonal properties of the modes, gives (assuming proportional damping)

$$M_i\ddot{\eta}_i + C_i\dot{\eta}_i + K_i\eta_i = 0 \tag{7-11}$$

where

$$M_i = \rho\int_0^L X_i^2\,dx$$

and

$$C_i = \gamma\int_0^L X_i^2\,dx$$

Substituting into Eq. (7-9) gives

$$\gamma = C = 2\zeta\rho\omega \tag{7-12}$$

Assuming 4½ in. drill pipe we obtain

$$\gamma = 2(0.0437)\frac{16.6}{32.2}2\pi(4) = 1.13 \text{ lb/fps per ft of pipe length}$$

This number represents a reasonable approximation for the modal damping coefficient. It includes the effects of all friction forces acting on the drill string that existed at the time of this measurement. A more accurate method would be to experimentally determine the transfer function for the axial modes. Modal damping could be taken from the imaginary component of each modal response. This could be done with current MWD technology.

7.1.3 Forced Vibration of Axial Modes

Drillstrings also resonate by frequency tuning. The driving force can excite one or several modes simultaneously. First consider the axial mode of vibration.

Axial vibrations have been observed at the surface through vertical vibrations of the Kelly for many years, especially while drilling hard formations. Accelerometer measurements of the swivel bounce indicated that the frequency of these vibrations were three cycles per rotation of the drillstring. The tops of core samples show a three-lobed pattern, which is consistent with frequencies observed at the surface (see Figure 7-7). Three lobed patterns in core tops are common.

How these lobed patterns are formed is unclear. Since they commonly occur in hard formations, it is clear that any slight reorientation of the axis of the drill bit would cause the WOB to be concentrated off of well bore center. This would start the process of forming the three lobes. Field measurements of bit force and drill collar bending directly above the bit (Figure 7-8) supports this theory.

Figure 7-7. Top of core taken from dolomite limestone formation (Courtesy of A. Scovil Murray, Imperial Oil Company, Calgary, Alberta, Canada).

The data also shows that bit force ranges from zero to near 150,000 lbs with an indicated WOB of 40,000 lbs [3]. Obviously, the bit disengages from the rock over part of the bit loading cycle so the displacement model applied to the drill bit is not exactly true.

These simultaneous measurements show a clear pattern of bending at the beginning of a buildup of axial bit force vibrations. In this case, BHA stabilization my help alleviate axial vibrations. Once the lobes in the hole bottom form they can cause boundary movement and corresponding boundary forces, which can drive longitudinal vibrations. The severity of the vibration response would depend on frequency tuning. Non-homogeneous formations may also play a role in the formation of these lobes.

Dareing and Livesay [4] modeled drillstrings as a system of two continuous pipes vibrating in a linearly damped environment. The analysis considers

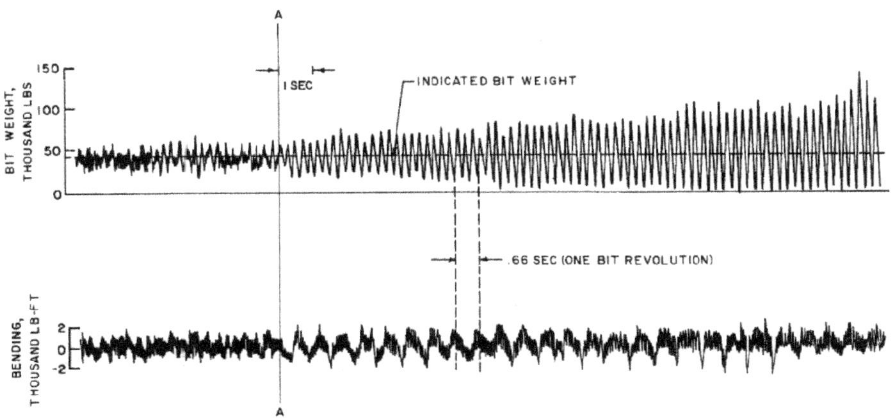

Figure 7-8. Transition into periodic bit force.

vibration response caused by boundary displacements induced by the three lobes associated with roller cone drill bits (Figure 7-7).

Summing forces on a differential element and equating them to the inertia term gives

$$AE\frac{\partial^2 u}{\partial x^2} = \gamma\frac{\partial u}{\partial t} + m\frac{\partial^2 u}{\partial t^2} \tag{7-13}$$

which applies to each tubular section. Variable tension caused by the weight is a static load and does not affect the dynamic analysis. The damping coefficient, γ (having units of lbs per ft/second per ft of pipe length), is best determined experimentally.

Damping distribution is needed, to make vibration response predictions. The mass and elastic can easily be calculated. Damping normally has to be determined experimentally because of the many factors that affect it.

The solution to Eq. (7-13) applies to both the drill pipe and drill collar sections. Axial displacements in each section are identified by

u_1 = drill collar section

u_2 = drill pipe section

These functions are linked together with appropriate boundary conditions.

The boundary condition at the surface considers the mass of swivel and traveling block, M and stiffness of the cables in the draw works, k. From a free body of the top mass, M

$$A_2E_2\frac{\partial u_2}{\partial x} + ku_2 + M\frac{\partial^2 u}{\partial t^2} = 0 \tag{7-14}$$

The connection between the drill pipe and drill collars must satisfy

$$u_1(L_1,t) = u_2(L_1,t) \tag{7-15}$$

$$E_1A_1\left.\frac{\partial u_1}{\partial x}\right|_{x=L_1} = E_2A_2\left.\frac{\partial u_2}{\partial x}\right|_{x=L_1} \tag{7-16}$$

Finally, bit displacement (caused by the three-lobed pattern found in core tops) is defined by

$$u_1(0,t) = u_0\sin\omega t \tag{7-17}$$

In terms of complex numbers

$$u_1(0,t) = \mathrm{Re}\left[-iu_0e^{i\omega t}\right]$$

The solution to Eq. (7-13) can be expressed in terms of the complex function

$$u(x,t) = \bar{u}(x)e^{i\omega t} \tag{7-18}$$

After substitution, Eq. (7-13) becomes

$$\frac{d^2\bar{u}}{dx^2} + \eta^2\bar{u} = 0 \tag{7-19}$$

where

$$\eta^2 = \frac{\omega^2}{c^2} - i\frac{\gamma\omega}{AE}$$

The solution to Eq. (7-19) is

$$\bar{u} = B\sin(\eta x + b)$$

where B, η and b are complex numbers. B and b depend on the boundary conditions formulated earlier. The constants B_1 b_1 B_2 b_2 are complex numbers determined by computer calculations. The final solution gives dynamic response in both drill pipe and collars in complex form. The real solution to the drillstring vibration prediction is the imaginary part of the complex solution.

The solution gives the overall shape of the axial amplitudes (Figure 7-9) and the frequency response (Figure 7-10) corresponding to various levels of assumed damping for both drill pipe and drill collars. The phase angle diagrams have been omitted but are included in ref [4]. The damping coefficients used in these calculations should be considered reasonable approximations. Notice the multiple resonant spikes, indicating multiple mode participation possibilities to the applied bit motion.

Even though higher modes have been observed through measurements in shallow test wells, it is more difficult to develop the higher modes in typical drillstrings because of the narrow bandwidth of the frequency response as shown in Figure 7-10. It takes time to put energy into the higher modes and this has to be done at a fairly constant rotary speed. This suggests that drill collars dictate the axial vibrations in the drill pipe.

Note that the frequency response curves (Figure 7-10) are bounded along the lower edge by the response of the drill collars. The level of damping along the drill collars is represented by γ_1 having units of lb/ft per fps. γ_2 represents the level of damping along the drill pipe. The vibration of the drill collar section alone dictates the magnitude of dynamic forces on the drill bit and other tools near the bit.

7.1.4 Drillstring Response to a Force Vibration Device

A creative applications of the Moineau mechanism is its use as an actuator in a pressure pulsing tool [5]. The vibration mechanism is based on a fluid valve, which is caused to vary by the Moineau rotor. The variable fluid valve creates a time dependent pressure drop, which causes a variable axial force to excite a drillstring. An objective of the pressure pulsating tool is to create dynamic force on the drill bit to enhance rate of penetration. It can also be used to reduce friction between drill pipe and well bore as the tool can be located anywhere in a drillstring.

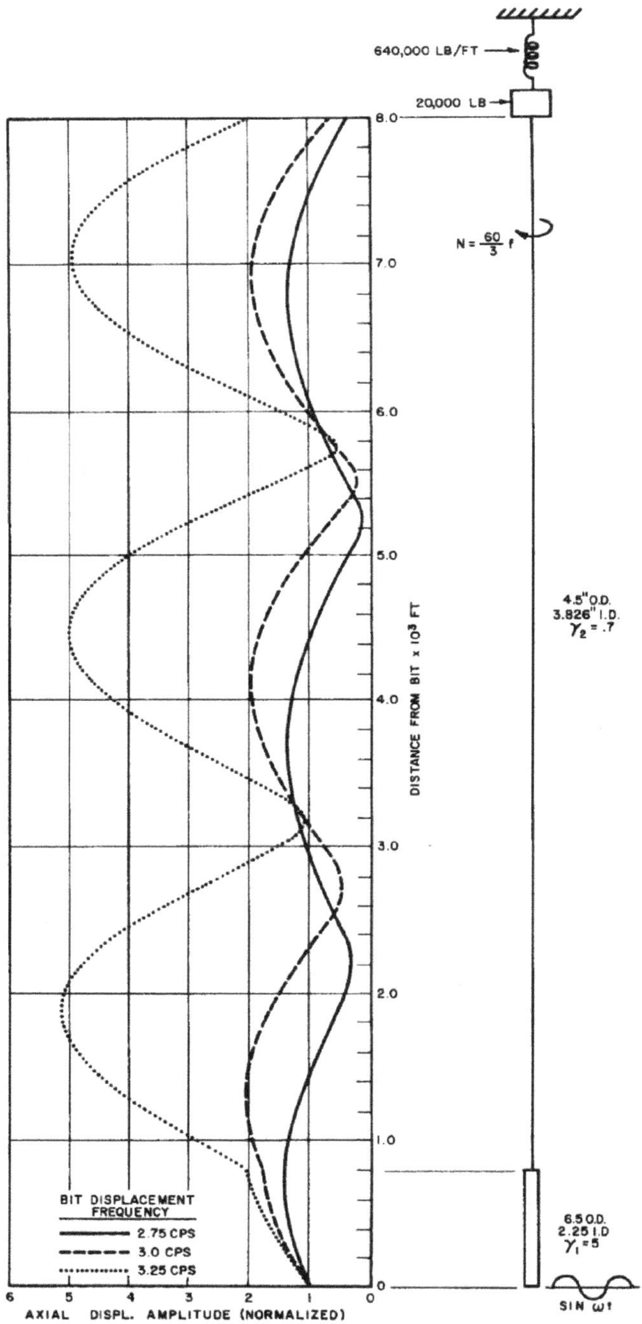

Figure 7-9. Frequency response of a drillstring.

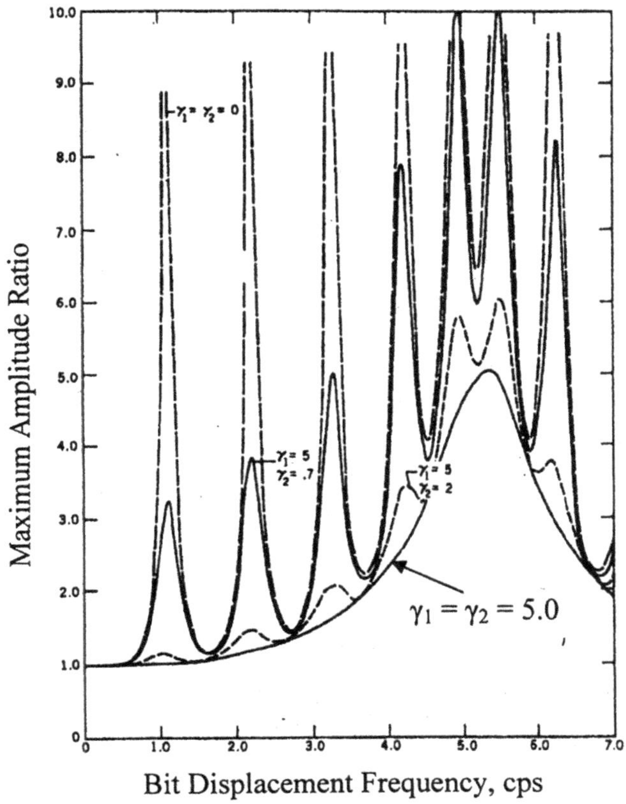

Figure 7-10. Frequency response of a drillstring.

The vibration tool uses the straight line rotor motion to activate a movable disc, creating time varying fluid flow area across a fixed disc. The movable valve member is rigidly attach to the lower end of the rotor. As the moving valve opening moves across the fixed valve opening, the effective valve (or nozzle) varies from a minimum to a maximum. This variation in flow area creates a pressure drop across the valve while drilling mud moves at constant flow, established by the mud pumps.

The vibration device (Figure 7-11) is typically located directly above a flexible tool. The variable opening valve is shown along with a pressure pulse, P(t). The simplified model shows the drill collars as a lumped mass; however, drill collars could be viewed as a distrusted mass having multiple degrees of freedom. The device can be located anywhere in the drillstring to excite axial modes. This application is especially useful in directional wells where pipe friction is a problem.

Consider the dynamic response of the lump mass indicated by WOB. The freebody diagram of the valve section gives

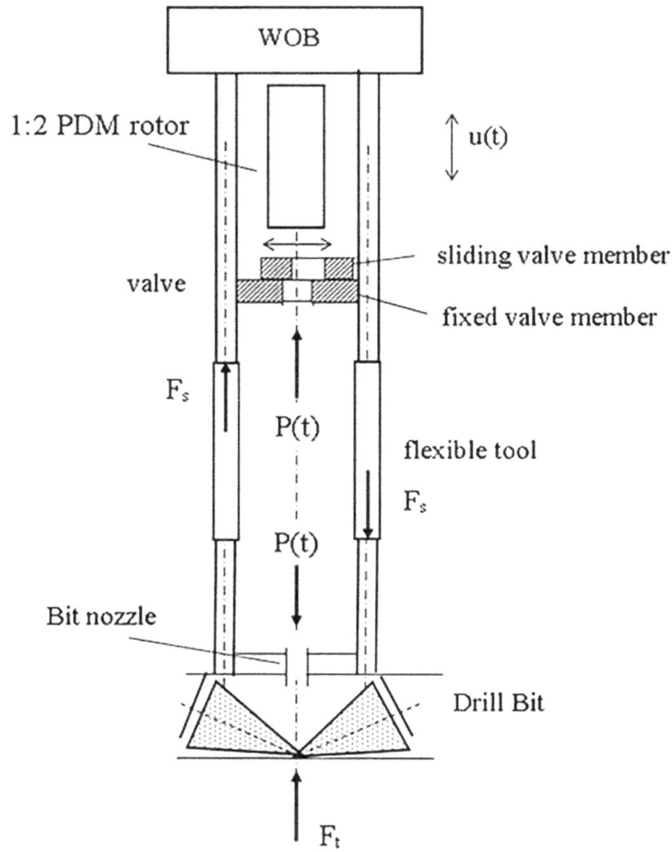

Figure 7-11. Mathematical model of vibration actuator.

$$M\frac{d^2u}{dt^2} = P(t) - WOB + F_s$$

The spring force is

$$F_s = k\left(\delta_{st} - u\right)$$

where

δ_{st} = static compression of the shock tool.
u = dynamic displacement of the mass.

By substitution

$$M\frac{d^2u}{dx^2} = P(t) - WOB + k\delta_{st} - ku$$

But WOB = δ_{st} k, so

$$M\frac{d^2u}{dt^2} + ku = P(t)$$

Assuming P(t) is a harmonic function, $P(t) = P_0 \sin \omega t$, the solution to the differential equation is

$$u(t) = \frac{P_0}{k}\frac{1}{1-\left(\frac{\omega}{\omega_n}\right)^2}\sin \omega t$$

This displacement translates into a dynamic force through the compression of the flexible tool. The greatest force coming when the frequency (ω) of excitation from the pressure pulses is tuned to the natural frequency of the spring-mass system.

If the pressure force, P(t) frequency is close to a natural frequency of any axial mode, then drill collars cannot be modeled as a lump mass. In this case the boundary condition at the lower end of the drill collar is modeled as shown Figure 7-12a. The vibration tool can be located anywhere within the drillstring, with or without a shock tool, and provide a source of excitation the a drillstring. If a pressure force is applied within a drill pipe section then boundary conditions at the point of application of P(t) must satisfy the conditions shown in Figure 7-12b. Drillstring response in either case can be determined by following the procedure explained previously [4].

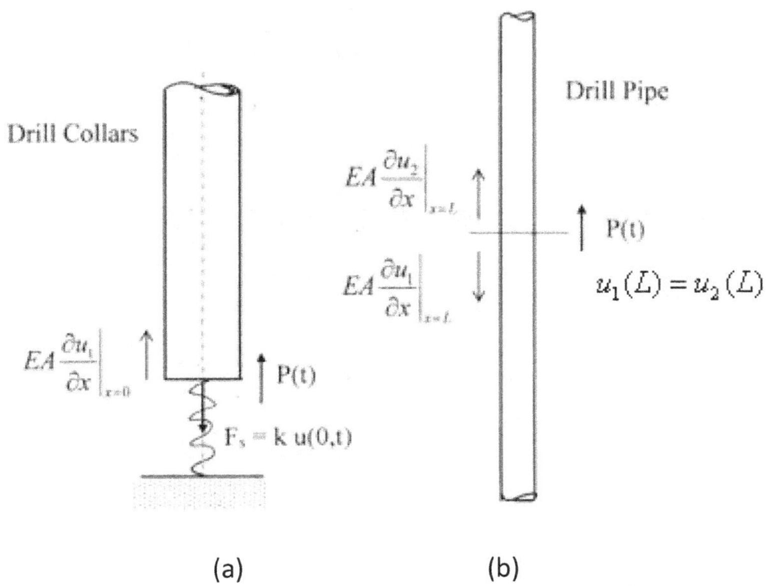

Figure 7-12. Two applications of the pressure forcing device.

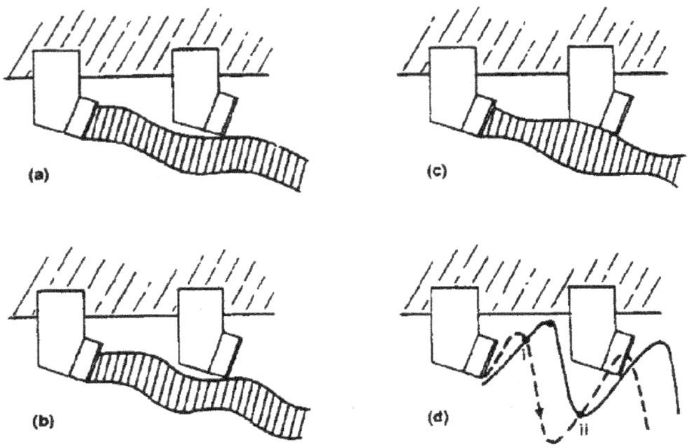

Figure 7-13. Self excited vibrations related to PDC drill bits.

7.1.5 Excitation by PDC Drill Bits

The self-exciting mechanism is based on the regeneration of surface undulations, which each cutter edge leaves behind if vibration is superimposed over the otherwise steady motion of the bit. The following edge cuts into the wavy surface and the corresponding variable cutting force excites the vibration, etc. The phase difference between the previous wave and the current wave gives rise to a variable instantaneous cutting depth and this variation account for the variable force, which drives the vibration (see Figure 7-13).

Whether self-excitation occurs (unstable cutting) or not (stable cutting) depends on the dynamic flexibility of the drillstring as expressed by the transfer function G(f) between the force acting on the bit, length of the cutting edges and on the rotational speed of the bit. The theory behind this mechanism of self-excitation is beyond the scope of this manuscript. For a complete discussion of this self-exciting mechanism and its application to drillstrings see ref [6].

7.2 TORSION VIBRATIONS IN DRILLSTRINGS

The differential equation for torsion motion has the same form as for axial motion. Adjusting for the torsion mode Eq. (7-13) becomes

$$JG\frac{\partial^2\theta}{\partial x^2} = \gamma_\theta\frac{\partial\theta}{\partial t} + j_m\frac{\partial^2\theta}{\partial t^2} \tag{7-20}$$

where

J = polar moment of inertia of pipe cross section, in.^{4}
G = modulus of rigidity, psi
θ = local pipe rotation, rad
γ_θ = torsion damping per unit length
j_m = mass polar moment of inertia per unit length

7.2.1 Coupling With Axial Bit Force

Downhole measurements indicate torsion vibrations are generated through force coupling with longitudinal vibrations at the drill bit (Figure 7-14). This data shows the torsion vibration is in step with the axial force, which can be related to three lobed undulations in core tops. Once longitudinal vibrations stop, torsion vibrations also stop.

The data was taken near Edmonton while drilling in a sandstone and shale formation with a roller cone drill bit. Well depth was 4370 ft and rotary speed was 110 rpm. This data shows the dynamic axial forces varied from zero (bit came off the bottom) to a maximum value of 150,000 lbs when the indicated (static) bit weight was 40,000 lbs [3]. At the same time bit torque ranged from 8000 ft-lbs in one direction to 1000 ft-lbs in the opposite directions; typical drilling torque is around 1500 ft-lbs. In each case, the frequency was three cycles per bit rotation.

7.2.2 Cutter Impulsive Force

Tooth brakeage on PDC drill bits is most likely due to large impulsive forces caused by momentum changes.

Consider a disc rotating at a constant angular velocity about its geometric center (Figure 7-15). Over a very short time interval, $\Delta t = t_2 - t_1$, the edge of the disc is caught by a fixed pin as shown. At time t_2 the disc sudden rotates about its instantaneous center, point “a”.

Over this short time interval (Δt) angular momentum with respect to point “a” is conserved. However, linear momentum with respect to the center of gravity is not conserved. Based on the conservation of angular momentum

$$H_{a2} = H_{a1} \tag{7-21}$$

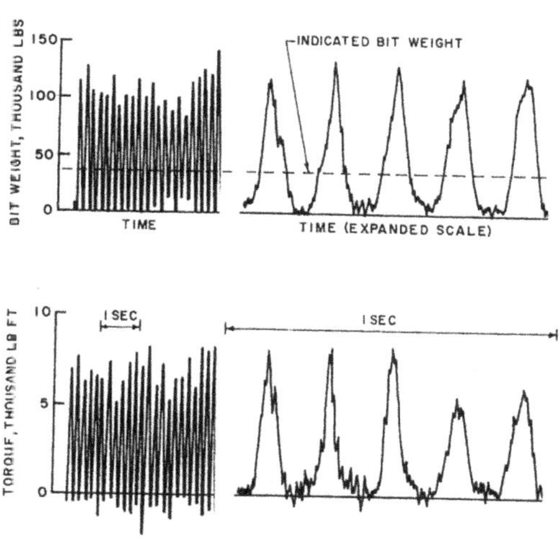

Figure 7-14. Variations in bit forces and torque; ESSO Production Research data taken near Edmonton, Alberta, Canada (1963).

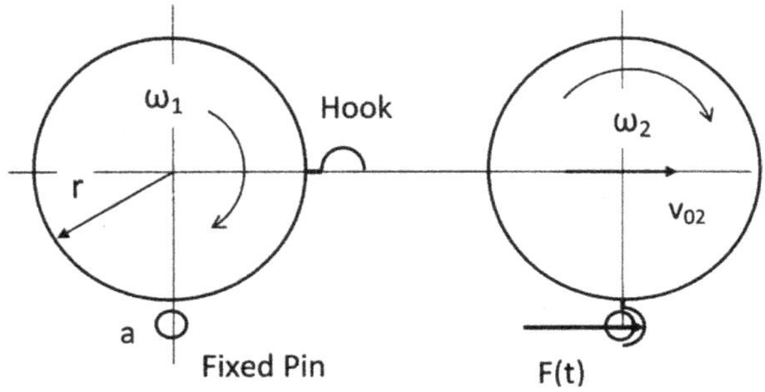

Figure 7-15. Cutter impulsive force.

from which

$$\left(I_G + Mr^2\right)\omega_2 = I_G\omega_1 \tag{7-22}$$

and

$$\omega_2 = \frac{I_G}{I_G + Mr^2}\omega_1 = \frac{1}{3}\omega_1 \tag{7-23}$$

assuming a solid disc. The velocity of the center of gravity is now $v_{02} = r\omega_2$. Considering the component of linear impulse to the right

$$\text{Im}_x = G_{x2} - G_{x1} \tag{7-24}$$

giving

$$\int_1^2 F(t)dt = v_{o2}M - 0 = r\omega_2 M \tag{7-25}$$

This change in linear momentum takes place over a very short period of time. The magnitude of the maximum impulsive force can be approximated by

$$\int_1^2 F(t)dt = F_{ave}\Delta t \tag{7-26}$$

Dividing the impulse by Δt gives the average impact force.

$$F_{ave} = \frac{r\omega_2 M}{\Delta t} \tag{7-27}$$

Consider, for example, a rigid disc weighing 40,000 lbs, having a 7-in. diameter and rotating at a speed of 100 rpm. These numbers were chosen to simulate drill collars. Adjusting the units for substitution into Eq. (7-27)

$$r = \frac{3.5}{12} = 0.2917 \text{ ft}$$

$$M = \frac{40{,}000}{32.2} = 1{,}242 \text{ slugs}$$

$$\omega_2 = \frac{1}{3}\frac{2\pi 100}{60} = 3.49 \text{ rad/second}$$

Putting these numbers into Eq. (7-27) gives

$$F_{ave} = \frac{0.2917(3.49)1{,}242}{0.05} = 79{,}496 \text{ lbs}$$

Here, Δt is arbitrarily chosen to be 0.05 second. A smaller Δt increases the predicted average force. The peak impact force on the pin would be much higher. Cutter damage and failure is possible from this level of force.

7.2.2.1 Accounting for Torsion Flexibility in Drill Collars

Now consider the torsion flexibility of drill collars of length, L (Figure 7-16). Assume the collars rotate clockwise as in drilling. Once a cutter makes contact with an obstruction, creating impact force, F(t), a shear wave starts to travel to the right with acoustic velocity c = 10,600 ft/second $\left(c = \sqrt{\frac{G}{\rho}} = 10{,}600\, fps\right)$. Over a short span of time the leading edge of the shear wave has travel a distance, $x = ct$. The shape of the impulse force, F(t), will be sustained within the collars as illustrated. The space interval Δx is related to the time interval, Δt by $\Delta x = c\Delta t$. As the shear wave travels to the right, linear momentum is changed according to

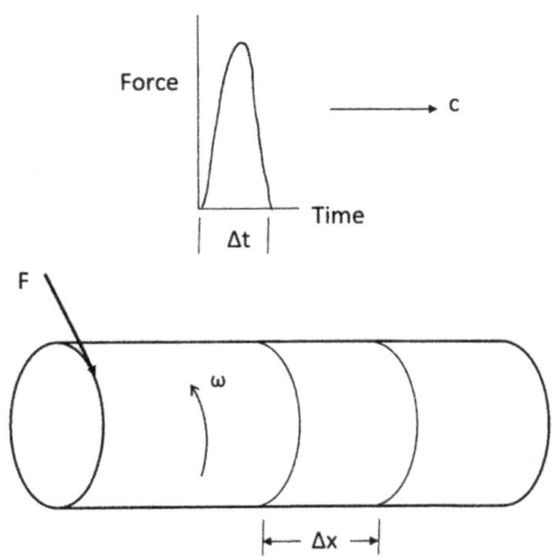

Figure 7-16. Torsion shear wave setup by cutter impact.

$$\int_0^{\Delta t} F(t)dt = r\omega_2 m\Delta x \quad (\text{m = mass per length}) \tag{7-28}$$

In terms of the average impulse force

$$F_{ave}\Delta t = r\omega_2 mc\Delta t \tag{7-29}$$

$$F_{ave} = r\omega_2 mc \tag{7-30}$$

Applying the previous numbers to Eq. (7-30) gives

$$F_{ave} = 0.291(3.49)(4.068)(10{,}600) = 43{,}894 \text{ lbs}$$

This number is substantially less than for the rigid disc because not all of the linear momentum of the collars is changed simultaneously.

7.2.3 PDC Gear Tracking

Another type of PDC drill bit motion is backward whirl. Laboratory testing by Bret, Warren, Behr [9] show a star shape pattern in rock samples. The resulting whirling motion is similar to a planetary gear train with the bit being the planet gear and well bore being the annular sun gear. In this case individual cutters on the bit would follow a cycloid path. Backward whirl of these bits creates centrifugal forces, which encourages backward whirl and well bore enlargement.

The whirling frequency in this case is

$$\frac{\omega_{bit}}{\omega_{whirl}} = 1 - \frac{r_{hole}}{r_{bit}} \tag{7-31}$$

$$\omega_{whirl} = -\frac{r_{bit}}{e}\omega_{bit} \tag{7-32}$$

where "e" is the radial location of the bit center. The equation indicates that whirl frequency is quite high initially.

7.3 LATERAL VIBRATIONS IN DRILLSTRINGS

Lateral vibrations may occur along any portion of drillstrings, however, they should be evaluated over the lower portion of the drill collar section where bottom-hole tools are located. Most of the drillstring is partially stabilized by well bore contact; however, local vibrations between contact points, such as stabilizers, could be severe. Potential lateral vibrations in drill collar sections containing instrumentation and other mechanical components should be analyzed for potential resonance. Lateral vibrations are typically detected by MWD downhole as they are not manifested at the surface as with axial and torsion vibrations.

7.3.1 Natural Frequencies of Lateral Modes (Short Beams)

Natural frequencies of beams under different boundary conditions are well documented (Table 7-2) [10]. In each case, the vibration model assumes there is no load applied to the ends of the beams.

The nomenclature is as follows:

$$\beta^4 = m\frac{\omega^2}{EI}$$

m = mass per unit length of beam
L = length of beam
EI = flexural stiffness

Bring everything together gives

$$\omega_n = (\beta_n L)^2 \sqrt{\frac{EI}{mL^4}}$$

where $(\beta_n L)^2$ is taken from Table 7-2. In the case of simply supported beams $\beta_1 L = \pi$ giving

$$\omega_n = \left(\frac{n\pi}{L}\right)^2 \sqrt{\frac{EI}{m}} \quad \text{n = 1, 2, 3--- (same as synchronous whirl)} \tag{7-33}$$

Timoshenko [11, p454] gives natural frequencies for simply supported beams under an axial tension load.

$$\omega_n = \frac{a(n\pi)^2}{L^2}\sqrt{1+\frac{SL^2}{(n\pi)^2 EI}} \tag{7-34}$$

where

$$a^2 = \frac{EI}{m}$$

EI = bending stiffness
L = length of simply support beam

Table 7-2. Natural frequencies of beams having different boundary conditions.

Beam Support	First Mode $(\beta_1 L)^2$	Second Mode $(\beta_2 L)^2$	Third Mode $(\beta_3 L)^2$
Simply supported	9.87	39.5	88.9
Cantilever	3.52	22.0	61.7
Free-free	22.4	61.7	121.0
Clamped-clamped	22.4	61.7	121.0
Camped-hinged	15.4	50.0	104.0
Hinged-free	0	15.4	50.0

m = mass per unit length
S = axial tension force in beam; also $(Q_{eff})_{ave}$
n = mode of vibration (n = 1, 2, 3, etc.)

For drill collars under compression

$$S = (Q_{eff})_{ave} = -\left(WOB - \frac{1}{2} w\ell BF\right)$$

Equation (7-34) is modified for internal pressure and applied tension as shown below.

$$\omega = a\left(\frac{n\pi}{L}\right)^2 \left[1 + \frac{L^2(F + P)}{(n\pi)^2 EI}\right]^{\frac{1}{2}} \tag{7-35}$$

Plus (+) F is tension and $P = -p_i A_1$ accounts for internal pressure. If the axial forces, F + P, are dropped

$$\omega_n = \left(\frac{n\pi}{L}\right)^2 \sqrt{\frac{EI}{m}} \tag{7-36}$$

which agrees with Eq. (7-33).

7.3.1.1 *Calculation of Natural Frequency Between Stabilizers*

Consider a 50 ft span of drill collars (7" × 3", 107 lb/ft) between two stabilizers and ignoring the compressive force for the moment. Noting that I = 113.88 in.^{4} and applying Eq. (7-33) (assuming simply support at the stabilizers)

$\omega_{n1} = 10.54$ rad/second

$f_{n1} = 1.68$ cps

$N_{cr1} = 101$ rpm

The natural frequency of the second mode is

$\omega_{n2} = 42.16$ rad/second

$f_{n2} = 6.71$ cps

$N_{cr2} = 403$ rpm

The critical speed for the first mode is within normal rotary speed range. The second mode could be excited by the rotor of PDMs.

Including the effects of compression, say WOB = 45,000 lbs, and buoyancy (BF = 0.8168) and applying Eq. (7-34)

$$S = -\left(45{,}000 - \frac{1}{2}(107)(50)0.8168\right) = -42{,}815$$

with

$$\frac{SL^2}{\pi^2 EI} = -\frac{42,815(50)^2}{\pi^2 30\times10^6(113.88)}\left|\frac{144\ \text{in}^2}{1\ \text{ft}^2}\right| = 0.4571$$

$$\sqrt{\frac{EI}{m}} = \sqrt{\frac{30\times10^6(113.88)32.2}{107}\left|\frac{1\ \text{ft}^2}{144\ \text{in}^2}\right|} = 2,672\ \text{ft}^2/\text{sec}$$

$$\omega_{n1} = \left(\frac{\pi}{L}\right)^2\sqrt{\frac{EI}{m}}(1-0.4571)^{\frac{1}{2}} \quad \text{(first mode)}$$

$$\omega_{n1} = \left(\frac{\pi}{50}\right)^2 2,672(0.7368) = 7.77\ \text{rad/second}$$

$$f_{n1} = 1.24\ \text{cps}$$

$$N_{cr1} = 74.4\ \text{rpm}$$

a somewhat lower critical rotary speed. The difference is due to the average effective compression in the collars.

7.3.2 Excitation of Lateral Modes

Because of well bore confinement, excitement of lateral modes is more likely to occur between stationary points of contact within sections of drill pipe or drill collars. A span of drill collars between stabilizers is an example.

7.3.2.1 Excitation by Positive Displacement Motors (PDM)

Lateral excitation is generated within positive displacement motors, which are in tandem with drill collars. The exciting force comes from the planetary movement of the rotor within the stator (Figure 7-17). This motion creates internal forces between rotor and stator, which can cause bottom hole assemblies to

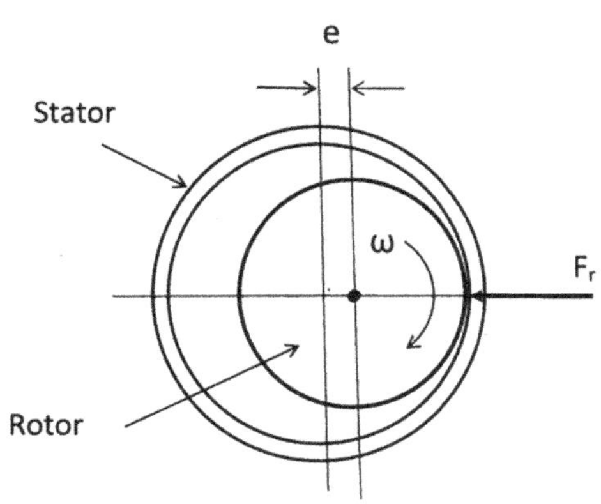

Figure 7-17. PDM rotor dynamics.

vibrate laterally, especially if the driving frequency is near a natural frequency. Drill collar can response with impact forces against the well bore and create high accelerations in instrumentation.

Assuming planar motion (all forces lie within a transverse plane), the equations of motion give

$$\sum F_r = M_r a_{Gr} \tag{7-37}$$

where a_{Gr} is the radial component of the acceleration of the center of gravity of the rotor and F_r is the radial force that produces this acceleration component. This radial force also pushes against the stator to excite the drill collars. This force is significant and cyclic.

Consider the following parameters:

Lobe ratio $= \dfrac{n_s}{n_r} = 2$

N = 200 rpm (bit speed)

$\omega = \dfrac{2\pi N}{60}$ rad/second (bit speed)

$W_r = 200$ lbs (rotor weight)

r_r = pitch radius of rotor (= 2 in.)

From the kinematics of the stator/rotor (Table 7-3)

$$\omega = \left(1 - \frac{n_s}{n_r}\right)\omega_e \tag{7-38}$$

meaning that vector "e" rotates counterclockwise with angular velocity of ω_e. The acceleration of center of the rotor is $a_{Gr} = \omega_e^2 e = \omega_e^2 r_r$; in this particular PDM $\omega_e = \omega$. The force $F_r = Ma_{Gr}$ or

$$F_r = \frac{200}{32.2}\left(\frac{2\pi 200}{60}\right)^2 \frac{2}{12} = 454 \text{ lbs}$$

The force vector rotates with $\omega_e = 20.94$ rad/second or a frequency of 3.33 cps, which is proportional to bit rotary speed. Force magnitude and frequency are both important in determining possible resonance and response of lateral modes between stabilizers. These numbers depend on stator/rotor ratios, output speed, and rotor mass.

Table 7-3. Kinematics of hypo cycloid gear train.

	Arm (e)	Rotor (r)	Stator (s)
with e	1	1	1
relative to e	0	$-r_s/r_r$	−1
total motion	1	$1 - r_s/r_r$	0

7.3.2.2 Unbalanced Drill Collars

Dykstra, Chen, Warren, Azar [12] consider the effect of unbalanced drill collars on lateral vibrations in BHA. Drill collars can be unbalanced due to bore misalignment, initial curvature, and gradual wear during service. Severe lateral vibrations were predicted for unbalanced inertia forces under resonant conditions. Natural frequencies and mode shapes determined from a finite element analysis. Measurements indicated lateral accelerations reached 50 g and higher. Under resonant conditions, surface torque increased from around 500 ft-lbs to 3000 ft-lbs.

7.3.3 Natural Frequencies of Lateral Modes (Long Beams)

Buckling of structures can be predicted from a static analysis or a dynamic analysis. The static analysis approach is based on seeking eigenvalues from static equations of bending. The principle of minimum potential energy is useful in this regard. The dynamic analysis approach predicts buckling from natural frequency predictions; when natural frequency is zero, buckling occurs. The following discussion approaches drill pipe buckling and lateral modes of vibration from the dynamic perspective following Huang and Dareing [13]. The physical model is illustrated in Figure 7-18.

The application of this analysis is the possible lateral vibration of drill pipe directly above the drill collar section. The drill pipe in this area experiences lowest tension and can be excited by drillstring rotation provided rotary speeds are tuned to any of the natural lateral modes as described below.

The lower end of the drill pipe is modeled as a fixed end with zero slope because of the relative bending stiffness of the drill collars. The top end is pinned. The axial force applied to the pipe at x = 0 is the internal force between the pipe and collars. This force is designated as F_B in the expression for effective tension. Plus F_B means tension and minus means compression.

The analysis is based on vertical pipe under static conditions and unconstrained laterally. Vibration and buckling modes occur in a vertical plane. The basic equation of motion is developed from Eq. (7-40), noting that fluid density is the same both inside and out. In this case, effective tension is

$$T_{eff} = (F_B + wx) + (L - x)w_m \tag{7-39}$$

where F_B is the actual tension at x = 0.

$$EI\frac{\partial^4 y}{\partial x^4} - \frac{\partial}{\partial x}\left\{\left[\left(F_B + w_m L\right) + \left(w - w_m\right)x\right]\frac{\partial y}{\partial x}\right\} + m\frac{\partial^2 y}{\partial t^2} = 0 \tag{7-40}$$

Consider the case where weight-on-bit puts the neutral point at the top of the drill collars. This means that the internal stress at the lower end of the drill pipe is equal to the local hydrostatic pressure and the force in the drill pipe at this interface is,

$$F_B = -Lw_m \tag{7-41}$$

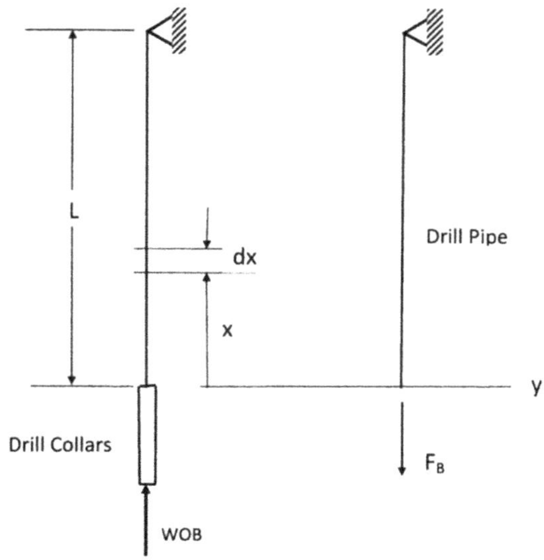

Figure 7-18. Effective tension in drill pipe.

Under this condition, the pipe will not buckle as explained in Ch 5. However, if WOB is increased further by F, then the force at the bottom of the drill pipe becomes

$$F_B = -Lw_m - F \tag{7-42}$$

This additional force can cause drill pipe buckling. The critical buckling force, F_{cr}, applied to the bottom of the drill pipe will be discussed next.

Substituting Eq. (7-42) into Eq. (7-40) gives

$$EI\frac{\partial^4 y}{\partial x^4} - \frac{\partial}{\partial x}\left\{\left[-F + \left(w - w_m\right)x\right]\frac{\partial y}{\partial x}\right\} + m\frac{\partial^2 y}{\partial t^2} = 0 \tag{7-43}$$

Here, F is a plus number and represents added compressive force beyond hydrostatic compression. The minus sign accounts for compression. The boundary conditions at the lower and upper ends are

$$y(0,t) = 0 \tag{7-44a}$$

$$\frac{\partial y}{\partial x}(0,t) = 0 \tag{7-44b}$$

$$y(L,t) = 0 \tag{7-44c}$$

$$\frac{\partial^2 y}{\partial x^2}(L,t) = 0 \tag{7-44d}$$

Assuming the solution to Eq. (7-43) is of the form

$$y(x,t) = Y(x)\sin\omega t \tag{7-45}$$

then Eq. (7-43) converts to

$$\frac{d^4Y}{d\zeta^4} - \alpha\zeta\frac{d^2Y}{d\zeta^2} + \beta\frac{d^2Y}{d\zeta^2} - \alpha\frac{dY}{d\zeta} - \lambda^4 Y = 0 \tag{7-46}$$

where

$$\varsigma = \frac{x}{L}$$

$$\alpha = \frac{(w - w_m)L^3}{EI}$$

$$\beta = \frac{FL^2}{EI}$$

$$\lambda^4 = \frac{m\omega^2 L^4}{EI}$$

Boundary conditions become

$$Y(0) = 0 \tag{7-47a}$$

$$\frac{dY}{dx}(0) = 0 \tag{7-47b}$$

$$Y(L) = 0 \tag{7-47c}$$

$$\frac{d^2Y}{dx^2}(L) = 0 \tag{7-47d}$$

7.3.3.1 Natural Frequencies of Drill Pipe (Lateral Mode)

Eigenvalues for this problem are determined from the characteristic equation derived from the expansion of a coefficient matrix [13]. Values of λ that satisfy the characteristic equation are given in Table 7-4.

Mode shapes and their bending moments are of interest because they show the intensity of bending at the interface between drill pipe and drill collars. Mode shapes are indicative of the effects of variable tension along drill pipe. Bending at the top is low because of high tension and high at the collars because of low tension and the structural constraint at the interface. Figure 7-19 gives mode shapes for the first two vibration modes and corresponding modal moments for both modes (corresponding to $\alpha = 4000$ and $\beta = 800$. Bending moments are greatest at the interface between the drill pipe and collars even though as shown above they are relatively small. Fatigue damage is most likely

Table 7-4. Frequency parameters, λ.

α	$\frac{\beta}{\alpha}$	λ_1	λ_2	λ_3
100	0	5.409	8.336	11.232
	0.1	5.230	8.147	11.063
	0.2	5.029	7.944	10.886
	0.3	4.801	7.724	10.700
250	0	6.354	9.458	12.325
	0.1	6.070	9.130	12.001
	0.2	5.733	8.761	11.648
	0.3	5.315	8.336	11.260
600	0	7.534	11.010	14.009
	0.1	7.100	10.488	13.463
	0.2	6.531	9.863	12.838
	0.3	5.667	9.073	12.104
1000	0	8.373	12.147	15.310
	0.1	7.807	11.470	14.589
	0.2	6.999	10.613	13.731
	0.3	5.335	9.416	12.659
2500	0	10.178	14.664	18.277
	0.1	9.262	13.579	17.109
	0.2	7.390	11.952	15.547
	0.3	....	8.585	13.155
6000	0	12.341	17.726	21.963
	0.1	10.809	15.980	20.093
	0.2	....	12.236	16.980
	0.3	....	....	10.859

cause by drillstring rotation with the drill pipe/drill collar interface being located in a dog leg under conditions of high tension.

7.4 VIBRATION CONTROL

The primary cause of severe vibrations is frequency tuning between forcing frequency and a natural frequency. This is true of all vibration types, whether axial, torsion or lateral and for each of their multi modes.

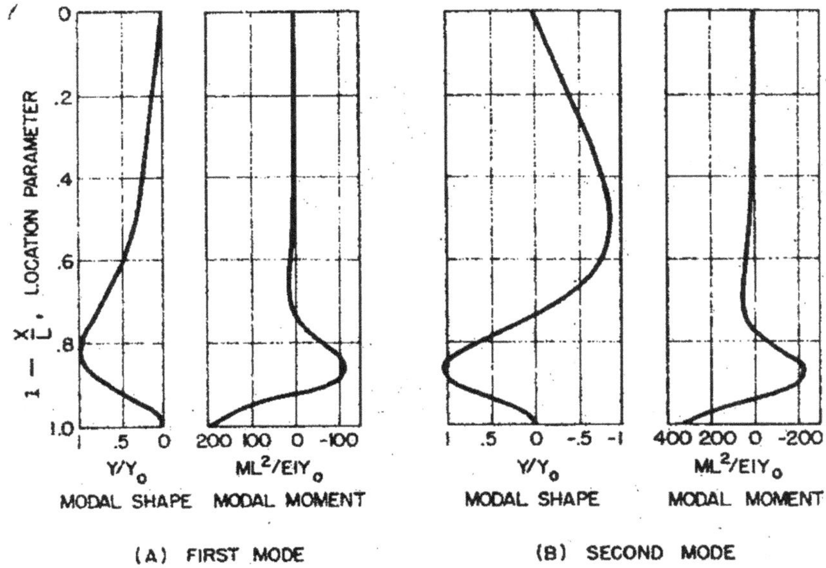

Figure 7-19. Vibration modes for α = 4000, β = 800.

In most cases, vibrations are driven by applied forces that vary with time. It is useful to identify the possible sources of excitation and quantify their magnitude and frequency. It is also useful to have knowledge of the natural frequencies of each type (axial, torsion, lateral) of the system. Each vibration type will have many vibration modes. Even with this information there are many possibilities for frequency tuning. In practice, active modes will identify themselves during operation. MWD measurements can be most useful in this regard. Vibration control is a matter of identifying active modes and detuning them from driving frequencies.

Mode vibration in drilling operations is a reality that has been shown by field measurements. Jogi, Macpherson, Neubert [14], measured vibration responses for axial, torsion and lateral modes using a downhole tool in a shallow test well at Amoco's test facility in Catoosa, Oklahoma. A 526 .8 ft drilling assembly containing a roller cone bit, drill collars (~8.0 in. OD), MWD, and stabilizers. The objective of these tests was to validate analytical predictions of natural frequencies for all three modes. Sources of excitation were inherent in collar whirl, imbalance, stabilizer/hole contact and drill bit forces. In nearly every case, natural frequencies predicted by vibration theory agreed with measured frequencies of each of the three modes of vibration. These tests show that severe vibrations in drillstrings are caused by frequency tuning.

Drillstring vibration control is fundamental. The basic options are:

1. Eliminate the source of excitation
2. Change the frequency of the force of excitation (usually linked to rotary speed)
3. Reduce the magnitude of the driving force

4. Change the natural frequencies of the drillstring through design noting that natural frequencies continually change as the well bore advances
   a. It is important to know natural frequencies of all modes of vibration
5. Apply damping
   a. Some damping is inherent
   b. Creating damping is usually costly and should be a last resort

Controlling severe vibrations driven by a self-excited mechanism is subtler. Here, the driving mechanism and elastic response are interdependent and feed energy into the vibration on a continuous basis. Self-excited vibrations are caused by PDC bits as explained earlier.

The vibration response caused by self-excitation depends on the feedback mechanism. The forcing mechanism depends on the vibration response; when one stops, the other stops. It is not always obvious how to untangle the two. A mathematical analysis, based on a proven feedback mechanism, can give useful guidance. A reliable analysis can identify operating regions of stability and instability. PDC chatter is a good example. Rotational stick-slip due to negative damping is another.

Solutions to self-excited vibration problems are:

- Eliminate the source
- Change the limit of dynamic stability for self-excited vibrations or operate within the stable region.

Usually the design of engineering structures is based on static analyses followed by a dynamic analysis. Once the structure is built and tested. Vibration measurements are used to check for unforeseen vibration problems. If a vibration problem occurs, then one of the control methods listed above can be implemented.

7.4.1 Controlling Axial Vibrations

Prior to 1960 a field method used to eliminate Kelly bounce was to lift the bit off bottom, spin the string, and touch back down for a few minutes [15]. It was envisioned that this would “clean off these high spots”. This approached worked until new lobes developed. The mechanics of this practice is scientifically sound because vibration motion is 180 deg. out of phase with the force when operating on the high side of resonance.

A force/phase analysis of a drillstring (see Figure 7-20) [16] shows a similar effect for this particular distributed system; the analysis assumes the bit does not separate from the formation. Note the effects of both drill pipe and drill collars. As the drill collars begin to resonate, the phase change becomes more pronounced.

Roller cone bits are known to produce axial and torsion vibrations. The driving frequency is simply 3 cycles per bit rotation. Care should be taken to avoid tuning natural frequencies of axial and torsion modes to this driving frequency.

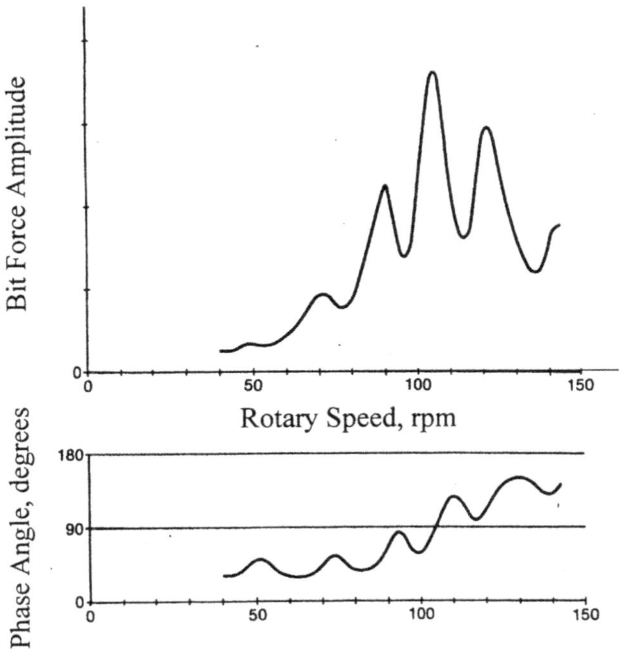

Figure 7-20. Phase angle between bit force and bit amplitude.

7.4.2 Use of Shock Absorbers

Shock absorbers are commonly used in hard rock areas where drill bit vibrations are often severe. These tools provide a mechanical spring effect and are place directly above the bit. The elastic element in these tools typically has a spring constant in the range of 50,000 lb/in. They have been made of

1. Rubber
2. Bellville springs
3. Steel springs of various designs

In hard rock drilling areas, more drill collars are needed and this factors into the vibration problem. Shock absorbers basically change the natural frequency of drill collars. A simple, but practical model, is a lump mass setting on a spring (Figure 7-21). The natural frequency of this SDOF model is simply $f = \frac{1}{2\pi}\sqrt{\frac{k}{M}}$. The differential equation of motion assuming base motion is

$$m\ddot{x} + c\dot{x} + kx = ky + c\dot{y} \tag{7-48}$$

The solution to Eq. (7-48) is shown in Figure 7-22 [10].

The frequency response shows that internal damping in shock absorbers is counterproductive at operating frequency ratios above $\sqrt{2}$, where these tools normally operate.

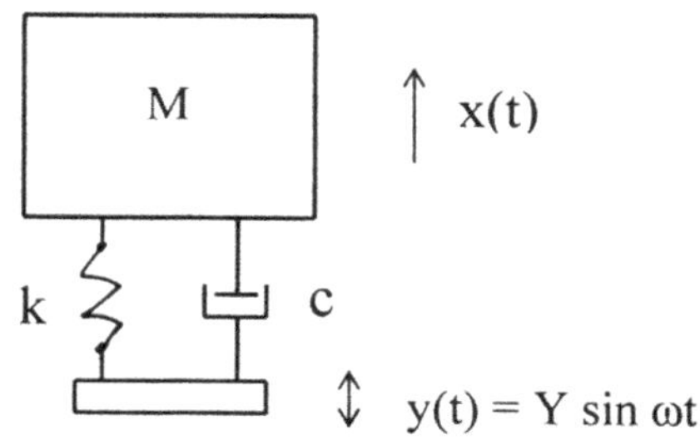

Figure 7-21. SDOF model of drill collars and shock absorber.

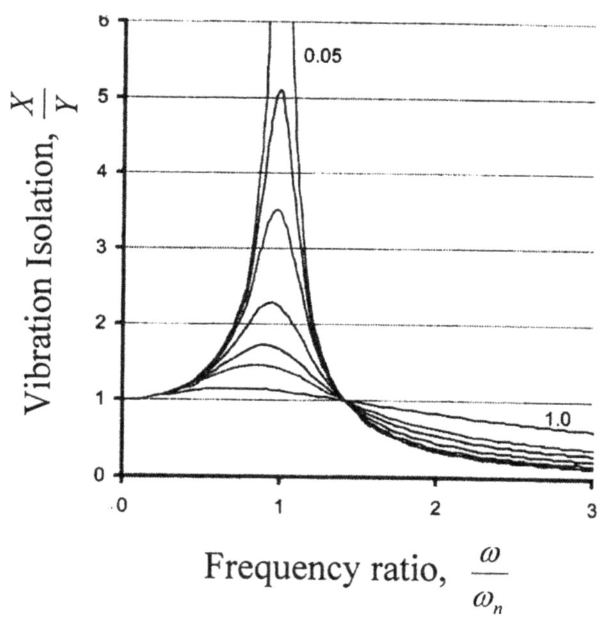

Figure 7-22. Frequency response of a SDOF system to base motion.

A shock absorber located directly above a drill bit changes the frequency response curve from the ones shown in Figure 7-10 to the ones shown in Figure 7-23. The response of the drill collar section can still be seen but it has been shifted to the left. On the right side of collar resonance, amplitudes of vibration are greatly dimensioned.

Note that the drill pipe may participate in these vibrations, but at a lower level than those shown in Figure 7-10. The shift in the frequency response is due mainly to the response of the drill collar section.

For deeper wells the frequency band width of drill pipe response becomes small. This makes it difficult to excite the drill pipe section because of the time required

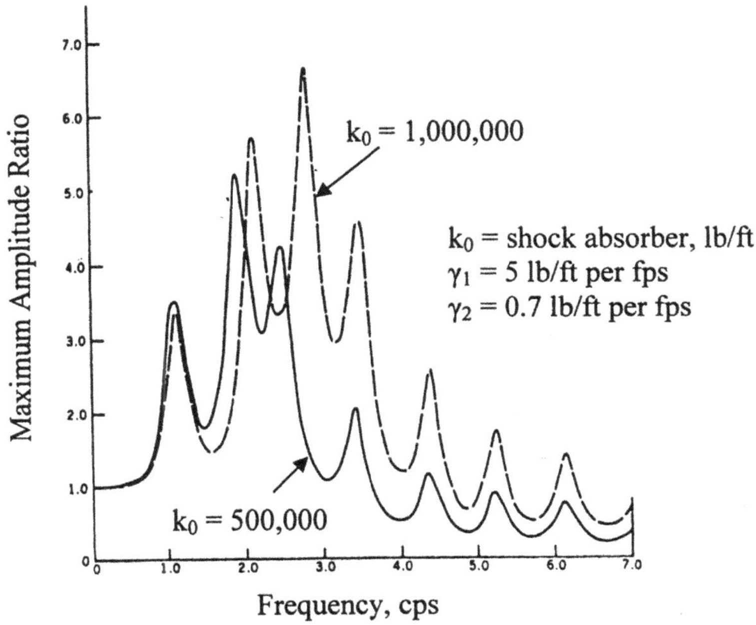

Figure 7-23. Frequency response with shock absorber [4].

to feed energy into the pipe and the necessary precision of rotary speed. In this case, the drill collar section controls the axial and torsional vibrations of the entire drillstring. Therefore, the focus in deep wells should be in controlling vibrations in the drill collars section alone. This greatly simplifies the vibration model.

7.4.3 Effect of Drill Collar Length

For many years drill collar length has been based on static analysis alone as discussed in Ch 5. Dareing [17] shows that axial modes of drill collars are often tuned inadvertently to critical rotary speeds and this condition should be considered in drill collar design and operation. This condition is especially critical in hard formations requiring long drill collar sections.

Consider, for example, the frequency response of just drill collars (see Figure 7-24). The effect of the drill pipe section has been stripped out. The top boundary condition is assumed to be free. This is reasonable because the cross-sectional area of drill pipe is much smaller that the cross-sectional area of drill collars.

The development of the natural frequency of axial mode is reviewed below.

$$\frac{\partial^2 u}{\partial x^2} = \frac{1}{c^2}\frac{\partial^2 u}{\partial t^2} \tag{7-49}$$

Assume a solution of the form

$$u(x,t) = X(x)\sin\omega t$$

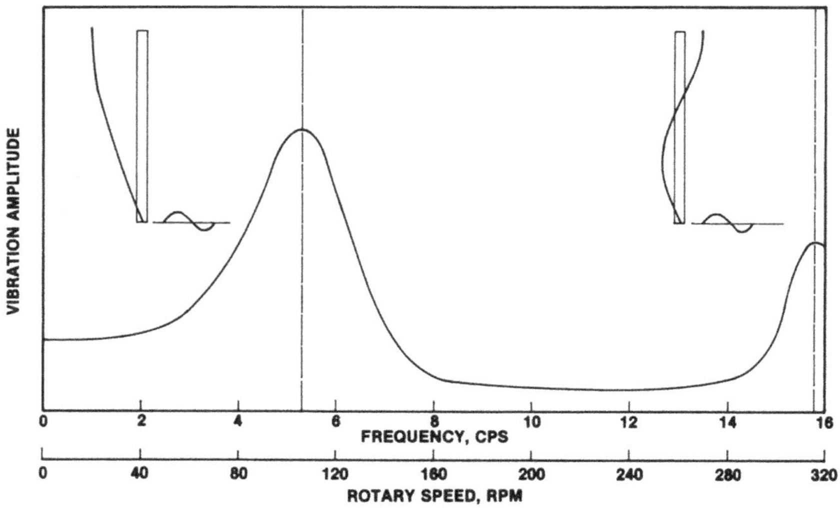

Figure 7-24. Frequency response of drill collar section.

By substitution

$$\frac{d^2X}{dx^2} + \frac{\omega^2}{c^2}X = 0$$

whose solution is

$$X(x) = A\sin\frac{\omega}{c}x + B\cos\frac{\omega}{c}x \tag{7-50}$$

Imposing boundary conditions

$$X(0) = 0 \quad \mathrm{B} = 0$$

$$\left.\frac{dX}{dx}\right|_{x=L} = 0 \quad \cos\frac{\omega}{c}L = 0 \quad \frac{\omega}{c}L = \frac{n\pi}{2}, \ \mathrm{n} = 1, 3, 5, ---$$

Natural frequencies of axial modes are defined by

$$\omega = \frac{n\pi}{2}\frac{c}{L} \text{ rad/second}$$

$$f_n = \frac{\omega}{2\pi} = \frac{nc}{4L}$$

Natural frequency of the first mode is

$$f_n = \frac{16{,}850}{4L}$$

Vibration analyses show how drill collars respond to bit motion [4]. Drill pipe responds too but it is their higher modes that respond to typical rotary speeds.

It is easier to resonant the drill collar section than to develop the higher drill pipe modes. Therefore, the drill collar section is a critical area in the drillstring to monitor [17, 18].

Drill collar length is dictated by required WOB. In hard rock drilling areas, more drill collars are carried to generate a high bit force. The irony of this practice is the natural frequency of axial modes of vibration becomes close to frequencies of excitation associated with normal drilling speeds. This situation creates frequency tuning almost automatically. The numbers for this scenario are given below.

The natural frequencies of a drill collars (pinned at the bottom and free at the top) are determined from

$$f_{na} = \frac{n16,850}{4L}, \text{ n} = 1, 3, 5, \text{----} \quad \text{(axial mode)} \tag{7-51}$$

Note that collar size is not a factor, only length. For the first mode, n = 1 and

$$f_{1a} = \frac{4212}{L}$$

If, for example, L = 600 ft, f_{1a} = 7.02 cps.

Now compare this frequency to that associated with a three-roller cone bit rotating at, 140 rpm. The drill bit frequency in this example would be

$$f = \frac{3N}{60} \text{ cps}$$

$$f = \frac{140}{20} = 7 \text{ cps}$$

Actual drilling cases may vary but the reality remains. Both natural frequency and driving frequency are close enough to excite the first mode of the drill collars. The same condition could exist for higher drill collar modes when PDMs and turbines are used. Drill collar response under this assumption, i.e., ignoring drill pipe participation, is shown in Figure 7-25.

7.4.3.1 *Vibration Absorbers (Axial Mode)*

The effect of shock absorbers is easier to explain using this model. Figure 7-26 illustrates how a shock absorber shifts the frequency response to the left and allows a much smoother response at the 120-rpm rotating speed. In general, it is best to operate any structure on the high side of response where vibrations are minor. The second mode could come into play at higher speeds and should be consider when operating at high speeds, especially when motors are used.

7.4.3.2 *Natural Frequencies With Vibration Absorbers*

Vibration absorbers offer a practical solution for changing natural frequency of drill collars. Drill collar weight (length) can remain the same for a required WOB, while lowering natural frequency.

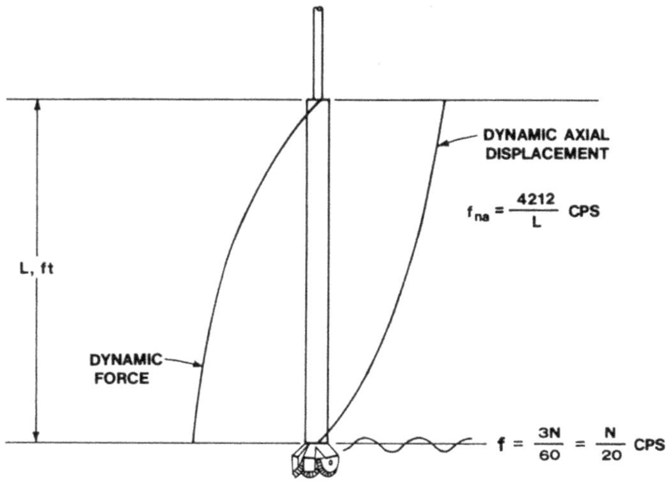

Figure 7-25. Vibration mode shape of drill collars.

The natural frequencies of drill collars with vibration absorber are developed as follows. Starting with the basic equation of motion for the distributed model.

$$EA\frac{\partial^2 u}{\partial x^2} = m\frac{\partial^2 u}{\partial x^2}, \text{ noting that } c^2 = \frac{AE}{m} = \frac{E}{\rho} \tag{7-52}$$

where m is mass per unit length, ρ is mass density, and c is acoustic velocity of a compression in the collars (16,850 ft/second).

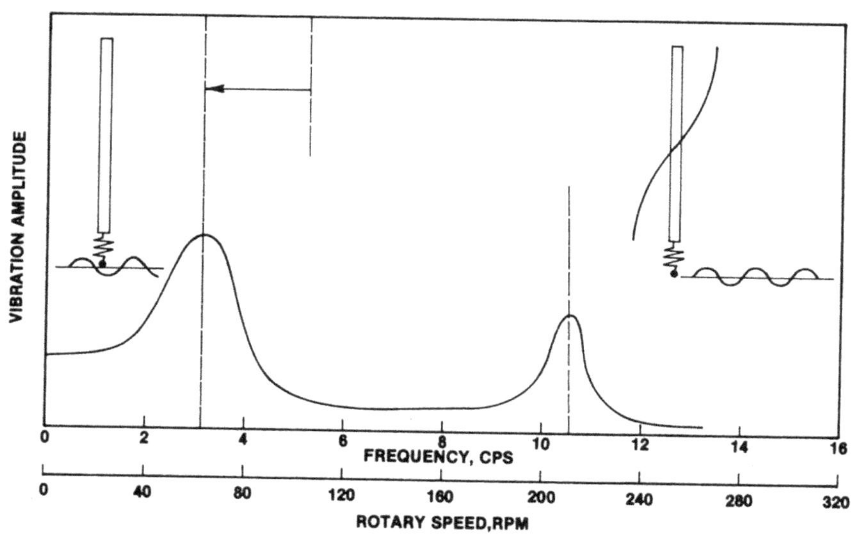

Figure 7-26. Effect of shock on drill collar response.

The formulation for determining natural frequencies of drill collars with a vibration absorber is given below. The math model is shown in Figure 7-27.

$$\frac{\partial^2 u}{\partial x^2} = \frac{1}{c^2}\frac{\partial^2 u}{\partial x^2} \tag{7-53}$$

For the solution, assume

$$u(x,t) = X(x)\sin \omega t \tag{7-54}$$

By substitution

$$\frac{d^2X}{dx^2}\sin \omega t = -\frac{\omega^2}{c^2}X \sin \omega t \tag{7-55}$$

$$\frac{d^2X}{dx^2} + \left(\frac{\omega}{c}\right)^2 X = 0 \tag{7-56}$$

General solution is

$$X(x) = A_1 \sin\frac{\omega}{c}x + B_1 \cos\frac{\omega}{c}x \tag{7-57}$$

$$\frac{dX}{dx} = A_1\frac{\omega}{c}\cos\frac{\omega}{c}x - B_1\frac{\omega}{c}\sin\frac{\omega}{c}x \tag{7-58}$$

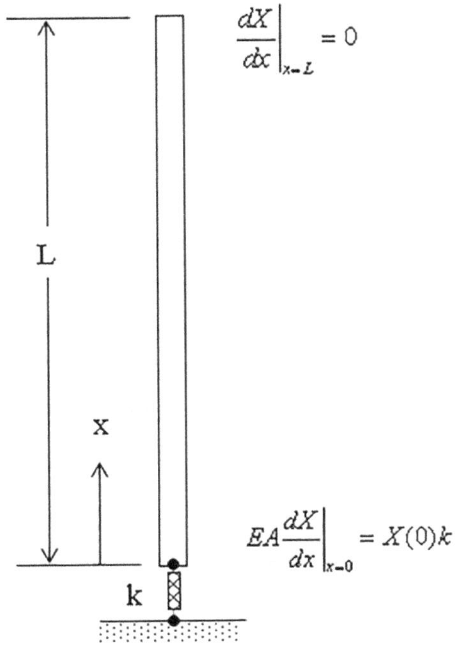

Figure 7-27. Math model for natural frequencies with vibration absorber.

Imposing boundary conditions

$$\left.\frac{dX}{dx}\right|_{x=L} = 0 \quad 0 = A_1 \cos\frac{\omega}{c}L - B_2 \sin\frac{\omega}{c}L$$

and

$$EA\left.\frac{dX}{dx}\right|_{x=0} = kX(0) \quad EA\left[A_1\frac{\omega}{c}\right] = kB_1$$

Combining these two equations gives

$$\left(\frac{\omega}{c}L\right)\tan\left(\frac{\omega}{c}L\right) = \frac{kL}{EA} \tag{7-59}$$

Let

$$\alpha = \frac{kL}{EA} \text{ and } \beta = \frac{\omega}{c}L$$

Then

$$\beta \tan\beta = \alpha \tag{7-60}$$

The eigenvalue we seek is $\beta = \frac{\omega}{c}L$, which is implicit in the characteristic equation. This equation was programmed to obtain the relation between the two dimensionless numbers. The relationship α and β is shown in Figure 7-28 and Table 7-5.

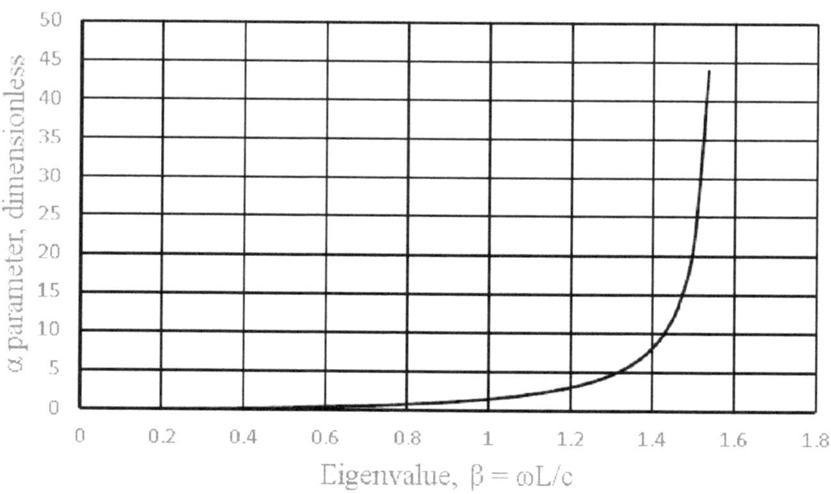

Figure 7-28. Eigenvalue of first mode with vibration absorber.

Table 7-5. Eigenvalues for first mode.

β	α
0.314159	0.102076
0.349066	0.127049
0.383972	0.155135
0.418879	0.186497
0.453785	0.221326
0.488692	0.259842
0.523598	0.302299
0.558505	0.348992
0.593411	0.400261
0.628318	0.456499
0.663225	0.518167
0.698131	0.585801
0.733038	0.660029
0.767944	0.741594
0.802851	0.831375
0.837757	0.930422
0.872664	1.039999
0.90757	1.161635
0.942477	1.297206
0.977384	1.449028
1.01229	1.62

Consider the collar-vibration absorber arrangement

- (6½ × 2, 102 lb/ft) collars
- k = 50,000 lb/in.

Lump mass model

$$f\,(cps) = \frac{1}{2\pi}\sqrt{\frac{k}{m}} = \frac{1}{2\pi}\sqrt{\frac{50{,}000}{700(102)/385}} = 2.61 \text{ cps}$$

Distributed mass model

$$\alpha = \frac{kL}{EA} = \frac{50000(700)12}{30 \times 10^6(30)} = 0.467$$

From table

$$\beta \sim 0.628$$

$$\frac{\omega}{c}L = 0.628$$

$$\omega = \frac{0.628(16,850)}{700} = 15.12 \text{ rad/second}$$

$$f = 2.4 \text{ cps}$$

Beta for the *second* mode is 3.28.

$$\frac{\omega}{c}L = 3.28$$

Natural circular frequency for second mode is

$$\omega_{n2} = \frac{3.28(16,850)}{700} = 78.95 \text{ rad/second}$$

$$f_{n2} = \frac{78.95}{2\pi} = 12.57 \text{ cps}$$

7.4.4 Torsion Vibration Mode

Torsion modes are predicted in a similar. The form of the equation of motion is the same on the parameters change. For example

$$\frac{\partial^2\theta}{\partial x^2} = \frac{1}{c^2}\frac{\partial^2\theta}{\partial t^2} \tag{7-61}$$

The acoustic velocity, c, represents the velocity of a shear wave, which for steel pipe is 10,650 ft/second. Therefore,

$$f_{nT} = \frac{n10,650}{4L} \quad \text{(torsion mode)} \quad n = 1, 3, 5 \tag{7-62}$$

Again, collar size is not a factor. For the first torsion mode

$$f_{1T} = \frac{2662}{L} \tag{7-63}$$

Torsion frequencies are somewhat lower than the frequencies of the axial modes as shown in Figure 7-29. Dynamic torque, however, follows the fluctuations in axial bit force as indicated in Figure 7-14.

7.4.5 Control of Lateral Vibrations

Lateral vibrations are especially critical in the lower portion of BHA where motors and MWD tools are located. The most likely source of lateral vibrations is a

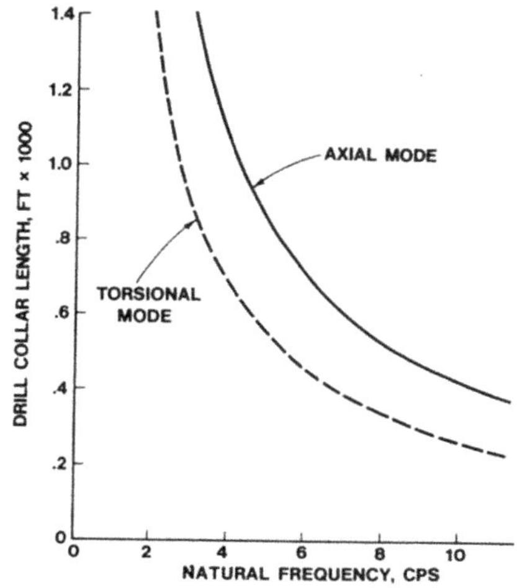

Figure 7-29. Natural frequency of drill collars.

positive displacement motor (PDM), which can excite modes between stabilizers as explained earlier. Frequency tuning (lateral modes) of drill pipe directly above the drill collar section should be checked for possible frequency tuning with rotary speeds.

7.4.6 Closing Remarks

Many years ago, I had the honor to discuss mechanical vibrations, and in particular, drillstring vibrations with Professor J. P. den Hartog, author of the first comprehensive treatise (1934) on mechanical vibrations [19]. Our discussion came as Jersey Production Research Company (JPRCo) was beginning to measure and analyze vibration data collected (for the first time ever) at drill bits with a downhole recorder [3].

As a young engineer, I was fascinated with his approach to data interpretation. Together we looked at the raw data and he commented .... "The information we need is here in front of us". He began counting frequencies and discussed possible sources of excitation. I came away from our visit realizing that while mechanical vibrations can be overwhelming, vibration analysis and control are rather simple ... determine natural frequencies and establish sources of excitation. If there is a vibration problem, it is almost certain to be frequency tuning.

Self-excited vibrations can also occur in drillstrings. Dareing, Tlusty, and Zamudio [6] explain the feedback mechanism developed at PDC drill bits. This phenomenon has been verified experimentally by Elsayed [20]. Limits of stability can be identified and provide guidance to bit design and operation.

Friction can also be a source of self-excited vibrations. Large torsion vibrations at the surface are not uncommon. These vibrations may damage surface equipment and cause safety concerns. Reference [21] gives a review of stick-slip friction math models along with recent studies in the application of time-delay methods of analysis for drillstring vibration control.

One hypothesis for the trigger mechanism of torsion self-excited vibrations, focuses on the combination of whirl, torque and axial instabilities. Using the analyses of Ch 6, lateral instability may occur when

$$\frac{Q}{Q_{Euler}} + \left(\frac{M_t}{M_{t,Greenhill}}\right)^2 + \left(\frac{\omega}{\omega_{crit}}\right)^2 \geq 1$$

While this equation was proposed for short pipe, it gives insight as to what might happen in very long pipe.

While any one component may not be enough to cause instability in drillstrings, the combination of the three might. Analysis show that in each case, (axial, whirl, torque), lateral deflections are greatest at the *same location*, namely, directly above the drill collars where pipe tension is lowest. Cyclic instability, accompanied by lateral deflections from all three factors, could accelerate torsional stick–slip to almost an uncontrollable level. In support of this hypotheses, drill pipe in the past has shown significant wear on one side, suggesting instability by axial and torsion buckling. Heavy weight drill pipe and pipe stabilizers directly above drill collars may be worth considering.

For sure three possible types of vibration (axial, torsion, and lateral) occur in drillstrings. They can be coupled or exist independently during drilling operations. In many cases each of these three vibration types can be isolated and analyzed separately to obtain reasonable engineering results. This has been my philosophy in writing this chapter.

7.5 SUMMARY OF KEY EQUATIONS

Natural Frequency of Lateral Vibrations

$$\omega_n = \left(\frac{n\pi}{L}\right)^2 \sqrt{\frac{EI}{m}} \quad \text{(same as synchronous whirl)}$$

$$\omega_n = \frac{a(n\pi)^2}{L^2}\sqrt{1 + \frac{SL^2}{(n\pi)^2 EI}}$$

Natural Frequency of Axial Modes in Drill Collars

$$f_{na} = \frac{n16{,}850}{4L} \quad \text{n = 1, 3, 5, ----} \quad \text{(axial mode)}$$

$$f_{1a} = \frac{4212}{L} \quad \text{(first mode)}$$

Frequency of Excitation

$$f = \frac{3N}{60} \text{ cps} \quad \text{(drill bit frequency, roller cone bit)}$$

Natural frequency of axial modes are determined from

$$\frac{\partial^2 u}{\partial x^2} = \frac{1}{c^2}\frac{\partial^2 u}{\partial t^2}$$

Natural frequency of torsion modes are determined from

$$\frac{\partial^2\theta}{\partial x^2} = \frac{1}{c_\theta^2}\frac{\partial^2\theta}{\partial t^2}$$

Equation of motion for axial response with damping

$$AE\frac{\partial^2 u}{\partial x^2} = \gamma\frac{\partial u}{\partial t} + m\frac{\partial^2 u}{\partial t^2}$$

7.6 PROBLEMS

1. The force-displacement data for a given shock absorber is

Force, lbs	Displ. in.
10,000	⅛
20,000	¼
30,000	¾
40,000	1

Determine the tool's spring rate.

2. If the spring rate for a shock absorber is 50,000 lb/in. How much bit weight causes the tool to bottom out over a maximum allowable displacement of 1½ in?

3. Determine the natural frequency of 695 ft (7 × 2½, 114 lb/ft) of drill collars place on top a shock absorber having a spring constant of (k = 50,000 lb/in). A roller cone drill bit is directly below the vibration absorber. What is the critical rotary speed for this BHA? Use the lump mass model.

4. Develop the frequency response curve for 800 ft (117 lb/ft) drill collars separated from a roller cone drill bit by a 45,000 lb/in vibration absorber. Assume no damping within the vibration absorber. Use the lump mass model.

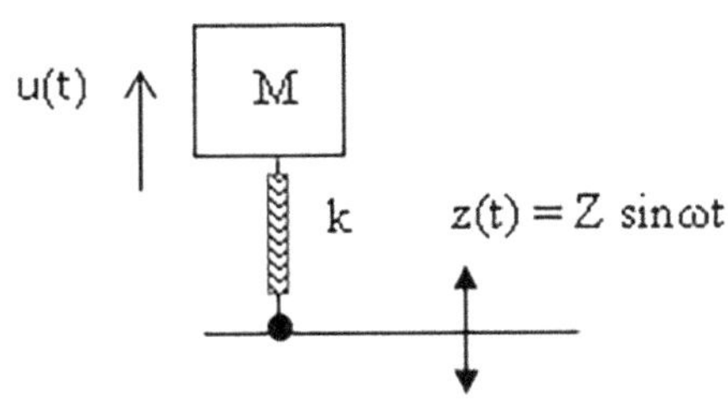

5. Determine the natural frequency of the second mode (axial mode) of a drill collar 6½ × 2 (102 lb per ft), 700 ft long for the following two cases:

a. No vibration absorber
b. A vibration absorber having a spring constant of 50,000 lb/in
What is the shift in natural frequencies?

6. Develop the equation of motion defining response of drill collars to a periodic forcing function applied to the lower end as shown in the figure. A vibration absorber separates the collars from the drill bit.

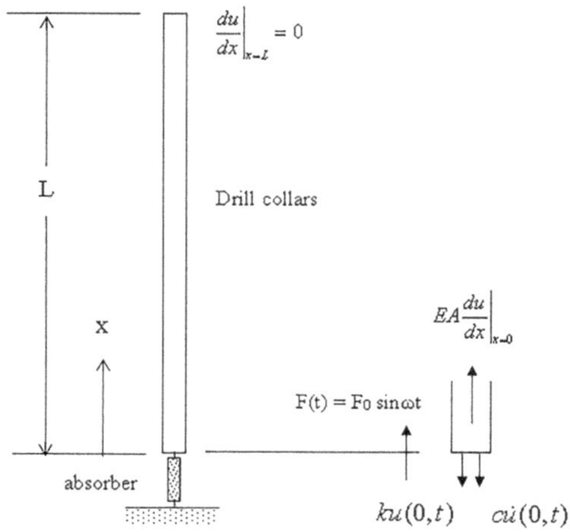

7. Drill collars are subjected to cyclic torque at the drill bit. A torsional dynamic absorber is attached at the top to isolated drill pipe from these oscillations. Develop the mathematical expression for drill collar response to these cyclic torques.

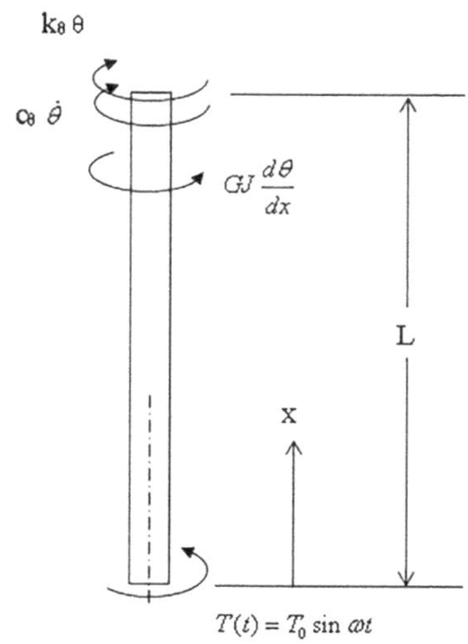

8. The following conditions are apply to a given BHA:
   a. Bit rotation speed is 300 rpm.
   b. Shock Absorber with a spring rate of 50,000 lb/in
   c. Drill collars — 9 × 3 in. (192 lb/ft)
   d. Maximum expected bit weight is 50,000 bs

   Determine:
   1. drill collar length such that the neutral point such that the neutral point is 85% of drill collar length
   2. natural frequencies of first and second axial modes
   3. critical bit speeds for both first and second modes
   4. compare with designated bit speed of 300 rpm

9. A 60 ft span of drill collars between two stabilizers is free to vibrate laterally. Assuming the following DC conditions:
   a. Drill Collar — 6½ × 2½ (96 lb/ft)
   b. WOB = 40,000 lb
   c. Mud weight is 11 ppg

   Determine:
   a. Natural frequency of the 1st mode of lateral vibration
   b. Critical rotary speed.

10. Maintaining the same operational parameters of problem 9 except a 3:2 PDM motor is producing a bit rotary speed of 350 rpm, what is the frequency of excitation produced by the mass of the rotor? What lateral mode is this frequency most likely to excite?

7.7 LIST OF REFERENCES

1. Garrett, W. R.; "Effect of a Downhole Shock Absorber on Drill Bit and Drill Stem Performance, Paper ASME 62-Pet-21 presented at the ASME 1962 Petroleum and Mechanical Engineering Conference, Dallas, Sept 23–26, 1962.
2. D. W. Dareing; *Mechanics of Drillstrings and Marine Risers*, ASME Press, 2012.
3. Dareing, D. W.: "Downhole Measurements of Drill-String Forces and Motions", Esso Production Research Report, No. EPR65-PR12, October 28, 1965.
4. Dareing, D. W. and Livesay, B. J.; "Longitudinal and Angular Drillstring Vibration with Damping," Trans. ASME *Journal of Engineering for Industry*, Vol. 90, Series B, No. 4, 1968, 671–679.
5. Ronnie Hardie and Anthony R. Lambert: DOWNHOLE APPARATUS, Canadian Patent No. CA 2 255 065, 1996.
6. Dareing, D. W., Tlusty, J., Zamudio, C.; "Self-Excited Vibrations Induced by Drag Bits," ASME, *Journal of Energy Resources Technology*, Vol. 112, March 1990, pp. 54–61.
7. Yao-Qun Lin and Yu-Hwa Wang; "Stick-Slip Vibration of Drill Strings", ASME *J. of Engineering for Industry*, Vol. 113, February 1991, pp. 38–43.

8. Yigit, Ahmet S. and Andreas P. Christoforou; "Stick-Slip and Bit Bounce Interaction in Oil-Well Drillstrings", *J. of Energy Resources Technology*, December 2006, Vol. 128, pp. 268–274.
9. J. Ford Brett, Thomas M. Warren, Suzanne M. Behr; "Bit Whirl — A New Theory of PDC Bit Failure", SPE Drilling Engineering, December 1990.
10. William T. Thomson and Marie Dillon Dahleh; *Theory of Vibrations with Applications*, 5th Ed. Prentice Hall, 1998.
11. S. Timoshenko, D. W. Young, and W. Weaver; *Vibration Problems in Engineering*, 4th Ed., John Wiley & Sons; 1974.
12. W. Dykstra, D. C.-K. Chen, T. M. Warren, J. J. Azar; "Drillstring component Mass Imbalance: A Major source of Downhole Vibrations", SPE Drilling & Completion, December 1996, pp. 234–241.
13. Huang, T. and Dareing, D. W.; "Buckling and Lateral Vibration of Drill Pipe," Trans. ASME, *Journal of Engineering for Industry*, Vol. 90, Series B., No. 4, 1968, 613–619.
14. Pushkar N. Jogi, John D. Macpherson, and Michael Neubert "Field Verification of Model-Derived Natural Frequencies of a Drill String", (ASME) *J of Energy Resources Technology*, Vol. 124 Sept 2002.
15. Garrett, Bill; Personal Discussions, 1980.
16. Dareing, D. W.; "Vibrations Increase Available Power at the Bit", ASME, *J. for Energy Resources Technology*, Vol. 107, March 1985, pp. 136–141.
17. Dareing, Don W.; "Drill Collar Length is a Major Factor in Vibration Control", *J. of Petroleum Technology*, April 1984, pp. 637–644.
18. Dareing, D. W.; "Guidelines for Controlling Drill String Vibrations," Trans. ASME, *Journal of Energy Resources Technology*, Vol. 106, June 1984, 272–277.
19. J. P. den Hartog, *Mechanical Vibrations*, McGraw-Hill Book Co.
20. Elsayed, M. A. and David W. Raymond, "Measurement and analysis of chatter in a compliant model of a drillstring equipped with a PDC Bit:, ETCE2000/ER-10280", Proceeding of ETCE/OMAE2000 Joint Conference, February 14–17, 2000, New Orleans, LA.
21. Belem Saldivar, Islam Boussaada, Hugues Mounier, Silviu-Iulian Niculesci: *Analysis and Control of Oilwell Drilling Vibrations*, Springer, 2015.

Chapter 8

DRILLING ECONOMICS

As discussed earlier, there are three types of drill bits: roller cone, PDC, and natural diamond. Each bit type has multiple design features targeting different geologic formations. Making choices among so many drill bits is based on *performance, cost, availability,* and *risk* as relates to technology. Bit performance is weighted highest among the four-decision metrics with footage cost being a key factor.

8.1 ECONOMICS OF DRILL BIT SELECTION

Expected bit performance is estimated from a data base of past performance in a given area. Lowest cost per foot is the basis of bit selection. Footage cost is calculated using

$$C = \frac{C_{rig}(t_T + t_d) + C_{bit}}{ROP^* t_d} \tag{8-1}$$

where

C = footage cost ($/ft)
C_{rig} =rig rate ($/hour)
C_{bit} = bit cost (S)
t_T = trip time (hours) (in and out)
t_d = drilling time (hours)
ROP = average rate of penetration over time, t_d(ft/hour)

This formula does not include costs related to hole problems, casing, etc. It is used to evaluate the cost effectiveness of drill bits. Equation (8-1) is displayed graphically (Figure 8-1) showing how footage cost steadily reduces with on-bottom drilling time (t_d) and rate of penetration (ROP), which depend directly on bit life and reliability of downhole tools.

Multiple cost curves can be drawn on the same diagram where each curve corresponds to a given *average* rate of penetration. The cost curves shaft upward as average rate of penetration reduces, indicating an increase in footage cost. Conversely, if average rate of penetration increases, footage cost reduces.

From this presentation of the cost equation, we see that drill bits that can sustain high rates of penetration (ROP) over a long drilling interval, t_d, give lowest footage cost.

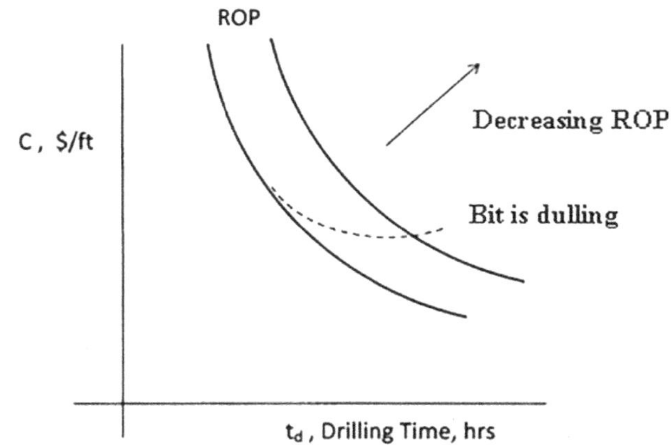

Figure 8-1. Footage cost versus drilling time and ROP.

For example, if

C_{rig} = \$300/hour
C_{bit} = \$8000
t_T = 10 hours
t_d = 100 hours
ROP = 15 ft/hour

then

$$C = \frac{300(10+100)+8000}{15(100)} = \$27.33/\text{ft}$$

Rig rate, trip time and bit cost are constant values in this plot. Using the numbers in the above example

$$C = \frac{300(10+t_d)+8000}{ROP(t_d)} \tag{8-2}$$

By choosing an ROP, footage cost is plotted against on-bottom drilling time as shown in Figure 8-1.

With this type of diagram, one can track footage cost by dividing footage drilled by drilling time and locate the corresponding footage cost on the diagram. The footage cost at each time step is the footage cost if the bit were tripped out of the hole at that particular time. If ROP is constant, footage cost would reduce along a given ROP curve. If average ROP reduces with drilling time, the footage cost point moves from one ROP curve to a lower ROP curve.

8.2 MONITORING DRILL BIT PERFORMANCE

The following example shows the effect of keeping a drill bit on bottom after it shows signs of reduced drilling performance as measured in terms of rate of penetration (ROP).

Consider the following scenario

Rig Rate = \$500/hour
Trip Time = 12 hours
Bit Cost = \$7000

The initial *average* rate of penetration with a new bit is ROP = 5 fph. However, after drilling for 100 hours or a distance of 500 ft the average ROP begins to reduce, and while drilling an additional 500 ft, the average rate of penetration drops to ROP = 3 fph. During the second 500 ft interval, footage cost moved from one curve (ROP = 5 fph) to another (ROP = 3 fph) as shown in Figure 8-2.

Lower Curve	Upper Curve
Rig Cost = \$500/hour	Rig Cost = \$500/hour
ROP_{ave} = 5 fph	ROP_{ave} = 3 fph
t_T = 12 hours	t_T = 12 hours
Bit Cost = \$7000	Bit Cost = \$7000

The footage cost using a *single* bit to drill the 1000 ft is

$$C = \frac{500(12 + 333.3) + 7000}{1000} = \$179.60/\text{ft}$$

where $t_d = \dfrac{1000\text{ ft}}{3\text{ fph}} = 333.3\text{ hr}$

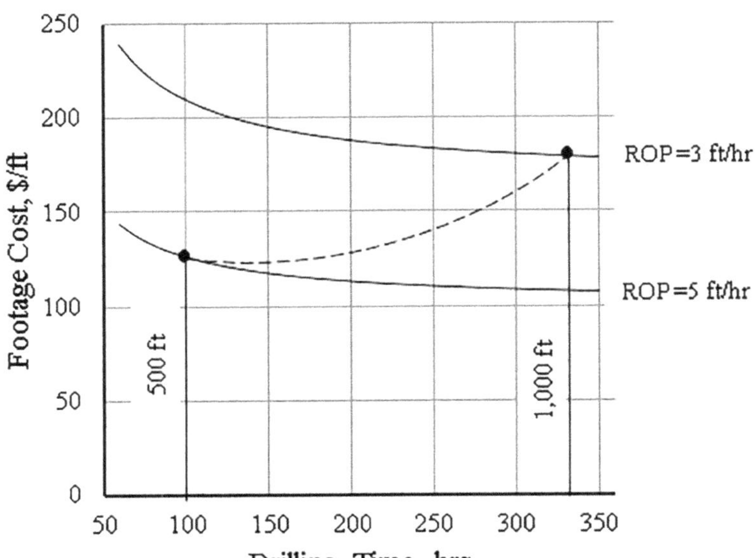

Figure 8-2. Effect of bit wear on footage cost.

When *two* drill bits are used to drill the 1000 ft, a second round-trip is required to replace the first bit. The footage cost for each 500 ft interval is

$$C = \frac{500(12 + 100) + 7000}{500} = \$134.00/\text{ft}$$

where $t_d = \dfrac{500 \text{ ft}}{5 \text{ fph}} = 100$ hr

The savings in footage cost using two bits instead of one is \$45.60/ft or over the 1000 ft interval is (45.60)(1000) = \$45,600. This example shows how footage cost increases by riding a bit too long. As the drill bit dulls, *instantaneous* ROP reduces, however, footage cost is based on average ROP over t_d.

This type of graphical presentation of the footage cost equation is commonly used to track the performance (ROP and footage cost) of drill bits.

8.3 BREAKEVEN CURVE

A bit that drills a given formation at the lowest cost per foot is the best choice. Estimated performance of drill bits is based on past performance in similar formations. Statistical data are useful for predicting bit performance and for economic comparisons of various bit options.

One method used to compare the performance of two, say, competitive drill bit bits is by use of the *breakeven curve*. The breakeven curve is based on the footage cost equation [Eq. (8-1)].

Since the average rate of penetration is

$$ROP = \frac{D}{t_d} \tag{8-3}$$

The drilling cost formula can also be written

$$C = \frac{C_{rig}(t_T + t_d) + C_{bit}}{D} \tag{8-4}$$

Rearranging Eq. (8-4),

$$D = \frac{C_{rig}}{C} t_d + \frac{C_{rig} t_T + C_{bit}}{C} \tag{8-5}$$

Equation (8-5) predicts the drilling interval required to achieve footage cost, C, during on-bottom drilling time, t_d.

A breakeven curve is developed from Eq. (8-5) as follows.

$$D = K_2 t_d + K_3 \tag{8-6}$$

where

$$K_2 = \frac{C_{rig}}{C}$$

$$K_3 = \frac{C_{rig} t_T + C_{bit}}{C}$$

Combining Eqs. (8-3) and (8-6) by eliminating t_d gives

$$ROP = \frac{K_2 D}{D - K_3} \tag{8-7}$$

where K_2 and K_3 are constants as defined above. This equation predicts the average rate of penetration required to achieve a given footage cost over a drilled interval D. A breakeven curve is a plot of this relationship (Figure 8-3). The ROP curve is bounded by K_2 and K_3. K_2 is parallel to the D axis, while K_3 is parallel to the ROP axis.

This type of presentation is useful for comparing the economic performance of one drill bit against another. The breakeven curve is developed and used as follows:

a. Determine the footage cost based on performance and price of Bit A, using Eq. (8-4).
b. Using the footage cost of Bit A, develop a breakeven curve based on the price of Bit B [use Eq. (8-7)].
c. Track the performance of Bit B on the breakeven curve based on its actual ROP.
d. Bit B breaks even with Bit A after footage D when the horizontal line intersects the breakeven curve. Beyond this footage, drilling cost is reduced and Bit B begins to out-perform Bit A.

This diagram is called a breakeven curve because it graphically shows the drilling interval, D at which Bit B breaks even ($/ft) with Bit A.

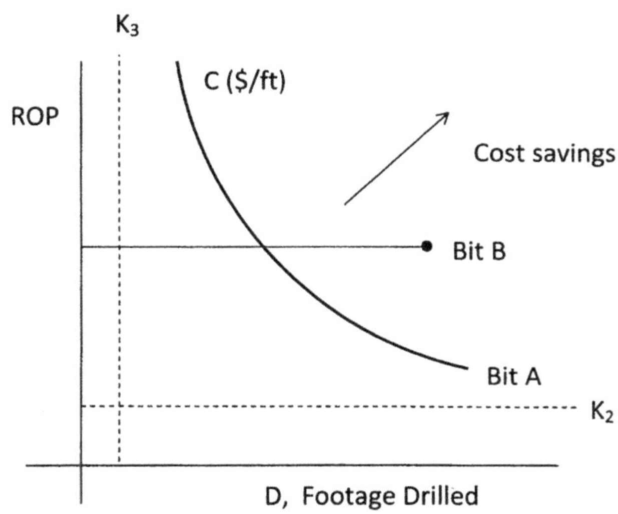

Figure 8-3. Breakeven cost curve.

8.3.1 Prices of Bit A and Bit B are Equal

As an example, consider

C_{rig} = \$300/hour
t_T = 9 hours
$C_{bit,A}$ = C_{bitB} = \$7000
ROP = 30 fph
t_d = 20 hours

Also assume that offset bit records for Bit A show that on an average, Bit A drills 600 ft of the formation in 20 hours. The footage cost using Bit A is

$$C = \frac{300(9 + 20) + 7000}{600} = \$26.17/\text{ft}$$

The cost curve for Bit B corresponding to \$26.17/ft is developed as follows.

$$K_2 = \frac{300}{26.17} = 11.46$$

and

$$K_3 = \frac{300(9) + 7000}{26.17} = 370.65$$

The footage cost curve for \$26.17/ft is shown in Figure 8-4 with the performance of Bit A on this curve. Assuming that the cost of Bit B is the same (or comparable) with Bit A, it will breakeven with Bit A once it has drilled 518.64 ft. The footage cost with Bit B reduces with additional footage drilled.

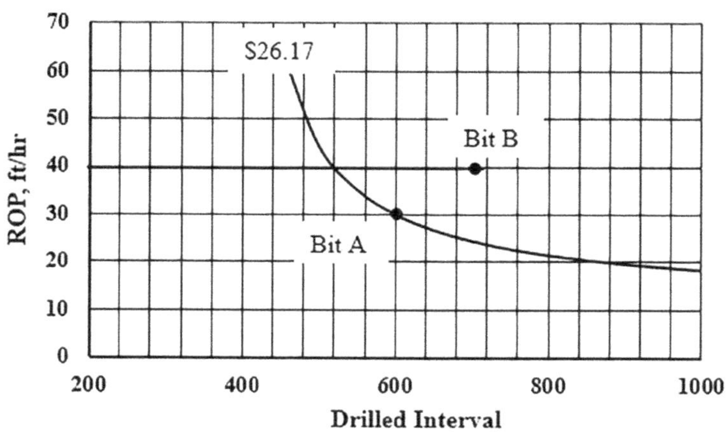

Figure 8-4. Breakeven curve for example case.

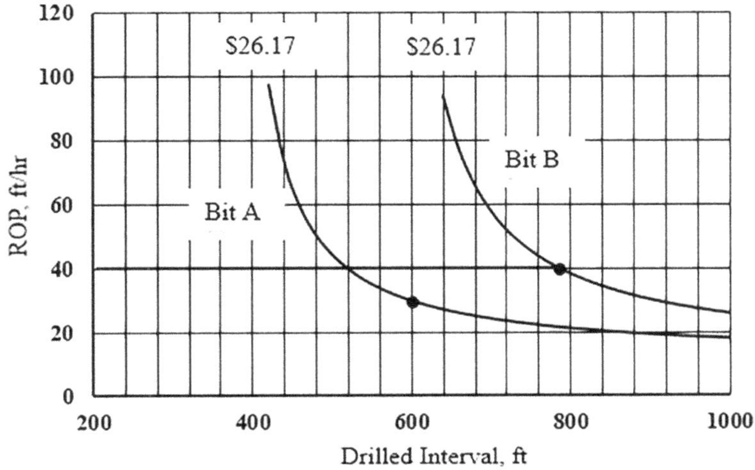

Figure 8-5. Breakeven curve for higher Bit B cost.

8.3.2 Prices of Bit A and Bit B Are Different

If the cost of drill bit B is higher than the cost of drill bit A, say $12,000, then K_3 = 561.71 and the breakeven curve for Bit B changes (see Figure 8-5).

Bit B would have to drill a 787.26 ft interval to reach achieve a footage cost of $26.17/ft and breakeven with Bit A. Footage cost with Bit B reduces with more footage drilled beyond this point.

8.4 FOOTAGE COST USING DRILLING MOTORS

Downhole drilling motors (PDMs) or turbines are typically charged to a drilling interval only for the time they are on bottom drilling. They are not charged to the interval while tripping. The footage cost equation is modified accordingly.

$$C = \frac{C_{rig}\left(t_T + t_d\right) + C_m t_d + C_{bit}}{D} \tag{8-8}$$

where C_m is motor (or turbine) rate ($/hour). The previous cost diagram still applies where now

$$K_2^* = \frac{C_{rig} + C_m}{C} \tag{8-9}$$

The expression for K_3 is unchanged.

Downhole motors provide more mechanical power to drill bits but add cost to the bottom hole assembly. Motors (and turbines) have to increase rate of penetration to justify this additional cost. The ROP required to produce the same footage cost, C, as conventional rotary drilling is determined as follows.

Following the development of Eq. (8-7),

$$ROP_{motor} = \frac{K_2^* D}{D - K_3} \tag{8-10}$$

$$ROP_{rotary} = \frac{K_2 D}{D - K_3} \tag{8-11}$$

where

$$K_2^* = \frac{C_{rig} + C_m}{C}$$

$$K_2 = \frac{C_{rig}}{C}$$

Combining Eqs. (8-10) and (8-11) gives

$$\frac{ROP_{motor}}{ROP_{rotary}} = \frac{K_2^*}{K_2} = \frac{C_{rig} + C_m}{C_{rig}} \tag{8-12}$$

$$ROP_{motor} = \left(1 + \frac{C_m}{C_{rig}}\right) ROP_{rotary} \tag{8-13}$$

where

C_m = motor cost ($/hour)
C_{rig} = rig cost ($/hour)
C is the same for both motor and rotary (thus cancel out)
D is the same for both motor and rotary (thus cancel out)

Equation (8-13) shows that for the footage cost to be the same for both motor and rotary, ROP with motor must be $1 + \frac{C_m}{C_{rig}}$ times as much as ROP with conventional rotary.

Experience with downhole motors and turbines suggest that rate of penetration can be increased over conventional rotary drilling by a factor of 1.5 to 1.9 with downhole motors and increased by a factor of 2.0 to 2.9 with downhole turbines.

Consider

$C_{rig} = \$500/\text{hr}$

$C_m = \$200/\text{hr}$

Then for downhole motors to compete with rotary power

$$(ROP)_{motor} = \left(1 + \frac{200}{500}\right)(ROP)_{rotary}$$

$$(ROP)_{motor} = 1.4\,(ROP)_{rotary}$$

for the same footage cost. Since, in practice, motors drill 1.5 to 1.9 times faster than rotary, the motor is justified.

Assuming a rig rate of $5000/hour,

$$(ROP)_{motor} = 1.04(ROP)_{rotary}$$

In this case, motor cost is small compared to rig rate so the motor is easily justified.

8.5 EFFECT OF PREMATURE TOOL FAILURE ON FOOTAGE COST

Premature failure of any tool in the BHA adversely affects footage cost. The probability of premature failure is elevated by the use of mechanically and electrically complex equipment, which must perform in a hostile environment. Reliability and maintenance of these tools is critical to drilling operations especially offshore where rig rates are relatively high. The cost of premature failure, which is measured in terms of on-bottom drilling time is explained below.

Tool failure affects drilling time, t_d, and footage cost according to

$$C(\$/ft) = \frac{C_{Rig}\left(t_T + t_d\right) + C_{PDM}\,t_d + C_{Bit}}{ROP t_d} \qquad (8\text{-}14)$$

This equation was programmed with drilling time (t_d) and rig rate (C_{rig}) as variables. The other parameters were assumed to be

C_{PDM} = motor cost ($200/hour)
C_{bit} = bit cost ($10,000)
t_T = trip time (10 hours)
ROP = 50 ft/hour

Table 8-1 and Figure 8-6 compare footage cost for three rig rates corresponding to; drill ship, jack-up rig and land rig. The independent variable is on-bottom drilling time, t_d.

The impact of premature tool failure is greatest offshore. Consider the drill ship column. If there is a tool failure (any tool in the BHA) after 10 hours of drilling, then the footage cost over this time is $824/ft. If applied over 1000 ft of drilling, then cost to drilling this interval is $824,000. On the other hand, if the BHA is pulled after 100 hours of drilling, footage cost is $446/ft making total drilling cost over the 1000 ft, $446,000. The cost of premature failure is viewed as $378,000. Loss of future revenue due to delayed production is an added cost to the operator.

Table 8-1. Drilling Cost per Foot; $/ft

Drilling Time, hrs	Drill Ship $480,000/day $20,000/hr	Jack-up Rig $150,000/day $6,250/hr	Land Rig $72,000/day $3,000/hr
5	1244.00	419.00	224.00
10	824.00	274.00	144.00
15	684.00	225.67	117.33
20	614.00	201.50	104.00
25	572.00	187.00	96.00
30	544.00	177.33	90.67
35	524.00	170.43	86.86
40	509.00	165.25	84.00
45	497.33	161.22	81.78
50	488.00	158.00	80.00
55	480.36	155.36	78.55
60	474.00	153.17	77.33
65	468.62	151.31	76.31
70	464.00	149.71	75.43
75	460.00	148.33	74.67
80	456.50	147.13	74.00
85	453.41	146.06	73.41
90	450.67	145.11	72.89
95	448.21	144.26	72.42
100	446.00	143.50	72.00
105	444.00	142.81	71.62
110	442.18	142.18	71.27
115	440.52	141.61	70.96
120	439.00	141.08	70.67

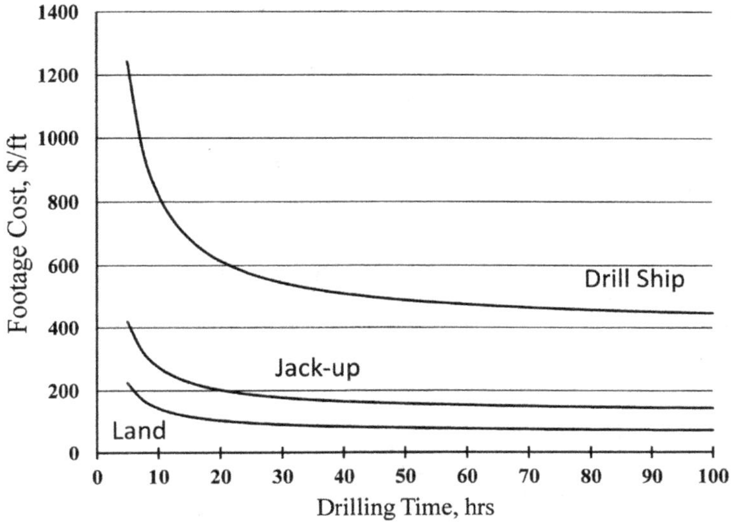

Figure 8-6. Effect of rig rate on footage cost.

8.6 SUMMARY OF EQUATIONS

Footage Cost

$$C = \frac{C_{rig}(t_T + t_d) + C_{bit}}{ROP^* t_d}$$

Breakeven Equation

$$ROP = \frac{K_2 D}{D - K_3}; \quad K_2 = \frac{C_{rig}}{C}; \quad K_3 = \frac{C_{rig} t_T + C_{bit}}{C}$$

Footage Cost with Drilling Motors

$$C = \frac{C_{rig}\left(t_T + t_d\right) + C_m t_d + C_{bit}}{D}$$

$$ROP_{motor} = \left(1 + \frac{C_m}{C_{rig}}\right) ROP_{rotary}$$

8.7 PROBLEMS

1. Drilling parameters are given as follows:

   Rig Rate = \$1000/hour
   Trip Time = 12 hours
   Bit Cost = \$10,000

The initial *average* rate of penetration with a new bit is ROP = 6 fph. However, after drilling for 100 hours or a distance of 600 ft the average ROP begins to reduce, and while drilling an additional 1200 ft, the average rate of penetration drops to ROP = 4 fph. During the second 600 ft interval, footage cost moved from one curve (ROP = 6 fph) to another (ROP = 4 fph) as shown in Figure 8-2.

Determine the cost savings by using two drill bits over the 1200 foot section of the hole as opposed to one drill bit.

2. Using the following drilling parameters calculate and plot (computer) the breakeven curve showing how ROP and drilling interval, D, vary for a constant footage cost. Operating parameters for Bit A are:
   - C_{rig} = \$500/hour
   - C_{bit} = \$9000
   - t_T = 8 hour (trip time: in and out)
   - t_d = 42 hours (on bottom drilling time)
   - D = 840 ft (drilled interval, to establish footage cost, \$/ft)
   - ROP = 20 ft/hour

   A competitive drill bit (Bit B) costs the same as Bit A, but can drill at a faster rate (ROP = 30 ft/hour). At what drilling interval does Bit B breakeven with Bit A?

3. Repeat problem 2 for the same operating parameters, except assume Bit B cost \$13,000. At what interval does Bit B breakeven with Bit A?

4. Develop and plot two breakeven curves; one curve corresponding to no motor in the BHA and the other corresponding a motor in the BHA. Base the curves on the following data.

   Rig rate = \$1000/hour
   Trip time = 10 hour
   Bit cost = \$10,000
   Motor cost = \$250/hour

   Develop both curves for a footage cost of \$300/ft and show that

$$ROP_{motor} = \left[1 + \frac{C_{motor}}{C_{rig}}\right] ROP_{rotary}$$

   Let D range from 75 to 400 ft in increments of 25 ft.

Chapter 9

MAINTENANCE AND RELIABILITY

The previous chapter underscores the importance of performance and duration of on-bottom drilling time (t_d) between bit runs. Tool reliability becomes especially important in offshore drilling operations where rig rates are relative high compared with on-land drilling. Premature failure of anyone of these tools effects footage cost by a lower on-bottom drilling duration (t_d).

Maintenance and reliability of downhole drilling equipment has become an important consideration as premature failure of any of the bottom hole tools will drastically affect footage cost. The performance of complex drilling tools such as, MWD, PDMs, Rotary Steerable Tools, and drill bits are especially important in directional drilling. Each of these tools has become more and more complicated mechanically and electronically. The failure of any tool may require removal and replacement before the hole can be advanced. Each must operate in a harsh downhole environment including; vibrations, possible drilling mud contamination and temperature. Design issues that relate to this harsh environment are: integrity of the electronic packages and components, dynamic seals, rolling contact bearing life and other mechanical components.

Three approaches to managing tool maintenance are [1]:

1. *Reactive maintenance* is based on running equipment until it fails and then fix it. The problem is that an unexpected failure could be costly, i.e., untimely downtime and total loss of equipment. For example, if a tool in the bottom hole assemble fails prematurely, footage cost increases substantially and future revenue is delayed. A premature failure of downhole equipment could also affect well control and safety.
2. *Preventive maintenance* requires action on a routine schedule usually based on past performance. The goal is to detect any degradation of the total assembly or component with the aim of extending useful life. This requires monitoring of performance and taking steps to prolong the life of system components. Monitoring the rate of penetration (ROP) of drill bits is a good example. If a drill bit dulls or is not performing, it may be necessary to check mud weight, reduce load, and speed or even pull the bit prior to total bit destruction. In some offshore drilling operations, drill pipe is not reused. It is simply not worth the risk and cost of a drill pipe failure.
3. *Predictive maintenance* relies on measurements that detect the onset of a sudden failure, thus allowing time to reduce or eliminate critical stresses by adjusting loads or system dynamics. Adjustments in speed, for example, may change frequency tuning effects. Also, pressure

and flow rate monitoring at the surface may reveal inadequate performance of PDMs. Measurement-While-Drilling (MWD) equipment become valuable tools in this regard.

Reliability refers to the probability that a tool will perform its function over a stated period of time. The bottom line is footage cost. If performance and life expectations are not met, footage cost will be substantially increased. In relief well drilling, lost time means loss of production and usually substantial environmental damage. Underlying principles of good design practice are:

a. Minimize the number of parts. Reliability varies inversely with number of parts
b. Keep the design simple, Complexity reduces reliability
c. Use standard parts when possible; there is plenty of statistical performance data

9.1 DOWNHOLE TOOL DESIGN

Maintenance and reliability start with good engineering design practice. There is a process, though not always understood or followed, that can improve tool reliability.

9.1.1 Objective of Design

A clear statement of the purpose of the design needs to be documented. A brief statement of strategy is also appropriate. These statements do not have to be long, but they need to be clear.

9.1.2 Setting the Design Specifications

Design specifications are an itemized set of constraints placed on the design of a product. They identify what the product is supposed to do and how the product should perform. They represent the initial "engineering baseline" from which all concepts are generated. This initial engineering baseline advances as the design progresses. Design specifications may become part of a legal contract.

Performance Requirements — Performance requirements identify specifics such as: loads, motions, flow rates, operating pressures, temperature limits, etc. In addition, technical specifications may include physical and chemical properties of materials to be used in the design. Performance expectations should be specific and may require preliminary testing to define design loads.

Codes and Standards — During the midst of the industrial revolution of the 19th century, it became apparent that mechanical components needed to be standardized to assure that, for example, bolts made in one place could fit together with nuts made in another place. ASME took a leading role in standardizing mechanical components such as pulleys, gears, key seats, etc. Even shop drawing symbols needed to be standardized.

Environmental — Environmental impact should be considered from over the products life cycle, i.e., from fabrication to product disposal. Even in the early stages of designing offshore platforms, engineers have to consider how to dispose of an offshore structure at the end of the production life (about 20 to 30 years) of an offshore oil and gas field.

Reliability — There are two methods of design, which relate to the safety and reliability of products: 1) factor of safety method and 2) statistical or reliability approach to design.

The *factor of safety* method is commonly used in engineering design. It is a time proven design tool and when used properly, sound and safe designs can be developed. Factor of safety (FS) is the ratio of failure stress to design stress. Failure stress depends upon criteria of failure of material used in the design. Designs are configured (size, dimension) according to the design stress or allowable stress. Factors of safety are intended to cover all uncertainties not identified in a set of specifications.

$$F.S. = \frac{F_{fail}}{F_{allow}} = \frac{\sigma_{yld}}{\sigma_a} \tag{9-1}$$

In general, the higher the factor of safety, the higher the product weight and cost. So it is important to keep the FS as low as possible. Factors of safety are now being substantially lowered because of stress computational advances brought about by high-speed computers and sophisticated computer software; uncertainty is thus lowered.

The *reliability* approach to design seeks to establish design parameters so that the product performs to an expected level of reliability [2]. Reliability is a statistical measure of performance and represents the probability a product that will perform to a desired goal. A product reliability of 0.9 means, that there is a 90 per cent chance that a given product will perform its proper function without failure. The method requires statistical data on all random variables, such as strength, size, weight, etc. With this approach, products can be designed to a required level of reliability. However, in many cases, product reliability is determined in the market place after the product is put into use.

Cost Considerations — Economics is considered during the early stages of concept generation. Production methods that allow for early return on investments may be critical. As the development costs continue to rise, oil companies want to begin production as early as possible, maybe prior to the completion of the overall production system. Time is money and if it takes 3 years to bring a reservoir on line that is 3 years without return on investment if early production is not achieved. The true savings on early production is determined by "discounted cash flow" calculations.

9.1.3 Creating Design Alternatives

Once design specifications have been set, possible design alternatives can be generated. Since design is open ended, i.e., there are many possible solutions, it is desirable to generate several design concepts and then evaluate them as a group before choosing the best possible design solution. The appropriateness of each concept is limited only by the imagination and experience of the design team. The objective at this point is to create design concepts that satisfy the specifications.

Each concept must be shown to be technically feasible before it can be properly evaluated. This means that engineering analyses and testing may be required and even drawings made to advance an idea beyond a hand sketch and a plausible alternative. A preliminary study may be needed to show an idea is workable, realistic and satisfies all design constraints or specifications, including cost constraints. In engineering practice, feasibility analyses may require extensive computer calculations and/or laboratory testing.

9.1.4 Evaluating Alternatives and Making a Choice

Decisions are typically based on four basic metrics: performance, cost, risk, and availability. Each of these metrics is defined as follows:

Performance — The capability to achieve needed operational characteristics, plus reliability.
Cost — The estimated cost of the design, including development and manufacturing costs.
Risk — The possibility that performance may not be met because of the design approach, absence of testing, or some specific technical consideration.
Availability — The availability of a design depending upon the stage of development.

A typical procedure for scoring several alternatives is to break each of these four categories into key sub elements and give each an appropriate weight.

Table 9-1. Evaluation scoring matrix.
(Hypothetical Numbers; sub metrics are given as a guide)

	Weight	Concept			
	%	A	B	C	D
PERFORMANCE (40)					
• Functional requirements	25	10	18	25	20
• Reliability/maintenance	15	9	12	15	17
RISK (30)					
• Accuracy of approach	20	10	12	14	7
• Experience of personnel	10	8	8	7	7
SCHEDULE (20)					
• Time of delivery (to market, production site)	20	12	15	18	18
COST (10)					
• Development, implementation	5	3	2	3	1
• Early production of reservoir	5	3	3	2	2
TOTAL SCORE	100	55	70	84	72

Each design alternative can then be scored under each sub element. The scoring is strictly a judgmental call so experience becomes an important factor. The more experience the better the judgment or scoring. Once each concept has been scored for each sub element, the numbers are totaled for a composite score for each concept. The total scores provide a means of comparing each concept against the others (see Table 9-1).

This numerical evaluation scheme has two key objectives. The first is to have a way to quantify one's judgment against a fixed scale so that each design concept can be rated in the same manner, thus, showing their truest level of merit in comparison with each other. The second objective is to give a way to examine the rationale of the final scores by looking at the sub elements of each concept in order to see the strong and weak features of each. Since all design selections are the result of trade-offs, the scoring system aides in selecting a preferred concept. The example given in Table 9-1 shows that concept C scored the highest.

9.1.5 Design Considerations

A preliminary design refines the conceptual design and more accurately describes the subsystems and mechanical elements through engineering calculations and material selection. Classic mechanics equations calculations (mechanics of materials, fluid mechanics, heat transfer, mechanical vibrations, etc.) will suffice. The goal at this point is to refine the design as close as possible to the products final configuration. The engineering design can be further refined by using finite element software. Quality calculation methods lead to a more reliable product. A few things to consider are discussed below.

9.1.5.1 Combined Stresses

Stress condition at any point in a design is usually the combined effects of different types of loads, such as; direct pull, bending, torsion, pressure, etc. Each of these loads produces stresses, which are superimposed to form a stress state at each point in a design. Local stresses are then expressed in terms of normal and shear stresses, which are then transformed into principal stress components and maximum shear.

Consider the case where pipe is loaded by torsional shear, internal pressure plus bending. Using the following pipe data and loads

- Pipe is 4-in. OD and 3.75 ID
- Internal pressure is 1000 psi
- Applied torque is 2000 ft-lb
- Bending moment is 5000 ft-lb
- Material yield is 60,000 psi

The challenge here is to determine the factor of safety against yielding. First, we calculate the three stress components.

Moments of inertia are needed to determine torsion shear and bending moment stresses.

$$I = \frac{\pi}{4}\left(r_0^4 - r_i^4\right) = \frac{\pi}{4}\left(2^4 - 1.875^4\right) = 2.8591 \text{ in.}^4$$

$$J = 2I = 5.7183 \text{ in.}^4$$

$$A = \pi\left(r_0^2 - r_i^2\right) = \pi\left(2^2 - 1.875^2\right) = 1.5217 \text{ in.}^2$$

Internal Pressure

$$\frac{t}{r_0} = \frac{0.125}{2} = 0.0625 \quad \text{or} \quad \frac{r_0}{t} = 16 \succ 10$$

The pipe must be considered a thick-walled cylinder. Using eq. (10-9)

$$\sigma_\theta(r = 2) = \frac{1000(1.875)^2 + 1.875^2 1000}{2^2 - 1.875^2} = 14{,}516 \text{ psi}$$

$$\sigma_\theta(r = 1.875) = \frac{1000(1.875)^2 + 2^2 1000}{2^2 - 1.875^2} = 15{,}516 \text{ psi}$$

Applied Torque

$$\tau = \frac{Tc}{J} = \frac{2000(12)2}{5.7183} = 8{,}394 \text{ psi}$$

Bending Moment

$$\sigma_b = \frac{5{,}000(12)2}{2.8591} = 41{,}971 \text{ psi}$$

Assuming the pipe is support axially so that the internal pressure does not create axial stress, the state of stress is

$\sigma_x = \sigma_b = 41{,}971$ psi
$\sigma_y = \sigma_\theta = 15{,}516$ psi
$\tau_{xy} = 8{,}394$ psi

These stresses are shown in Mohr's circle (Figure 9-1), which also indicate principal stresses and maximum shear for the combined stresses.

$\sigma_{p1} = 44{,}409$ psi
$\sigma_{p2} = 13{,}077$ psi
$\tau_{max} = 15{,}666$ psi

The orientation of the principal axis from the x axis is $\theta = 11.2$ deg. counterclockwise.

The combined effect of three different types of loading (internal pressure, bending, and torsion) produced a biaxial state of stress having principal stress as shown.

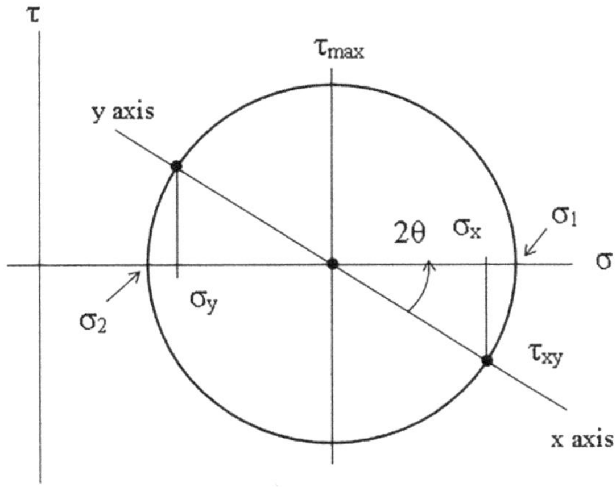

Figure 9-1. Mohr's circle for the example.

9.1.6 Stress Concentration

Classical means of calculating stress (bending, torsion, pressure, axial, and combined stress) are adequate in many cases to configure a reliable design considering factors of safety. However, in consideration of safety and cost, such as in aerospace and drilling equipment, it is desirable to refine stress calculations in critical areas of a design using finite element software or stress concentration factors that have been established analytically or experimentally.

Consider, for example, a small hole in the center of a long strap or plate (Figure 9-2). Using the theory of elasticity, Seely and Smith [3] show that local stress across a strap are defined by

$$\sigma_x = \frac{\sigma_0}{2}\left(2 + \frac{a^2}{r^2} + 3\frac{a^4}{r^4}\right) \tag{9-2}$$

This formula has been confirmed by strain measurements and photo-elastic method.

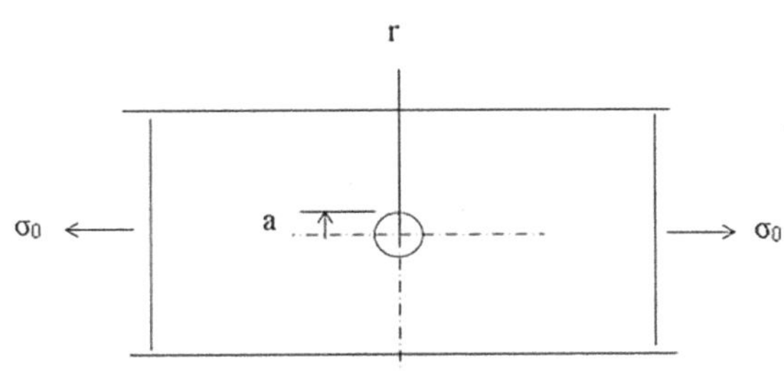

Figure 9-2. Plate with hole.

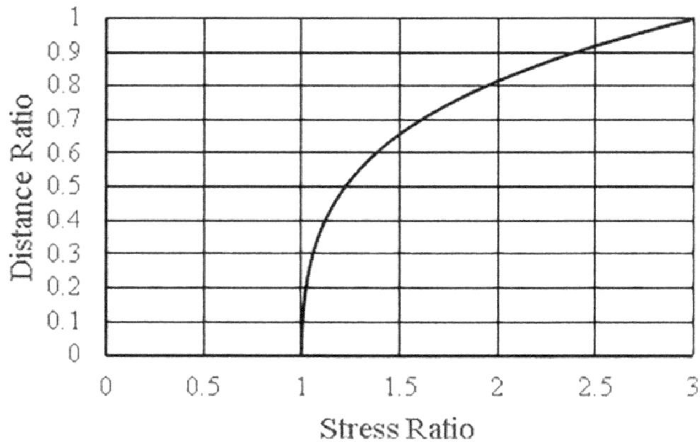

Figure 9-3. Stress distribution away from the hole.

Equation (9-2) is plotted in Figure 9-3 to show how local stress varies across the width of the plate starting at the edge of the hole and moving to the outside edge of the plate. The parameters in the drawing are

Stress Ratio — $\frac{\sigma_x}{\sigma_0}$

Distance Ratio — $\frac{a}{r}$

When r = a, the distance ratio is one and the stress ratio is 3, meaning that local stress at the hole is 3 times the nominal stress, σ_0. Also, as the distance ratio reduces to zero, meaning distance, r, is very large local stress approaches the nominal stress. The maximum stress is located at the edge of the hole. The stress concentration factor is 3.

Stress concentration factors for other geometric shapes have been established theoretically and experimentally [4]. Always avoid sharp geometric changes, such as fillets, openings, sharp thread roots, etc. as they are points of stress concentration.

9.1.6.1 Material Yielding

Material yielding is a state of stress-strain where the material stretches with a small increase in load leading to permanent distortion in a material. Yield strength is a common failure criterion in design. Once yielding is reached, plastic deformation occurs, and low cycle fatigue is possible. The criteria of failure commonly used to determine material yielding is the von Mises criteria based on energy of distortion [4]. For a biaxial state of stress this criterion shows, yielding occurs when

$$\sigma_A^2 - \sigma_A \sigma_B + \sigma_B^2 \geq \sigma_{yld}^2 \tag{9-3}$$

where σ_A and σ_B are local principal stresses. Brittle materials behave differently. Failure occurs suddenly without yield. The Coulomb-Mohr Criteria is commonly used in this case.

9.1.7 Impact Loading, Low and High Cycle Fatigue

There are three ranges of cyclic material fatigue failure [4]. These ranges are typically determined experimental by a bending test device developing stress reversals without an average stress component.

1. Low cycle fatigue (½ cycle to 1000 cycles). The one-half cycle refers to a static tension test.
2. Finite life cycle (1000 out to 10^6 cycles to 10^7 cycles for steel)
3. Infinite life cycle (no failure beyond 10^7 cycles. The stress level below, which no failure is expected is called the endurance limit. Tests indicate that the endurance limit for steel is about ½ of the ultimate strength up to an ultimate strength of 200 ksi. Beyond this strength level, the endurance limit is 100 ksi.

Impact loads are also possible and inertia effects should also be considered, especially in low temperature applications.

Since downhole tools are subjected to various modes of vibration as explained earlier, each of these zones of fatigue have to be considered in bottom hole tool design.

9.1.7.1 Fatigue

Metal fatigue results from cyclic stresses defined by

$$\sigma = \sigma_m + \sigma_A \cos \omega t$$

where

σ_m = mean stress
σ_A = stress amplitude
ω = frequency of stress cycle

Material failure depends on magnitude of both σ_m and σ_A. When $\sigma_m = 0$ there is complete reversal of stress. Endurance limits are determined experimentally under this condition. Tests are conducted on multiple specimens, each being cycles at different stress levels of σ_A until failure occurs. The number of cycles to failure is documented at each stress level. The endurance limit is defined as the stress level at which complete reversals do not cause failure beyond 10^6 cycles.

Test data for steel show that endurance limits depend on the ultimate strength.

$$S_e \sim 0.5 S_{ult} \quad \text{for} \quad S_{ult} \prec 200 \text{ ksi}$$
$$S_e \sim 100 \text{ ksi} \quad \text{for} \quad S_{ult} \succ 200 \text{ ksi}$$

In general, time dependent stresses cycle about a mean or average stress level. In this case, the Goodman diagram (Figure 9-4) is useful for determining

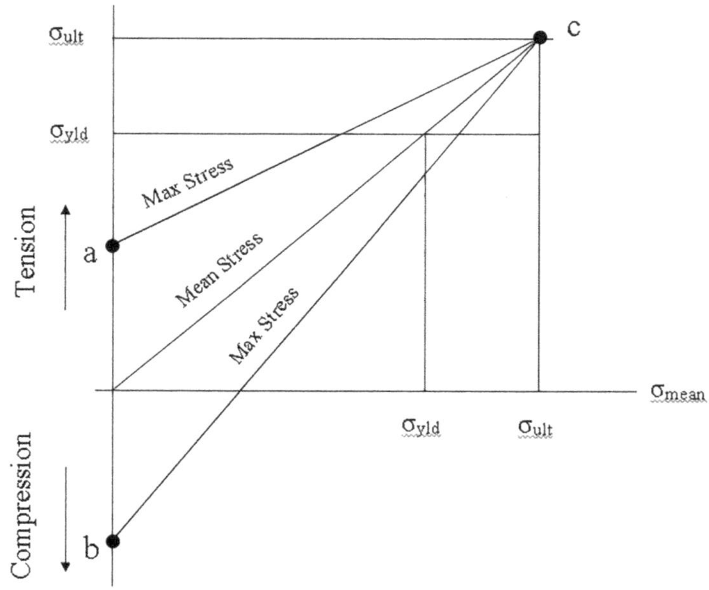

Figure 9-4. Goodman diagram.

fatigue limits for stress levels having a nonzero mean stress. Mean strength is limited by yield strength, σ_{yld}, of the material.

The Goodman diagram is constructed by marking the endurance limit on both the plus (a) and minus (b) directions on the vertical axis. The mean stress is zero and consistent with complete stress reversal. The ultimate strength sets the maximum possible value for the mean stress and amplitude stress coordinates. This point is represented by point (c). Straight lines are drawn from a-c and b-c. The lines establish the endurance limit for the material a different mean stress. Yield strength limits endurance stress. The two lines a-c and c-b establish the fatigue limits for different mean stress levels. Amplitudes and mean stress levels near yield strength levels usually lead to low cycle fatigue.

Cyclic fatigue limits in the compression range is insensitive to mean stress [4].

9.1.7.2 Stress Corrosion Cracking

Three factors must exist simultaneously for stress corrosion cracking to occur; 1) corrosive environment (such as the sea), 2) local tension stress (could be residual stress caused by forming) and 3) hard metal (base metal or produced by quenching). The corrosive environment causes any irregularity in the surface to form and deepen a crack. Localized tension encourages crack growth. Hard metal cannot blunt a crack and, therefore, is susceptible to sudden crack growth and brittle fracture.

This type of failure can occur in offshore pipelines. In one case, a failure analysis showed that an elbow in the pipeline had been formed by heating it to high temperatures and then quenched for convenience of handling. In the process of forming the elbow, residual tension stresses were left in the material. Local tension stresses could also have been generated during pipeline installation.

9.1.7.3 *Brittle Fracture*

Fracture mechanics deals with stresses at the tip of a crack. The crack could be visible at the surface or internal to the structural member. The major concern is a sudden, catastrophic failure — one that occurs without warning.

The Griffin theory [5] of brittle fracture states there are paths of weakness within the matrix of materials where cracks propagate. His work was based on experimental testing of glass. He showed that cracks form even though crystals within glass are much stronger than the nominal strength of glass.

There are two approaches for predicting brittle failure:

1. Experimental — Charpy V-Notch tests are conducted on several test pieces, having the same an initial crack, to determine impact energy levels required to cause the crack to suddenly extend and separate the material. Tests are conducted at different temperatures to determine a materials ability to support a crack. There is typically a temperature at which impact energy for failure substantially drops. This is the transition temperature, an important property of materials expected to operate at low temperatures. High strength materials ($\sigma_{yld} \geq 80{,}000 psi$) are particularly vulnerable to this type of failure, especially at low temperatures.
2. Theoretical — Stress predictions are based on the theory of elasticity, which predicts stresses at the end of an elliptical opening [4]. The elliptical opening (Figure 9-5) can achieve the appearance of a crack. Stress prediction show that normal stresses away from the tip of a crack are defined by

$$\sigma_y = \frac{\sigma\sqrt{a}}{\sqrt{2r}} \qquad (9\text{-}4)$$

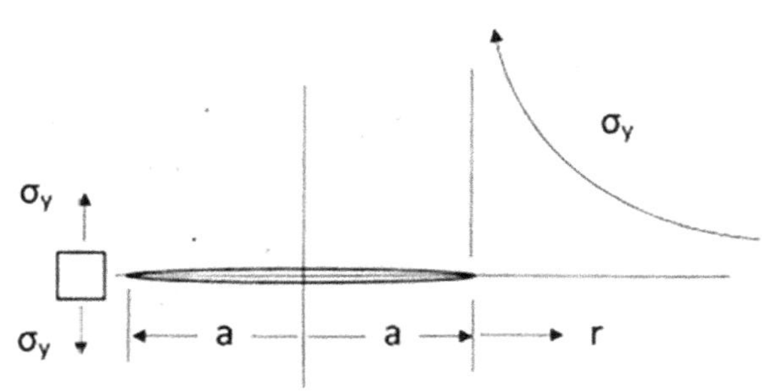

Figure 9-5. Stress at tip of crack.

This equation shows the normal stress is ∞ when r = 0. In many cases, failure does not occur in the form of a crack propagation. Ductile materials are able to create a plastic region at the tip of the crack, and thus blunt the crack inhibiting sudden propagation. Even in high strength materials, a crack may be stable. To quantify this phenomenon, the flowing equation is commonly used to determine critical crack length.

$$\sigma_y = \frac{K_0}{\sqrt{2\pi r}} \quad \left(K_0 = \sigma\sqrt{\pi a}\right)$$

where K_0 is the stress intensity factor. A crack propagates suddenly when K_0 = K_{cr}, a material constant determined experimentally.

9.1.8 Testing the Design

During the early stages of the Saturn/Apollo space project, system reliability was based almost completely on testing. This was the beginning of the computer age and prior to the availability of numerically based computer software. Engineering calculations were based primarily on classic mechanics. The Saturn rocket was powered by Rocketdyne's F-1 engine, the most powerful engine ever built, even by today standards. There were thousands of components. Overall system reliability depended to a large extent on component and system testing. Component and subsystem testing was critical to a successful launch. Testing results were fed back to design and more testing was conducted. Martin [6] stated that ... "testing was conducted until each team could say that everything that could possibly be done to assure success has been done." Later during the 1960's, computer capability expanded along with numerically based software allowing more precise engineering calculations and simulation.

Oil tool components and subsystems are typically tested for reliability and performance. Testing and analysis go hand in hand. One supplements the other. It could be argued that operating conditions for oil drilling are more complex than for space design.

It has been suggested that one way to determine the fitness of bottom hole tools is to chain them to a pick-up truck and drag it down a gravel road for a mile or so. A little drilling rig humor here, but the analogy does underscore the rugged operating conditions downhole.

9.2 THEORY OF STRESS

In general, local stress is defined by three stress components, $\sigma_x, \sigma_y, \tau_{xy}$, which are applied to this small element. This condition (Figure 9-6) defines stress at a point even though surfaces are required to yield forces, which must be in equilibrium. Stresses are shown in their "plus" direction.

From moments about point "a", $\tau_{xy} = \tau_{yx}$, showing that shear stresses on perpendicular planes have the same magnitude. These stress components may be established analytically or experimentally.

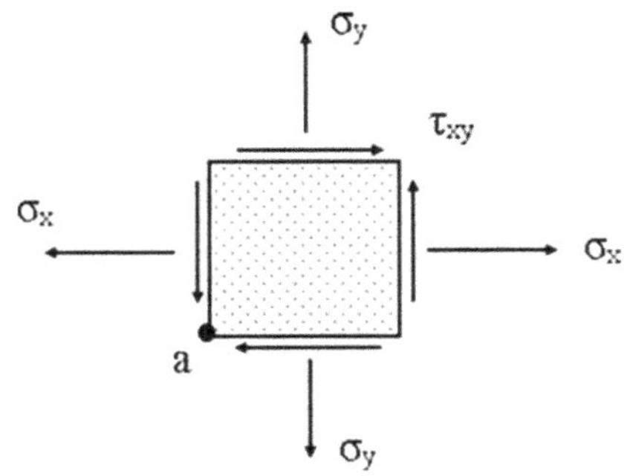

Figure 9-6. Stress components at a point.

9.2.1 Normal and Shear Stress Transformations

Normal and shear stresses on inclined planes vary as before. Forces associated with these stresses are in equilibrium. Equilibrium equations expressed in terms of the stress components are given below.

$$\sum F_n = 0 \tag{9-3}$$

$$\sigma_n = \sigma_x \cos^2\theta + \sigma_y \sin^2\theta + 2\tau_{xy} \sin\theta \cos\theta \tag{9-4}$$

$$\sum F_t = 0 \tag{9-5}$$

$$\tau_{nt} = -\frac{\sigma_x - \sigma_y}{2} \sin 2\theta + \tau_{xy} \cos 2\theta \tag{9-6}$$

The sign convention for each stress is plus as shown in Figure 9-7.

9.2.2 Maximum Normal and Maximum Shear Stresses

Maximum normal and maximum shear stresses and their orientation are of interest in design. Material failure may depend on both, depending on criteria

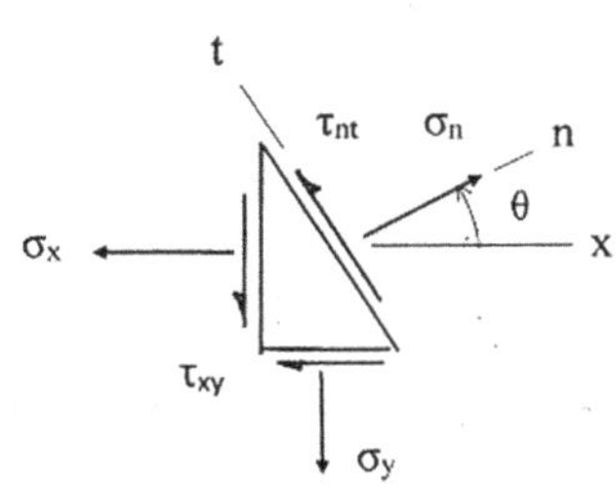

Figure 9-7. Stress on an inclined surface.

of failure. The location and magnitude of maximum stresses are determined by calculus. Starting with the normal stress,

$$\sigma_n = \frac{\sigma_x + \sigma_y}{2} + \frac{\sigma_x - \sigma_y}{2}\cos 2\theta + \tau_{xy}\sin 2\theta \tag{9-7}$$

$$\frac{d\sigma_n}{d\theta} = 0 \tag{9-8}$$

gives

$$\tan 2\theta_p = \frac{2\tau_{xy}}{\sigma_x - \sigma_y} \tag{9-9}$$

By substitution

$$\sigma_{p1,\,p2} = \frac{\sigma_x + \sigma_y}{2} \pm \sqrt{\left(\frac{\sigma_x - \sigma_y}{2}\right)^2 + \tau_{xy}^2} \tag{9-10}$$

Similarly, to determine maximum shear and starting with

$$\tau_{nt} = -\frac{\sigma_x - \sigma_y}{2}\sin 2\theta + \tau_{xy}\cos 2\theta \tag{9-11}$$

$$\frac{d\tau_{nt}}{d\theta} = 0 \tag{9-12}$$

$$\tan 2\theta_\tau = -\frac{\sigma_x - \sigma_y}{2\tau_{xy}}$$

$$\tau_{max} = \pm\sqrt{\left(\frac{\sigma_x - \sigma_y}{2}\right)^2 + \tau_{xy}^2} \tag{9-13}$$

Observations

1. Shear stresses are zero on planes of max and min normal stress (principal planes)
2. Principal planes are 45 deg from planes of max shear stress
3. The two principal planes are 90 deg apart; perpendicular to each other

Consider, for example, the stress condition given in Figure 9-8.

$\sigma_x = 10,000$ psi
$\sigma_y = -8,000$ psi
$\tau_{xy} = -4,000$ psi

Substituting these numbers into Eq. (9-10) gives, $\sigma_{p1,p2} = -8849$ and $+10,850$ psi.

$$\tan 2\theta_p = \frac{2\tau_{xy}}{\sigma_x - \sigma_y}$$

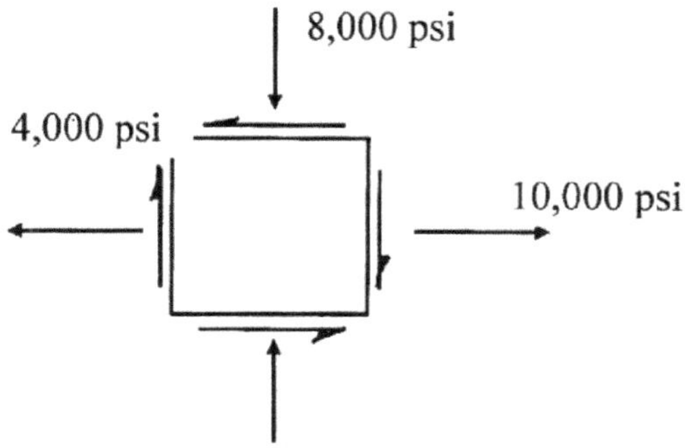

Figure 9-8. Example stress components.

$$\tan 2\theta_p = \frac{2(-4{,}000)}{10{,}000-(-8{,}000)} = -0.444$$ Two values of θ_p satisfy this equation

$$\theta_p = -12^o$$

and

$$\theta_p = 78^o$$

To determine which principal stress applies to which plane, substitute, say $\theta_p = -12$ deg. into

$$\sigma_n = \sigma_x \cos^2\theta + \sigma_y \sin^2\theta + 2\tau_{xy} \sin\theta \cos\theta \tag{9-14}$$

Gives σ_p = 10,850 psi. Both principal stresses are shown on the element in Figure 9-9.

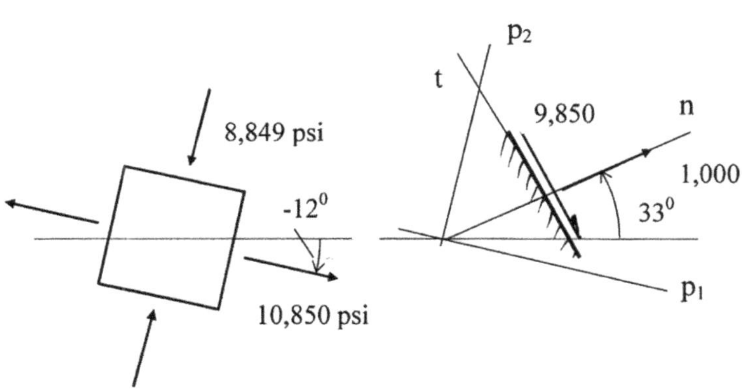

Figure 9-9. Maximum normal and shear stresses.

Now consider maximum shear stress.

$$\tan 2\theta_\tau = -\frac{\sigma_x - \sigma_y}{2\tau_{xy}}$$

$$\tan 2\theta_\tau = -\frac{10{,}000 - (-8{,}000)}{2(-4{,}000)} = 2.25$$

$$2\theta_{\max} = 66^o$$

$$\theta_{\max} = 33^o$$

$$\tau_{\max} = -9850 \text{ psi}$$

$$\sigma_n = 1000 \text{ psi}$$

9.2.3 Mohr's Stress Circle

In 1882, Otto Mohr, a German engineer, recognized that transformation of stresses at a point can be represented by a special circle [7]. Mohr's circle gives a graphical representation of all equations. Normal stresses are located on the horizontal axis (Figure 9-10). Shear stresses are located on the vertical axis. Stress components at a point are located on the circle as follows. σ_x and σ_y are marked on the horizontal axis. The average of these stresses is the center of the circle. Shear stress is measured from these two normal stress points. The coordinates, σ_x, τ_{xy} determine the radius of the circle.

$\sigma_x \succ \sigma_y$
τ_{xy} is plus and moment is CCW. It is plotted down in Mohr diagram

Both principle stresses can be formulated from the diagram. The reader is encouraged to produce the previous equations from Mohr's diagram.

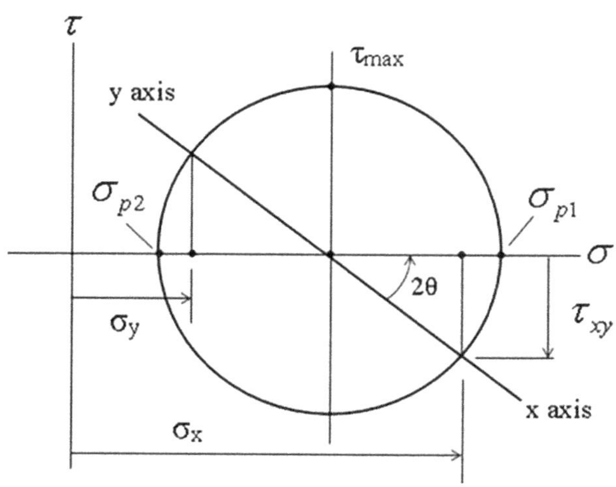

Figure 9-10. Mohr's stress circle.

$$\sigma_n = \frac{\sigma_x + \sigma_y}{2} + \frac{\sigma_x - \sigma_y}{2}\cos 2\theta + \tau_{xy}\sin 2\theta \tag{9-15}$$

$$\sigma_{p1,p2} = \frac{\sigma_x + \sigma_y}{2} \pm \sqrt{\left(\frac{\sigma_x - \sigma_y}{2}\right)^2 + \tau_{xy}^2} \tag{9-16}$$

$$\tan 2\theta_p = \frac{2\tau_{xy}}{\sigma_x - \sigma_y} \tag{9-17}$$

Maximum shear stress is formulated from the diagram.

$$\tau_{nt} = -\frac{\sigma_x - \sigma_y}{2}\sin 2\theta + \tau_{xy}\cos 2\theta \tag{9-18}$$

$$\tau_{max} = \pm\sqrt{\left(\frac{\sigma_x - \sigma_y}{2}\right)^2 + \tau_{xy}^2} \tag{9-19}$$

$$\tan\theta_\tau = -\frac{\sigma_x - \sigma_y}{2\tau_{xy}} \tag{9-20}$$

The use of Mohr's circle will be illustrated with the following example. In this example, assume the components of stress at a point are $\sigma_x = +8$ ksi, $\sigma_y = -6$ ksi and $\tau_{xy} = +4$ ksi (Figure 9-11). The objective is to determine the principal normal stresses, maximum shear stress, σ_{p1}, σ_{p2}, τ_{max}, and the orientation of the elements on which they act. Orientation of principal planes of stress and maximum shear are shown in Figure 9-12.

$$\tan 2\theta_p = \frac{4}{7} \tag{9-21}$$

$$2\theta_p = 29.74^0$$

$$\theta_p = 14.87^0$$

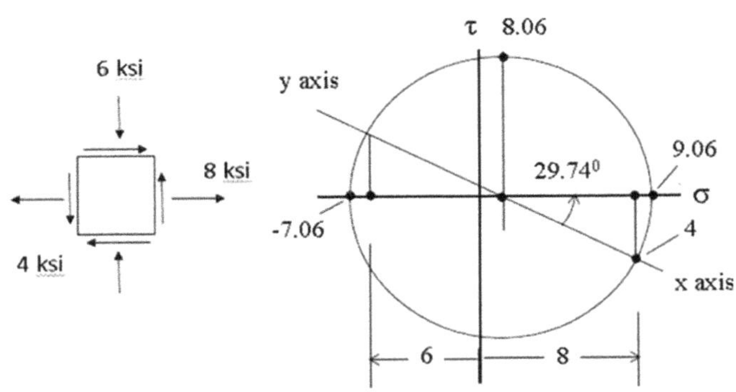

Figure 9-11. Solution by Mohr's circle (ksi Stress Units).

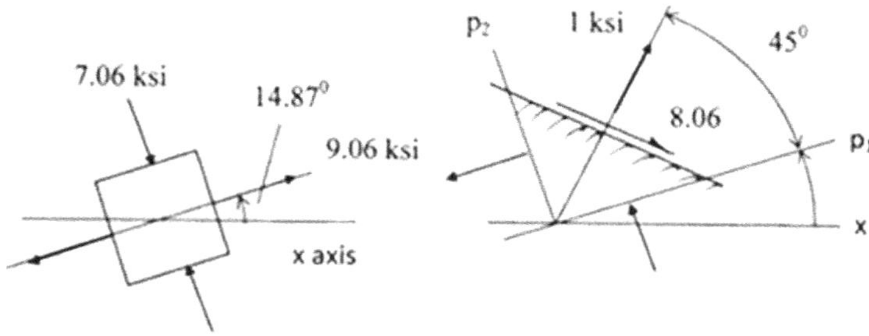

Figure 9-12. Plane of maximum shear stress.

9.3 THEORY OF STRAIN

The theory of strain is developed independently from the theory of stress. Both stress and strain are related by Hooke's Law, which will be cover later.

There are two types of strain, normal and shear, and they both relate to the two types of stress, normal and shear as previously discussed.

Normal strain is simple elongation (or compression) as shown in Figure 9-13. By definition

$$\varepsilon = \frac{\delta}{L} \quad \text{(dimensionless)} \tag{9-22}$$

In general

$$\varepsilon = \frac{du}{dx} \quad \text{(may not be uniform)} \tag{9-23}$$

Internal tension/compression may not be uniform.

The magnitude of normal strain is roughly

$$\varepsilon = \frac{\sigma_{yld}}{E} = \frac{50 \times 10^3}{30 \times 10^6} = 1.67 \times 10^{-3}$$

Normal strain is commonly expressed in micro units,

$\varepsilon = 1670\ \mu$ strain

$\varepsilon = 1670 \times 10^{-6}$

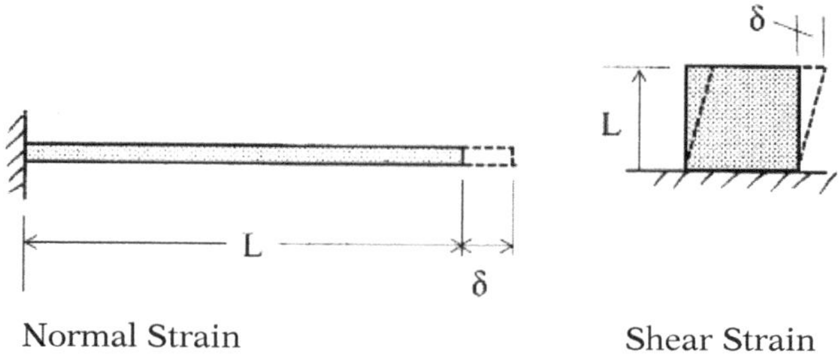

Figure 9-13. Two types of strain.

A normal strain of 2000 micro strain is a large strain and a good reference number.

Shear strain is a measure of the amount of distortion as opposed to stretching.

$$\gamma = \frac{\delta}{L} = \tan\alpha \approx \alpha \tag{9-24}$$

which shows that shear strain is a measure of distortion indicated by angle, α. A more general situation of distortion is shown in the side drawing. In this case

$$\gamma = \alpha + \beta \tag{9-25}$$

Both angles add to define the total distortion. Shear strain is also expressed by angle, θ.

$$\gamma = \frac{\pi}{2} - \phi \tag{9-26}$$

The sign convention for shear strain is

$\phi \prec \frac{\pi}{2}$ shear strain is plus

$\phi \succ \frac{\pi}{2}$ shear strain is negative

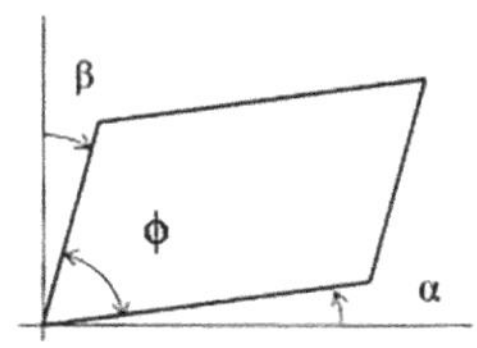

9.3.1 Strain Transformation

Strain at a point includes both normal and shear strains simultaneously. We will now transform these strains on an element oriented by θ from the x axis (Figure 9-14).

Normal strains, referenced from the n,t reference frame, can be expressed in terms of ε_x, ε_y, γ_{xy} by

$$\varepsilon_n = \varepsilon_x \cos^2\theta + \varepsilon_y \sin^2\theta + \gamma_{xy} \sin\theta\cos\theta \tag{9-27}$$

$$\varepsilon_n = \frac{\varepsilon_x + \varepsilon_y}{2} + \frac{\varepsilon_x - \varepsilon_y}{2}\cos 2\theta + \frac{\gamma_{xy}}{2}\sin 2\theta \tag{9-28}$$

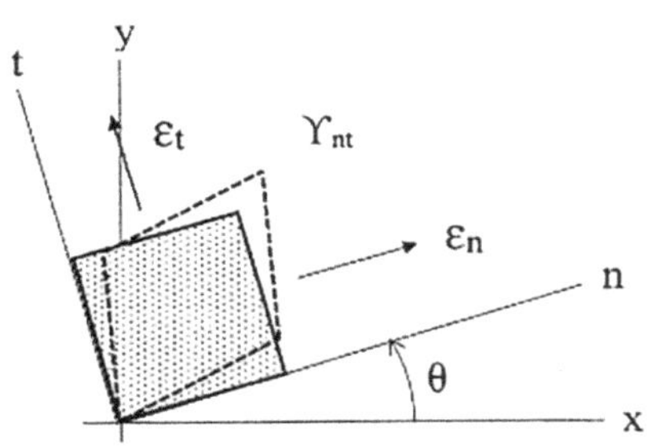

Figure 9-14. Strain transformation.

Maximum normal strains are determined by use of calculus.

$$\varepsilon_{p1}, \varepsilon_{p2} = \frac{\varepsilon_x + \varepsilon_y}{2} \pm \sqrt{\left(\frac{\varepsilon_x - \varepsilon_y}{2}\right)^2 + \left(\frac{\gamma_{xy}}{2}\right)^2} \tag{9-29}$$

$$\tan 2\theta_p = \frac{\gamma_{xy}}{\varepsilon_x - \varepsilon_y}$$

Shear strains, reference from the n,t coordinate reference frame, can be expressed in terms of ε_x, ε_y, γ_{xy} by

$$\frac{\gamma_{nt}}{2} = -\frac{\varepsilon_x - \varepsilon_y}{2}\sin 2\theta + \frac{\gamma_{xy}}{2}\cos 2\theta \tag{9-30}$$

Maximum strain at a point is determined from calculus.

$$\frac{\gamma_{max}}{2} = \pm\sqrt{\left(\frac{\varepsilon_x - \varepsilon_y}{2}\right)^2 + \left(\frac{\gamma_{xy}}{2}\right)^2} \tag{9-31}$$

$$\tan 2\theta_{\gamma\ max} = -\frac{\varepsilon_x - \varepsilon_y}{\gamma_{xy}} \tag{9-32}$$

Note the mathematical similarity between strain and stress transformations.

$$\sigma_n = \sigma_x \cos^2\theta + \sigma_y \sin^2\theta + 2\tau_{xy}\sin\theta\cos\theta \tag{9-33}$$

$$\varepsilon_n = \varepsilon_x \cos^2\theta + \varepsilon_y \sin^2\theta + 2\left(\frac{\gamma_{xy}}{2}\right)\sin\theta\cos\theta \tag{9-34}$$

$$\tau_{nt} = -\frac{\sigma_x - \sigma_y}{2}\sin 2\theta + \tau_{xy}\cos 2\theta \tag{9-35}$$

$$\frac{\gamma_{nt}}{2} = -\frac{\varepsilon_x - \varepsilon_y}{2}\sin 2\theta + \frac{\gamma_{xy}}{2}\cos 2\theta \tag{9-36}$$

The transformation equations for stress and strain are mathematically the same except that the shear strain is divided by 2. The means, analytical expressions for maximum normal and shear stresses also apply to strain. Mohr's circle also applies shear strain except the shear strain axis is divided by 2. The sign convention for both stress and stain are compared in Figure 9-15.

Consider for example

$\varepsilon_x = 800\ \mu$
$\varepsilon_y = -1000\ \mu$
$\gamma_{xy} = -600\ \mu$

What are the normal strains and shear strain on an element oriented –30 deg. from the x axis?

By substitution

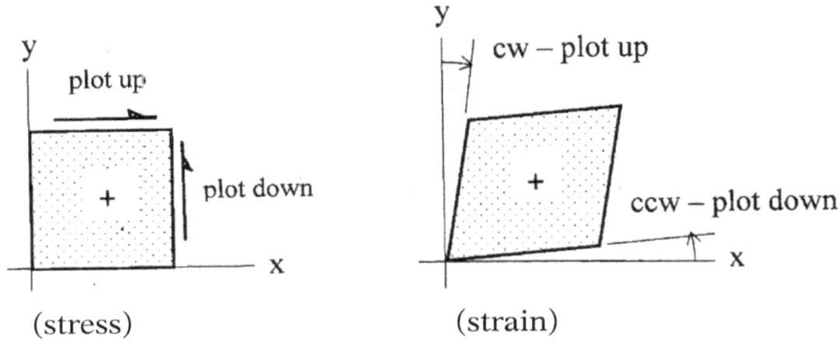

Figure 9-15. Sign convention for shear strain.

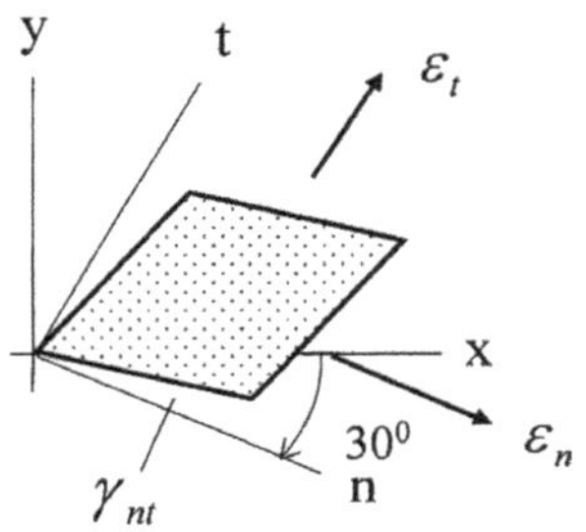

Figure 9-16. Strain transformation.

$\varepsilon_n = 609.8\ \mu$
$\varepsilon_t = -810\ \mu$
$\gamma_{nt} = 1259\ \mu$

These strains are shown in Figure 9-16.

9.3.2 Mohr's Strain Circle

Consider an example where

$\varepsilon_x = 1200\ \mu$
$\varepsilon_y = -600\ \mu$
$\gamma_{xy} = 900\ \mu$

Mohr's strain circle for this case is given in Figure 9-17. Planes of principal strain and maximum shear strain are shown in Figure 9-18.

Again, planes of maximum stress and strain are the same; No shear stress and strain.

The numbers for this example were chosen arbitrarily. It is interesting that the shear strain with this combination of numbers is a quite high. The shear stress, in this case, would be

$$\tau_{max} = G\varepsilon_{max} = \frac{E}{2(1+\nu)}\varepsilon_{max} = \frac{30 \times 10^6}{2(1+0.25)}(0.002012) = 24{,}144 \text{ psi}$$

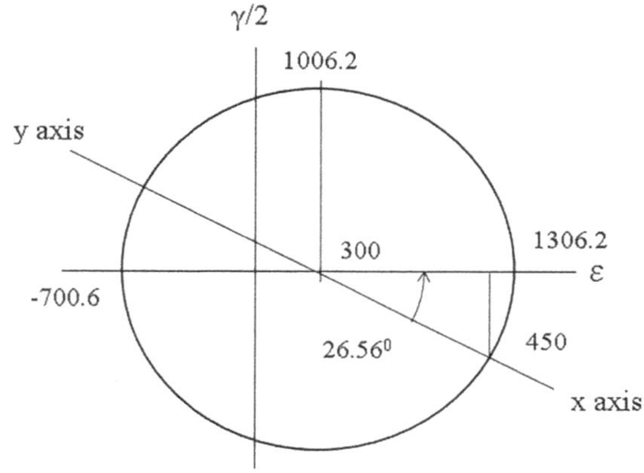

Figure 9-17. Mohr's strain circle.

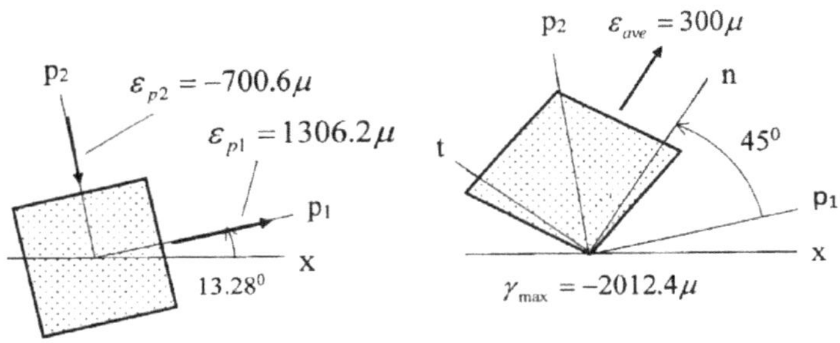

Figure 9-18. Planes of maximum normal and shear strains.

9.3.3 Principal Axes of Stress and Strain

From the theories of stress and strain, we know that there is no shear stress on the principal planes, therefore, there is no shear strain as shear stress and shear strain are related through

$$\tau = G\gamma \tag{9-37}$$

This means that the principal planes of normal stress are the same as the principal planes of normal strain (Figure 9-19).

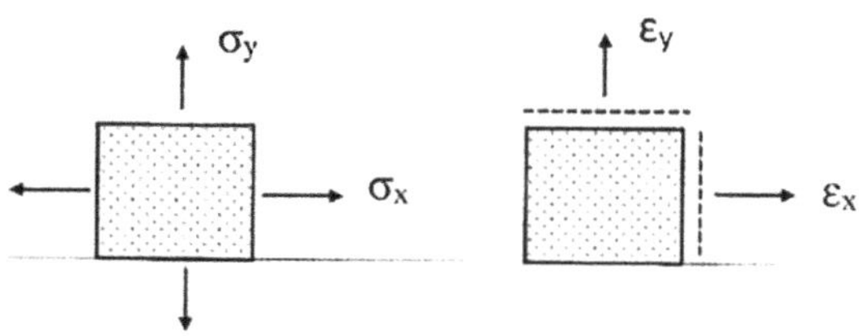

Figure 9-19. Comparison of normal stress and strain.

9.4 GENERALIZED HOOKE'S LAW

The theories of stress and strain were developed independent of each other. Both variables are related through Hooke's Law. Laboratory tests show that when a load is applied to a test rod, the rod stretches elastically. The relation between applied stress and resulting strain in a uniaxial state of stress is expressed by

$$\varepsilon = \frac{\sigma}{E} \tag{9-38}$$

The strain has the same direction as the stress. Tests also show that stress in one direction not only produces strain in the same direction but also a shrinking normal strain in transverse directions. The amount of shrinking is related to the strain in the direction of the load by

$$\nu = -\frac{\varepsilon_T}{\varepsilon_A} = -\frac{transverse}{axial} \tag{9-39}$$

This ratio is called Poisson's ratio and is a property of the material. Equations (9-38) and (9-39) are basic to the generalize Hooke's Law for a triaxial state of stress.

$$\varepsilon_x = \frac{1}{E}\left[\sigma_x - \nu\left(\sigma_y + \sigma_z\right)\right] \tag{9-40}$$

$$\varepsilon_y = \frac{1}{E}\left[\sigma_y - \nu\left(\sigma_x + \sigma_z\right)\right] \tag{9-41}$$

$$\varepsilon_z = \frac{1}{E}\left[\sigma_z - \nu\left(\sigma_y + \sigma_x\right)\right] \tag{9-42}$$

Be aware that in general the ratio of transverse strains is *not* equal to Poisson's ratio. The rules for finding Poisson's ratio must follow the procedure described above.

Hooke's Law includes the effects of direct stress plus Poisson's effect due to stresses in the other two transverse directions. Two common engineering situations well now be addressed; 1) plane stress and 2) plane strain.

9.4.1 Theory of Plain Stress

Under plane stress, $\sigma_z = 0$, however, $\varepsilon_z \neq 0$ (Figure 9-20a). An example is stress on a free surface. Stresses lie in the plane of the surface but stress normal to the surface is zero. Applying Hooke's Law gives

$$\varepsilon_x = \frac{1}{E}\left[\sigma_x - \nu\sigma_y\right] \tag{9-43}$$

$$\varepsilon_y = \frac{1}{E}\left[\sigma_y - \nu\sigma_x\right] \tag{9-44}$$

$$\varepsilon_z = -\frac{\nu}{E}\left[\sigma_x + \sigma_y\right] \tag{9-45}$$

By substitution gives

$$\varepsilon_z = -\frac{\nu}{1-\nu}\left(\varepsilon_x + \varepsilon_y\right) \tag{9-46}$$

Note that

$$\varepsilon_z \neq \nu\left(\varepsilon_x + \varepsilon_y\right) \tag{9-47}$$

Rearranging Eqs. (9-43) and (9-44) gives expressions for stress in terms of strain.

$$\sigma_x = \frac{E}{1-\nu}\left[\varepsilon_x + \nu\varepsilon_y\right] \tag{9-48}$$

$$\sigma_y = \frac{E}{1-\nu}\left[\varepsilon_y + \nu\varepsilon_x\right] \tag{9-49}$$

Shear stress is not involved in either equation, but relates to shear strain by

$$\tau_{xy} = G\gamma_{xy} \tag{9-50}$$

9.4.2 Theory of Plain Strain

Plain strain exists when an element is not allowed to move in the z direction ($\varepsilon_z = 0$), creating stress in the z direction (Figure 9-20b). Hooke's law in this case gives

$$\sigma_x = \frac{E}{(1+\nu)(1-2\nu)}\left[(1-\nu)\,\varepsilon_x + \nu\varepsilon_y\right] \tag{9-51}$$

$$\sigma_y = \frac{E}{(1+\nu)(1-2\nu)}\left[(1-\nu)\,\varepsilon_y + \nu\varepsilon_x\right] \tag{9-52}$$

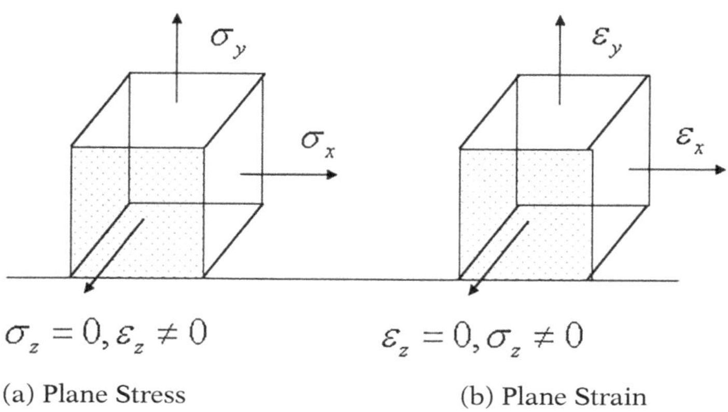

Figure 9-20. Elements of plain stress and plane strain.

$$\sigma_z = \frac{E}{(1+\nu)(1-2\nu)}\left[\nu\left(\varepsilon_x + \varepsilon_y\right)\right] \tag{9-53}$$

Still

$$\tau_{xy} = G\gamma_{xy} \tag{9-54}$$

As mentioned earlier, the modulus of elasticity is determined experimentally from a simple tension test. The modulus of rigidity, however, is *not* determined experimentally but is determined mathematically.

$$G = \frac{E}{2(1+\nu)} \tag{9-55}$$

9.4.3 Conversion of Strain Measurements to Strain and Stress

Strain gauges are used to experimentally determine local strain at a point. A single gauge may be used in a uniaxial loaded member. Only two gauges, 90 deg. apart, are required to determine strain at a point if aligned with known principal strain/stress directions. However, strain at a point usually requires a strain rosette containing three gauges because principal strain directions are unknown.

The basic construction of a strain gauge is shown in Figure 9-21. The "Grid" is composed of a metallic wire with a known electrical resistance. Manufactured strain gauges have a known relation between their change in elongation and associated change in electrical resistance. This relation is known as the "gauge factor," or GF. The "lead wires" may be sent directly to an Ohmmeter to measure resistance; however, the lead wires are more commonly used in a Wheatstone bridge configuration. The "backing" typically contains and adhesive for locking the gauge to the material of interest. Thus, as the material is strained, the wires in the grid are strained by the same amount.

Wheatstone bridge circuits are commonly used in conjunction with strain gauges, because they allow small changes in resistance to be measured accurately. This is important, because elongation-induced strains in the gauge result in very small resistance changes as compared to the gauge's initial resistance. The Wheatstone bridge is shown in Figure 9-22. Resistance R_1 (not shown) in the bridge is the strain gauge. The output voltage, E_0, may be measured to determine the change in resistance of the strain gauge. When the output voltage

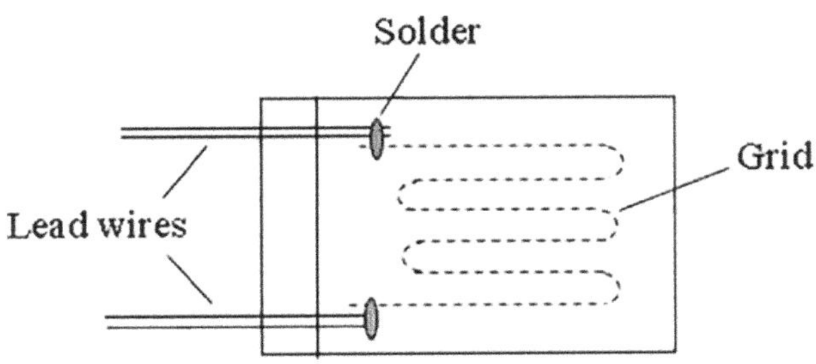

Figure 9-21. Strain gauge schematic.

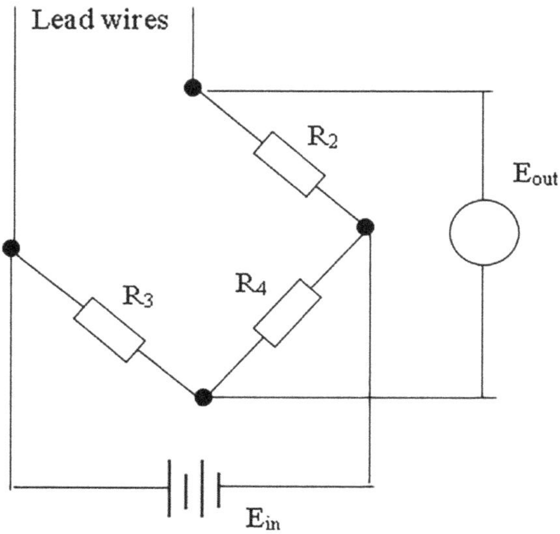

Figure 9-22. Strain gauge used as the fourth resistance in a Wheatstone bridge.

equals 0, the bridge is said to be *balanced,* meaning the resistances in opposite legs are equal and there is no measurable strain [3].

Strain rosettes may come in different gauge arrangements, such as shown in Figure 9-23.

They differ by the orientation of the gauges.

Rosette A $\theta_a = 0^0$ $\theta_b = 45^0$ $\theta_c = 90^0$
Rosette B $\theta_a = 0^0$ $\theta_b = 90^0$ (used when principal directions known)
Rosette C $\theta_a = 0^0$ $\theta_b = 60^0$ $\theta_c = 120^0$

Two things to remember about strain gauges: 1) they do not measure shear strain and 2) they do not measure stress. They can only measure normal strain in one direction. But, knowing three normal strains in three arbitrary directions allows us to *calculate* shear strain and principal stresses at a point.

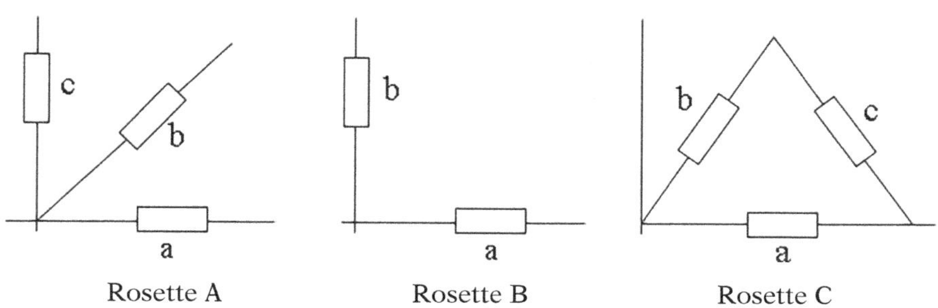

Figure 9-23. Strain rosette arrangements.

If normal strains are measured in three arbitrary directions, they can be used to find strains (ε_x, ε_y, γ_{xy}) associated with any xy coordinate system. From the theory of strain discussed earlier

$$\varepsilon_a = \varepsilon_x \cos^2\theta_a + \varepsilon_y \sin^2\theta_a + \gamma_{xy} \sin\theta_a \cos\theta_a \tag{9-55}$$

$$\varepsilon_b = \varepsilon_x \cos^2\theta_b + \varepsilon_y \sin^2\theta_b + \gamma_{xy} \sin\theta_b \cos\theta_b \tag{9-56}$$

$$\varepsilon_c = \varepsilon_x \cos^2\theta_c + \varepsilon_y \sin^2\theta_c + \gamma_{xy} \sin\theta_c \cos\theta_c \tag{9-57}$$

Theta, θ, is the angular orientation of each strain gauge.

Strain measurements on a free surface are useful for defining strain at a given location. Once the state of strain is determined, stresses are calculated from Hooke's law.

In matrix form

$$\begin{pmatrix} \varepsilon_a \\ \varepsilon_b \\ \varepsilon_c \end{pmatrix} = \begin{bmatrix} & & \\ & A & \\ & & \end{bmatrix} \begin{pmatrix} \varepsilon_x \\ \varepsilon_y \\ \gamma_{xy} \end{pmatrix} \tag{9-58}$$

Strain referenced from the xy frame is determined from inverting Eq. (9-58).

$$\begin{pmatrix} \varepsilon_x \\ \varepsilon_y \\ \gamma_{xy} \end{pmatrix} = \begin{bmatrix} & & \\ & A & \\ & & \end{bmatrix}^{-1} \begin{pmatrix} \varepsilon_a \\ \varepsilon_b \\ \varepsilon_c \end{pmatrix} \tag{9-59}$$

Once ε_x, ε_y, γ_{xy} have been determined, ε_{p1}, ε_{p2}, γ_{max} can be determined analytically or by using Mohr's strain circle as discussed above. Principle stresses are determined from Hooke's Law, according to the theory of plane stress as previously discussed.

9.4.4 Strain Transducers

Strain gauges are commonly used to determine stress at a point in a design. Stress is not measured directly, but is determined from measured strain through Hooke's law. One of the challenges is designing attachment element so that significant strains are developed without creating a point of weakness in the design. Strain levels of 2000 μ strains are considered high. For example from simple stress-strain

$$\sigma = E\varepsilon = 30 \times 10^6 (2000 \times 10^{-6}) = 60{,}000 \text{ psi}$$

a rather high stress. The challenge is to convert low strain to stress and still maintain instrument sensitivity. Wheatstone bridges are commonly used for this purpose.

Orientation of strain gauges is also important in designing a transducer area in a tool. Drilling parameters of interest are axial forces, bending moments and torque. Since strain gauges measure only normal strain, not shear, it is advantages to orient strain clusters in the direction of principal strains for maximum sensitive and focus. Consider the use of strain gauges to measure torque.

From our earlier discussion, principal normal strains, in the case of pure torque on a shaft, are 45 deg. from a longitudinal line (Figure 9-24). These principal strains are related to principal stresses by

$$\varepsilon_{p1} = \frac{1}{E}\left[\sigma_{p1} - \nu\sigma_{p2}\right]$$

$$\varepsilon_{p2} = \frac{1}{E}\left[\sigma_{p2} - \nu\sigma_{p1}\right]$$

But $\sigma_{p1} = \tau$ and $\sigma_{p2} = -\tau$, therefore,

$$\varepsilon_{p1} = \frac{\tau}{E}(1+\nu) \quad \text{and} \quad \varepsilon_{p2} = \frac{-\tau}{E}(1+\nu)$$

$$\tau = \frac{\varepsilon_{p1}E}{1+\nu}$$

Also, we know

$$\tau = \frac{Tc}{J}, \text{ so}$$

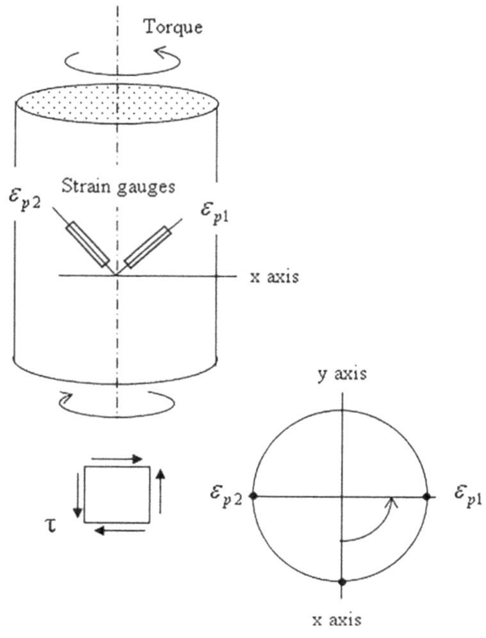

Figure 9-24. Torque transducer.

Applied torque relates to the principal strains by

$$T = \frac{J}{c}\frac{\varepsilon_{p1}E}{\left(1+\nu\right)}$$

If a similar strain rosette is attached to the opposite side of the shaft, and connected to the Wheatstone bridge circuit, temperature effects as well as bending and axial strains are eliminated from the torsional measurements [8]. Strain rosette arrangements for measuring axial and bending are discussed in ref [9].

Piezoelectric Transducer — Certain materials develop electrical potential when deformed by force, pressure, temperature and inertia effects. The electric voltage can be easily measure and calibrated to experimentally quantify various parameters. Materials that exhibit this effect are; quartz, Rochelle salt, polarized barium titanate, ammonium dihydrogen, ordinary sugar, etc.

The voltage output obtained from the materials due to piezoelectric effect is very small. To measure the output some amplifiers, auxiliary circuit and the connecting cables are required. Some advantages to piezoelectric transduces are: 1) High frequency response 2) can detect impact loads in microseconds, and 3) it is small and rugged. Limitations are: 1) output is low 2) It is difficult to construct the crystal into a directed shape with sufficient strength.

9.5 CHALLENGES IN DOWNHOLE TOOL DESIGN

A challenging constraint in downhole tool design is limited space. Every tool has to work within the diameter of a well bore, which may range roughly between 12 to 5 in. While the size is small, the loads are relatively high (~50,000 lb). Dynamic loads can reach 150,000 lbs or higher. The working environment (temperature, pressure, abrasive fluids) is also adverse. Formation properties add another level of uncertainty even though off-set wells provide a good reference for current drilling.

The vision for many years has been to navigate a drill bit to a given target several thousand feet away and reach the target within a few feet. This vision has been realized within the past few years through much research and development of complex tools (drill bit, motors/turbines, MWD, rotary stearable tools (RST). With complexity, brought about by multiple components (mechanical and electrical), comes uncertainty. Probability of achieving expected performance reduces with complexity of design and number of components. The challenge is to reduce uncertainty and increase reliability through design.

Two failure modes of importance in drilling tool design are wear and fatigue. Both are affected by the space constraint. The following gives an overview of these two failure modes.

9.5.1 Friction and Wear

Amontons in 1699 observed that friction force is independent of the area of contact and that friction force was proportional to the applied load. He also stated that the coefficient of friction for all materials is 0.3, a widely used number even today. These friction laws were further clarified by Coulomb, who in 1781 distinguished the difference between static and dynamic friction. The Amontons-Coulomb laws of dry friction are

1) The friction force is directly proportional to the load
2) The friction force is independent of the gross area of the contacting surfaces
3) The friction force depends upon the nature of the sliding surfaces

Over the years, grease (or animal fat) has been used to reduce the coefficient of friction in everything from water wheels to machinery. Grease was the main lubricant in railroad trains during the industrial revolution. Towers [10] discovered under certain conditions that thick films can actually separate the journal from the bearing surface thus reducing friction to fluid film shear. His discovery was followed (1886) by analytical studies by Reynolds [11] who developed the lubrication equations relating friction to film thickness, load, viscosity, and sliding speed.

9.5.1.1 *Thick Film Lubrication*

The smallest allowable film thickness in journal bearings depends on factors such as: 1) smoothness or finish of the sliding surfaces, 2) elastic rigidity of the bearing structure, 3) thermal distortion, and 4) size of contamination particles in the lubricant. Experiments conducted by Karelitz [12] and McKee [13] show that the smallest practical limit for fluid film lubrication is about 0.00005 in. (approximately 1 μm). However, for large commercial bearings, the minimum thickness should be kept above 0.0001 in. (2.54 μm), depending on size of contaminates in the lubricant.

Boundary lubrication is defined as a condition where the fluid film does not separate the surfaces greater than a distance equal to surface roughness. Under this condition, the surfaces come into intermittent contact with the microscopic peaks on each surface and the bearing surfaces are separated by a few layers of fluid molecules. Under this condition, friction and heat increase drastically; the bearing usually gets hot and surfaces often seize.

The study of boundary lubrication dates to Stribeck [14]. The objective of design is to operate at lowest friction while avoiding intermitted boundary contact. The primary focus of this work was lubrication in zone 2.

Fuller [15] identifies four regions of bearing friction, which are displayed in Figure 9-25. Coefficient of friction is plotted against the Hersey number, $\frac{ZN}{p}$, where Z is viscosity (cp), N is rotary speed (rpm), and p is average bearing pressure (psi).

1) Thick film region — Bearing surfaces are completely separated by a liquid film. When the lubricant is free of abrasive particles, wear is prevented. The coefficient of friction is low as 0.001 or even less. As a guide, this region begins at a Hersey number of in the range of 30 to 40. This number could be lower depending on surface finish and run-in time.

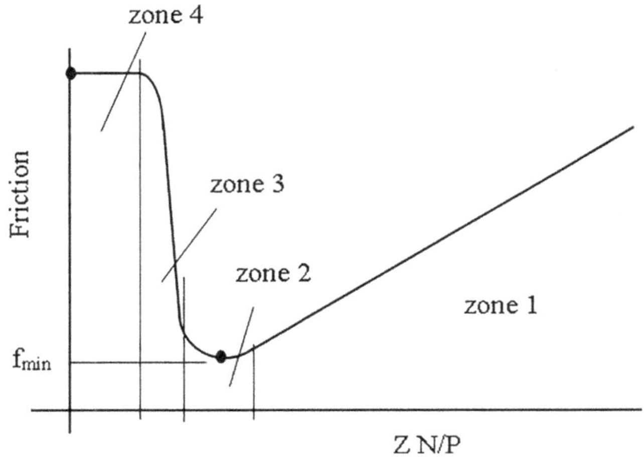

Figure 9-25. Plot of coefficient of friction versus ZN/P.

2) Thin film lubrication region — This region represents the lower limit of complete separation by a film. Film thickness varies from around 0.0002 in. 0.00005 in. Bearing rigidity and surface smoothness are important too.
3) Mixed film region — Loading is so severe that complete surface separation cannot be achieved by the lubricant. Most of the surface experiences rubbing without the benefit of complete separation of the peaks and valleys of the surface roughness. At best, lubrication is achieved by surface separation by a few layers of fluid molecules. The coefficient of friction under this condition may range from 0.02 to 0.08.
4) Boundary film region — There is no fluid film because of either low viscosity or velocity or because of high bearing load. Film thickness can only be described in terms of a molecules. The coefficient of friction may range between 0.08 and 0.15.

To the right of the minimum point, bearings operate under thick film conditions. Under this condition, a thick film completely separates the bearing surfaces. If the load on the bearing is increased, the operating point moves to the left until thermal equilibrium is again established. Additional increases in load will lower the coefficient of friction until the minimum point has been reached. It is best to operate journal bearings at this low friction point. Further increases in load will cause boundary film lubrication causing friction and heat to increase.

Two factors can move the minimum friction point to the left. One is surface smoothness and the other is the "oiliness" of the lubricant. Surface smoothness is sometimes achieves by "running in" bearings, i.e., initially running new machine at lower than specified operating speeds. This allows the bearing surfaces to polish themselves to some extent.

9.5.1.2 Rolling Contact Bearings

Roller contact bearings are used in nearly every industry as well as jet engines and rocket engines. They are reliable and forgiving. Failure usually comes in the form of pitting, but even then they still perform. Types of roller contact bearings are:

- Cylindrical
- Tapered cylindrical
- Ball
- Angular contact

There is a great deal of history on performance and a comprehensive data bank for making statistical predictions on their performance.

In a typical industrial setting rolling contact bearings are lubricated and, in many cases sealed. The life of these bearings, under steady load, is predicted by

$$L = \left(\frac{C}{P}\right)^q \tag{9-60}$$

where

L = bearing life in millions of revolutions (or hours for constant speed)
P = bearing load; assumed to be steady, lb
C = basic load rating of a given bearing size and type, lb
q = exponent (10/3 for roller bearings, 3.0 for ball bearings)

This empirical equation is based on laboratory testing and operating performance. Performance data is displayed on a log-log graph as illustrated in Figure 9-26.

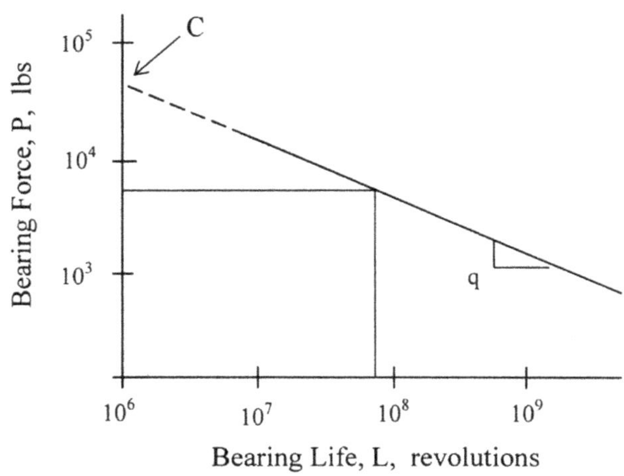

Figure 9-26. Bearing load rating.

The exponent, q, is determined as follows.

$$\frac{L}{10^6} = \left(\frac{C}{P}\right)^q \tag{9-61}$$

$$Log\,\frac{L}{10^6} = \log\left(\frac{C}{P}\right)^q = q\log\frac{C}{P} \tag{9-62}$$

$$q = \frac{\log L - \log 10^6}{\log C - \log P} \tag{9-63}$$

If we view "L" as life in millions of revolutions then Eq. (9-60) applies. The performance line is drawing through the data such that 90% of data is outside of the line and 10% is inside the line. This means that bearing performance based on this line has a reliability of 0.9. Rolling contact bearing failure is defined as fatigue pitting on either the roller or race. This type of bearing is forgiving in that they will still work even after pitting occurs.

Commercial bearings are rated by the value of "C", which corresponds to a life of 10^6 cycles. This coordinate is for convenience only as bearing loads are not expected to perform as this load level. Equation (9-60) is used to select bearings for specific applications.

$$C = PL^{\frac{1}{q}} \tag{9-64}$$

where P and L are design requirements and C is the bearing catalog load rating.

Coefficients of friction for each bearing type are given in Table 9-2. These friction data apply to a perfectly clean environment.

Equipment literally "wears out" because of wear. In most cases, friction and wear produce heat, which accelerates wear and if left unchecked can destroy equipment. Temperature is a measure of heat and a good indicator of friction and wear. The objective in design is to minimize friction and thus minimize mechanical wear and heat.

9.5.2 Effect of Vibrations on the Life of Rolling Contact Bearings

Radzimovsky and Dareing [16] predict the effect of dynamic forces on the life of rolling contact bearings. Dynamic forces are assumed to be of the form

$$P(t) = P_m + P_v \sin\frac{2\pi t}{T} \tag{9-65}$$

where

P_m = Mean value of load variation, lb
P_v = Variable component of load variation, lb
T = Period of cyclic load, second

Table 9-2. Comparison of friction.

	Coefficient of Friction (typical)
Dry friction	
Generic	0.3
Steel on Copper	0.15
Teflon on anything	0.1
Thick Film Bearings	0.001
Journal Bearings	
Slider Bearings	
Hydrostatic Bearings	
Squeeze Films	
Boundary Films	0.15
Sleeve Bearings	0.1
Roller Contact Bearings	0.002

To derive the equivalent load, it may be assumed that if a certain load acts for a certain fraction of the life duration, which the bearing expects to have under this load, then the same fraction of the life of the bearing is consumed. For another load, having different magnitude, only the remaining part of the bearing capacity maybe utilized. Experience supports this assumption. Therefore, over one load cycle

$$TP_e^q = \int_0^T P(\mathrm{t})^q dt \tag{9-66}$$

$$P_e = p_m \left[\frac{1}{T} \int_o^T \left(1 - \frac{P_v}{P_m} \sin \frac{2\pi t}{T} \right)^q dt \right]^{\frac{1}{q}} \tag{9-67}$$

From which

$$\beta = \left[\frac{1}{T} \int_o^T \left(1 - \alpha \sin \frac{2\pi t}{T} \right)^q dt \right]^{\frac{1}{q}} \tag{9-68}$$

where

$$\alpha = \frac{P_v}{P_m}$$

$$\beta = \frac{P_e}{P_m}$$

P_e is defined as a steady load that will cause fatigue failure in the bearing after the same number of revolutions (or hours, if speed is constant) as a given unsteady load defined by Eq. (9-65).

When contact roller bearings are properly lubricated, q = 3.33 for cylindrical roller bearings and q = 3 for ball bearings.

For ball bearings, the mathematical relation between these two parameters is [17]

$$\beta = \left[1 + 1.5\alpha^2\right]^{\frac{1}{3}} \tag{9-69}$$

For cylindrical bearings, the relationship has to be determined numerically.

The effect of dynamic loading on bearing life is defined by the life reduction coefficient, K.

$$K = \frac{L_{unsteady}}{L_{steady}} = \frac{L_e}{L} \tag{9-70}$$

K and β are related as follow.

$$\frac{L_e}{L} = \left(\frac{\frac{C}{P_e}}{\frac{C}{P}}\right)^q = \frac{1}{\left(\frac{P_e}{P}\right)^q} \tag{9-71}$$

$$K = \frac{1}{\beta^q} \tag{9-72}$$

The relation between K and β (L_e and P_e) is illustrated in Figure 9-27.

The results are given in Table 9-3. As an example, consider the following set of numbers:

C = 10,700 lb (load rating for deep groove *radial ball bearing*)
P_{min} = 900 lb
P_{max} = 5100 lb
N = 1800 rpm

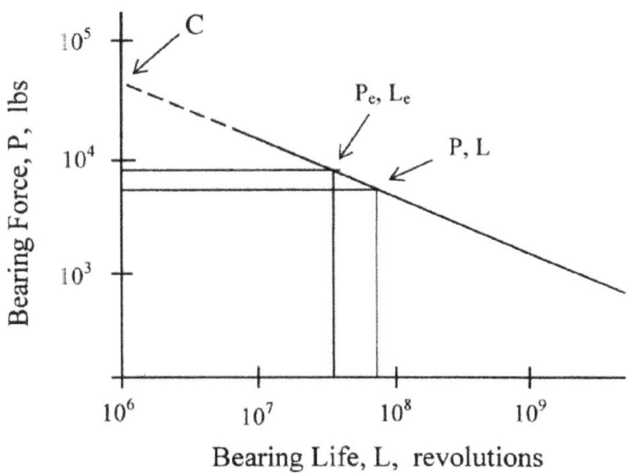

Figure 9-27. Effect of dynamic loading on bearing life.

Table 9-3. Equivalent load and life reduction factors.

	Roller Bearings		Ball Bearings	
α	β_r	K_r	β_b	K_b
0.0	1.0000	1.000	1.000	1.000
0.1	1.0042	0.986	1.005	0.985
0.2	1.0199	0.936	1.020	0.943
0.3	1.0458	0.861	1.043	0.881
0.4	1.9803	0.773	1.073	0.806
0.5	1.1216	0.683	1.112	0.727
0.6	1.1682	0.596	1.155	0.649
0.7	1.2187	0.518	1.202	0.576
0.8	1.2719	0.449	1.252	0.510
0.9	1.3270	0.390	1.303	0.451
1.0	1.3835	0.340	1.357	0.400

Therefore,

P_m = 3000 lbs
P_v = 2100 lbs

Expected bearing life under *steading* loading conditions is

$$L = \frac{(10{,}700/3000)^3 10^6 rev}{1800(60)\,\text{rev/hr}} = 421\ \text{hrs}$$

From this input data,

P_m = 3000 lb
P_v = 2100 lb

$$\alpha = \frac{2100}{3000} = 0.7$$

β = 1.202 (Table 9-3)
K = 0.576 (Table 9-3)

The equivalent load is

$$P_e = 1.202\ (3000) = 3606\ \text{lbs}$$

The bearing life under this dynamic load condition is

$$L_e = 421 \times 0.576 = 242\ \text{hour.}$$

Representing a 179/421 = 0.4252 or a 42.52% reduction in bearing life.

9.6 DRILL BIT DESIGN AND EXPECTATIONS

Initially, the drill bit was the only bottom hole tool. Since the first roller bit patented by Hughes [18], the roller cone bit has gone through many improvements. The design challenges are bearing life and cutter wear. Figure 9-28 illustrates a typical bit cone cross section showing the bearing arrangement. The ball bearing carries little or no load but is there to hold the cone in place. The balls are inserted through a hole that leads to the bearing cavity. Assuming WOB is 45,000 lbs, each of the three bearing legs supports 15,000 lbs. This load generates lateral, thrust and moment loads within the bearing arrangement. The small pin or sleeve bearing is required to support the thrust load with some help from the ball bearing. Lateral loads are carried by the cylindrical bearing and the sleeve bearing. This may be the best arrangement for the space but in an industrial application, cylindrical and sleeve bearings would not be use in parallel. Sleeve bearings require radial space, which in this case causes non-uniform loading to the cylindrical rollers.

For many years, these bearings were unsealed and operated in a corrosive and abrasive environment caused by drilling mud. Under this condition, bearing life is about 8 to 10 hours.

In recent years, roller bearing life has increased to over 100 hours due to reliable dynamic seals and lubricant reservoirs. In some designs, cylindrical roller bearings have been replaced with journal type bearings.

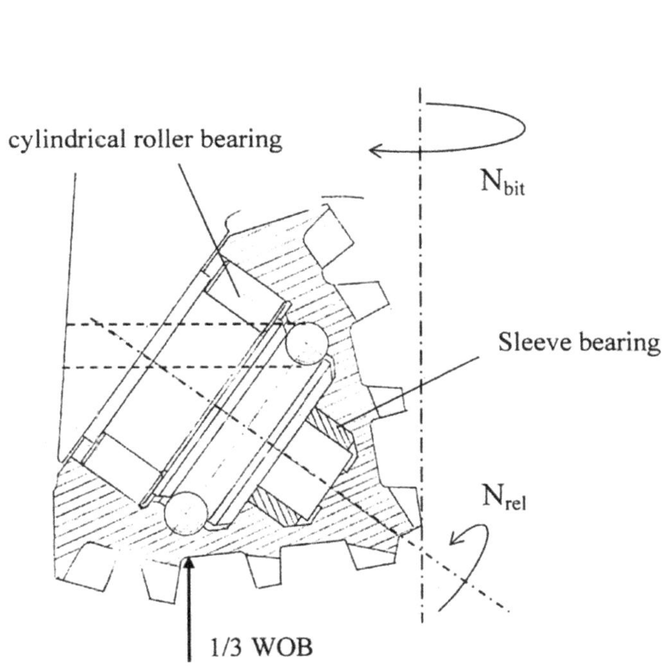

Figure 9-28. Bearings mounted inside a roller cone.

9.6.1 Effect of Environment on Drill Bit Bearing Life

Bearings in roller bit bearings have typically been open to the drilling mud plus high loads required for acceptable rates of penetration (ROP). Also, the overall bearing assembly comprised of cylindrical, ball and journal bearings within each bit cone is not conducive to prolonged bearing life. The cylindrical roller bearing usually fails first because of misalignment and abrasion.

Tests conducted by King [19] on 7" in. W7R drill bits in barite mud show that q = 1.4 while the same drill bit show q = 2.8 in air (Figure 9-29). The values of q are strikingly different from those for well lubricated ball and cylindrical bearings. These tests were conducted under steading external loading to each bit tested.

9.6.1.1 Effect of Vibration and Environment on Roller Cone Drill Bit Bearing Life

Dareing and Radzimovsky [16] expanded the analysis in ref. [17] to include other values of the exponent, q, in order to cover the effects of drilling mud. The definitions of the dimensionless parameters are still the same. Figure 9-30 relates, β,α and q.

The results are also presented as life reduction due to dynamic loading. A life reduction coefficient, K, is used to show reduction in bearing life caused by unsteady loading. K is the ratio of bearing life corresponding loading to bearing life corresponding to steady loading.

$$K = \frac{L_{unsteady}}{L_{steady}} \tag{9-73}$$

This ratio can also be related to β through

$$K = \frac{1}{\beta^{q}} \tag{9-74}$$

Figure 9-31 relates K, α, and q.

The relation between K and α is tabulated (Table 9-4) so values of K can be determined more accurately for small values of α.

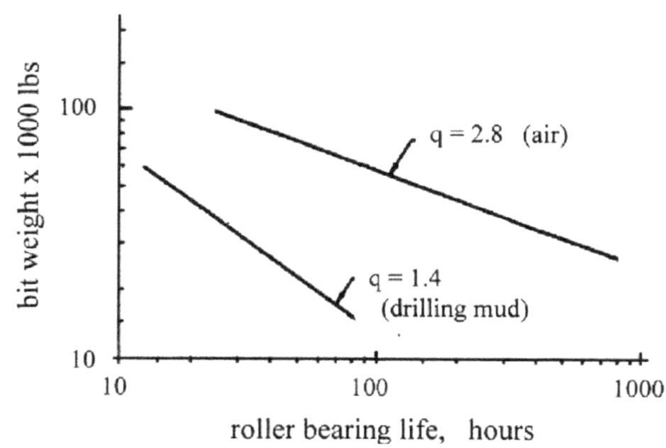

Figure 9-29. Life expectancy of unsealed bit bearing (N = 60 rpm).

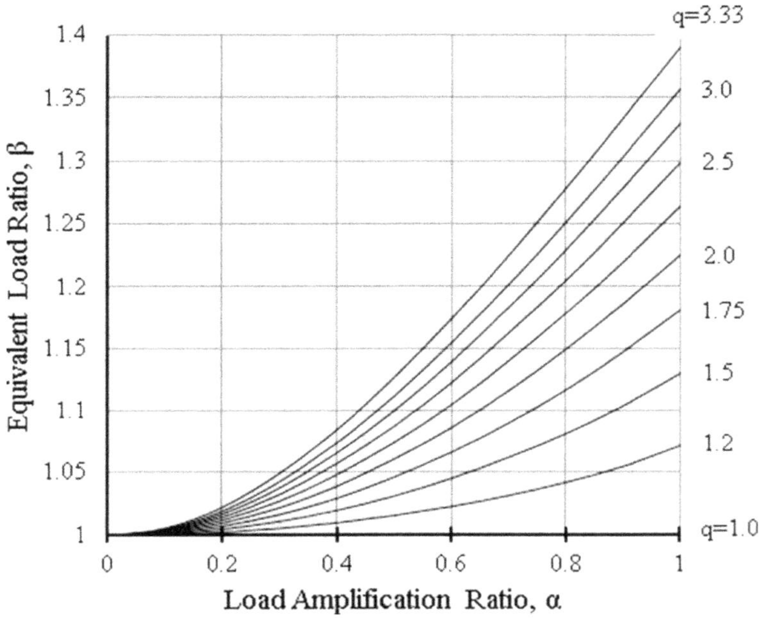

Figure 9-30. Equivalent steady loads yielding same bit life.

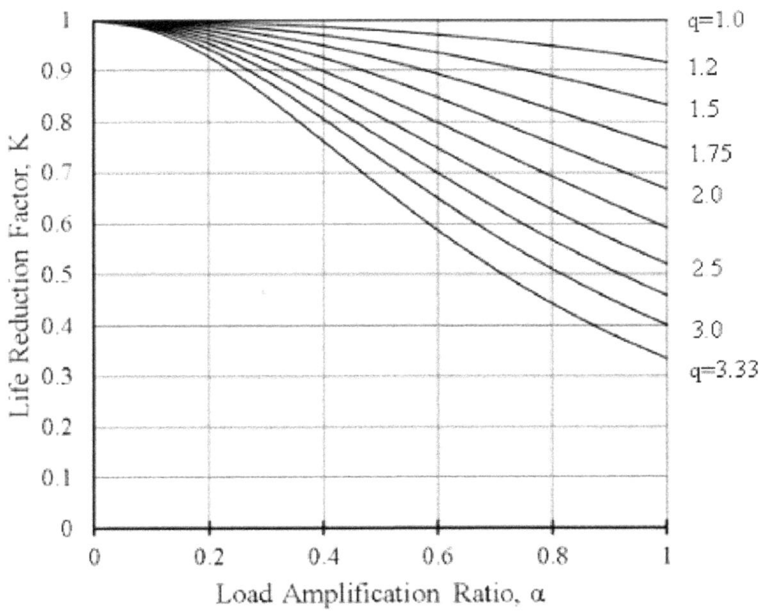

Figure 9-31. Bearing life reduction due to unsteady loading.

Table 9-4. Life reduction factors, K.

	K, Life Reduction Factors									
alpha, α	q = 3.33	3	2.75	2.5	2.25	2	1.75	1.5	1.25	q = 1
0	1	1	1	1	1	1	1	1	1	1
0.1	0.980968	0.985222	0.988113	0.990713	0.993018	0.995025	0.996729	0.998128	0.999219	1
0.2	0.927925	0.943396	0.954105	0.963878	0.972657	0.980393	0.987035	0.992542	0.996874	1
0.3	0.851058	0.881058	0.902392	0.922291	0.940545	0.956939	0.971265	0.983334	0.992965	1
0.4	0.762356	0.806452	0.838831	0.869853	0.899051	0.925925	0.949965	0.970653	0.987486	1
0.5	0.671915	0.727273	0.769301	0.810744	0.850884	0.888889	0.923827	0.954689	0.980427	1
0.6	0.586415	0.649351	0.698694	0.74877	0.798723	0.847458	0.893634	0.935668	0.971763	1
0.7	0.509306	0.576369	0.630517	0.686978	0.744934	0.803212	0.860209	0.913826	0.961448	1
0.8	0.441753	0.510204	0.56693	0.627543	0.691435	0.757575	0.82435	0.889389	0.949387	1
0.9	0.383566	0.451467	0.509022	0.571849	0.639658	0.711744	0.786777	0.862508	0.935369	1
1	0.333895	0.4	0.457135	0.52065	0.590592	0.666663	0.748039	0.833038	0.916869	1

As an example, consider the following case:

$\alpha = 0.25$ (variable force component is 25% of the average force component)
$P_m = 40{,}000$ lb
$q = 1.4$ (bearing cavity open to drilling mud)

According to Figure 9-27,

$$\beta = \frac{P_e}{P_m} = 1.01$$

This means that the effect of the 25% vibration amplitude is the same as increasing mean bit force by 400 lb, an insignificant affect.

Alternatively, the life reduction coefficient, K, for this example is

$$K = \frac{L_{unsteady}}{L_{steady}} = 0.99$$

Assuming the life expectancy for the steady case is 22 hours then the life expectancy for the unsteady force case is

$$L_{unsteady} = 22(0.99) = 21.78 \text{ hours}$$

The effect of the unsteady bit force is a reduction in bit life of 13 minutes. If the force variation is 100% instead of 25%, bit life would be reduced from 22 hours to 19 hours, a small but significant decrease in bit life.

Now assume that the bearing cavity is sealed with a wear exponent of $q = 2.5$ and a 50 hour life under a steady bit load of 40,000 lbs. If in addition a 40,000 lb variable force is applied, $\alpha = 1$ and $K = 0.52$. Bit life under this condition is

$$K = \frac{L_{unsteady}}{L_{steady}} = 0.52$$

$$L_{unsteady} = 50(0.52) = 26 \text{ hours}$$

In this case, bearing life has been cut in about half.

These calculations show that, the life of sealed bearings is more sensitive to dynamic forces than is the life of unsealed bearings. As bit bearing seals are improved, there is more incentive to eliminate bit force vibrations.

9.6.1.2 Dynamic Seals in Bearing Design

The key to the success of downhole tools is the ability to seal out drilling mud from internal components. In some cases, static seals are sufficient but, in most cases, dynamic seals are required. Dynamic seals refer to mechanical elements that allow relative motion between components while eliminating fluid flow across the seal elements. Fluid flow across dynamic seals is allowed if flow is outward.

The bearing cavity in early drill bits (roller cone) were not sealed so drilling fluid infiltrated the bearing cavity. The abrasive and corrosive environment caused by drilling mud impacted bearing life and typically drill bits lasted about 8 hours. The bearings simply wore out. In most cases, bit teeth still had plenty of wear capability left. The bearings were the weak link, which limited bit life and affected footage cost.

Recent improvements in bearing design are

1. Lubricant-reservoir within the legs of each bit body section (three legs)
2. Better dynamic seals
3. Replacing cylindrical bearing with journal bearings

Drill bits now perform over 100 hours as a result of these improvements.

A few parameters to monitor at the drill bit are:

1. Drop in rate of penetration (ROP) — a direct indicator of tooth wear.
2. Bearing temperature — an indicator of bearing distress and bearing wear rate.
3. Bit torque — an indicator of bearing or cone distress.
4. Axial force vibration magnitude — could causes reduction in bearing life.

9.7 POSITIVE DISPLACEMENT DRILLING MOTORS

These motors convert hydraulic power (pressure and flow rate) into mechanical power (torque and speed). The contour of the stator is made of flexible material while the contour of the rotor is steel. The planetary motion of the rotor creates lateral loads on the stator, which are cyclic and frictional. Both effects can damage the flexible stator lining. The pressure drop across the motor generates a downward thrust, which has to be support by rotary bearings. The power generated by the planetary motion of the rotor is brought back to the center by a universal joint. The universal joint creates side load to the motor housing, which are typically supported by hard metal sleeve bearings lubricated by the drill mud.

The eccentric motion of the rotor is also a source of excitation to the drill collars and bottom hole assembly.

Things to consider in monitoring drilling motors include:

1. Increase in pressure drop across a PDM — could indicate elastomer (stator) failure.
2. Drop in rate of penetration with rise in pressure — could indicate inefficiencies and component failure within a motor.

3. Temperature rise and high vibration frequency — could indicate distress in rolling contact bearings.
4. Monitor lateral vibrations — relates to PDM rotor speed and could excite lateral vibration modes between stabilizers.

9.8 DRILLING TURBINES

Drilling turbines operate at about twice the speed of PDMs and generate more drill bit power. Bearing support in both axial and lateral directions is critical. The axial bearing loads are due to the pressure drop and area of the rotor. Thrust loads can reach as high as 45,000 lbs. Appropriate bearing and dynamic seals become especially important in the reliability of these tools.

Rotor balance and internal stabilization become important as any rotating imbalance can possibly excite higher order lateral modes.

A by-pass valve at the top of the turbine provides a means to pump lost circulation material through the drillstring, thus by-passing the turbine blades.

Things to consider in monitoring drilling turbines are:

1. Operating at optimum power output is critical — a tachometer is essential.
2. Thrust bearing performance — could be monitored by temperature or vibration sensors.
3. Dynamic seals are critical to prolonged bearing life.
4. Rotor imbalance could indicate blade wear or failure.

9.9 MEASUREMENT WHILE DRILLING INSTRUMENTATION

MWD tools are the most complex tools in bottom hole assemblies. They contain multiple mechanical and electrical components and subsystems. They provide downhole information during the drilling process without disrupting drilling operations. MWD tools measure, capture and process and then transmitted drilling information back to the surface. MWD tools combined with rotary steerable tools provide a means of navigating drill bits accurately along a specified path to a target several miles into the earth.

It is not the purpose here to explain the components of this complex tool but to underscore the high level of engineering and research that has gone into developing this technology. Component design is challenged by the demanding operating environment of dynamics, fluid environment, temperature and pressures. Only one aspect of component design, i.e., the vibration isolation of electronic modules is discussed here.

9.9.1 Packaging Electronic Components

Vibration of bottom hole assemblies, no doubt, plays a role in the reliability of electronics. The functionality of electronic components and contact points could be affected along with module supports. Failure of electric contacts points due to fatigue is disruptive. Vibration isolation should be considered. In essence, the tool housing provides base motion related to drill collar vibrations. Electronic modules are discrete masses to be isolated.

9.9.1.1 Axial Support — Simple Harmonic Housing Motion

The response motion, say electronic modules, to motion of the housing is viewed at simple harmonic motion as depicted in Figure 9-32. The differential equation of motion for base motion for the two cases is the same.

$$m\frac{d^2x}{dt^2}+c\frac{dx}{dt}+kx=c\frac{dy}{dt}+ky \tag{9-75}$$

The solution to this equation shows that the ratio of the response amplitude, X, to the amplitude of the base motion, Y, is

$$\frac{X}{Y}=\sqrt{\frac{1+\left(2\varsigma r\right)^2}{\left(1-r^2\right)^2+\left(2\varsigma r\right)^2}} \tag{9-76}$$

This equation defines how much of base motion gets transmitted to the mass, m.

According to vibration isolation theory, avoidance of resonance of the electronic module is most desirable. A secondary approach is the use of damping in the support of electronic modules. This works for frequency ratios of $\frac{\omega}{\omega_n}\leq\sqrt{2}$. Outside of this frequency range damping becomes detrimental. Resonance occurs when the two frequencies are tuned. The practical application of this concept is the isolation of sensitive instruments or electronic components from tool housing motion.

Impact forces and accelerations between housing and well bore may be transmitted to the electronics, depending on the rigidity of the mounting.

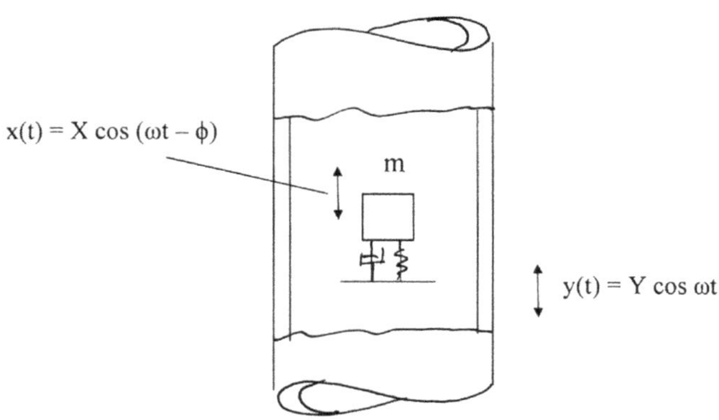

Figure 9-32. Isolation of axial vibration mode.

9.9.1.2 Lateral Support — Non-Periodic Acceleration of Housing

Though tools can vibrate laterally with simple harmonic motion, downhole measurements indicate large lateral accelerations are generated, from tool impact with the well bore. The packaging question is the same but the base motion is expressed in terms of acceleration instead of displacement (Figure 9-33).

$$m\ddot{x} + c\dot{x} + kx = ky + c\dot{y} \tag{9-77}$$

$$m\ddot{w} + c\dot{w} + kw = m\ddot{y} \tag{9-78}$$

where $\ddot{y} = a(t)$ and the relative displacement, w(t), is defined by

$$w(t) = x(t) - y(t) \tag{9-79}$$

Tool acceleration, a(t), is determined experimentally from downhole measurements.

In general lateral acceleration are non-periodic and the response is best determined by use of Duhamel's integral [19].

$$w(t) = \frac{1}{m\omega_n}\int_0^t F(\tau)e^{-\varsigma\omega_n(t-\tau)}\sin\omega_d(t-\tau)d\tau \quad \text{(damped case)} \tag{9-80}$$

$$w(t) = \frac{1}{m\omega_n}\int_0^t F(\tau)\sin\omega_d(t-\tau)d\tau \quad \text{(undamped case)} \tag{9-81}$$

where $F(t) = m\ddot{y}$

The total force on the module support is

$$P(t) = kw(t) + c\dot{w}(t) \quad \text{(damped case)} \tag{9-82}$$

$$P(t) = kw(t) \quad \text{(undamped case)} \tag{9-83}$$

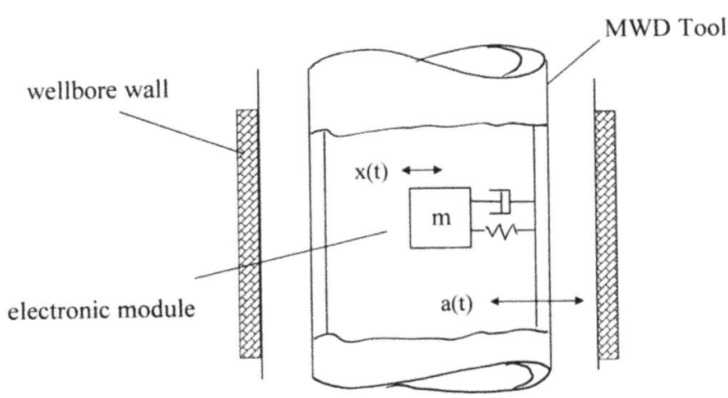

Figure 9-33. Effect of wall impact on electronic module.

Acceleration of the electronic module is

$$\ddot{x}(t) = \ddot{w}(t) + \ddot{y}(t) \tag{9-84}$$

This tool has the potential to monitor its own performance during actual drilling. Basic elements in the tool are: sensors, data storage and processing, binary code transmission, surface receiver and data processing at the surface, any one of which has multiple components. Reliability depends to a large extent on design, vibration control of all modes and packaging of the electronic modules.

Key factors to monitor are:

a. Dynamic and vibration motion of housing, especially impact forces — could affect performance of internal instruments.
b. Performance response of mechanical and electrical modules — magnitude module acceleration could affect life and performance.
c. Corresponding motion and stresses on key subsystems, such as mud pulsar, produced by tool housing motion.
d. Bearing performance and impending failure, such as high frequency noise signals produced by pitting contact surfaces.

9.9.2 Summary of Equations

$$\sigma_{p1,p2} = \frac{\sigma_x + \sigma_y}{2} \pm \sqrt{\left(\frac{\sigma_x - \sigma_y}{2}\right)^2 + \tau_{xy}^2}$$

$$\tau_{\max} = \pm\sqrt{\left(\frac{\sigma_x - \sigma_y}{2}\right)^2 + \tau_{xy}^2}$$

$$\varepsilon_{p1}, \varepsilon_{p2} = \frac{\varepsilon_x + \varepsilon_y}{2} \pm \sqrt{\left(\frac{\varepsilon_x + \varepsilon_y}{2}\right)^2 + \left(\frac{\gamma_{xy}}{2}\right)^2}$$

$$\frac{\gamma_{\max}}{2} = \pm\sqrt{\left(\frac{\varepsilon_x - \varepsilon_y}{2}\right)^2 + \left(\frac{\gamma_{xy}}{2}\right)^2}$$

$$\varepsilon_x = \frac{1}{E}\left[\sigma_x - \nu\sigma_y\right]$$

$$\varepsilon_y = \frac{1}{E}\left[\sigma_y - \nu\sigma_x\right]$$

$$\varepsilon_z = -\frac{\nu}{E}\left[\sigma_x + \sigma_y\right]$$

$$\varepsilon_z = -\frac{\nu}{1-\nu}\left(\varepsilon_x + \varepsilon_y\right)$$

$$\sigma_x = \frac{E}{1-\nu}\left[\varepsilon_x + \nu\varepsilon_y\right]$$

$$\sigma_y = \frac{E}{1-\nu}\left[\varepsilon_y + \nu\varepsilon_x\right]$$

Shear stress is not involved in either equation, but relates to shear strain by

$$\tau_{xy} = G\gamma_{xy}$$

Roller contact bearing life

$$\beta = \left[\frac{1}{T}\int_0^T \left(1-\alpha\sin\frac{2\pi t}{T}\right)dt\right]^{\frac{1}{q}}$$

$$\beta = \left(1+\alpha^2\right)^{\frac{1}{3}} \qquad \text{(ball bearings)}$$

$$K = \frac{1}{\beta^q}$$

9.10 PROBLEMS

1. A stress condition at a critical point in a drilling tool is calculated to be

   $\sigma_y = 40{,}000$ psi (tension in pipe)
   $\sigma_\theta = \sigma_x = 5{,}000$ psi (internal pressure)
   $\tau_{xy} = 10{,}000$ psi (torsion)

   Determine:
   a. principal stresses
   b. Factor of safety assuming the pipe is E75 and yielding based on von Mises' criteria of failure

2. Determine the principal strains for Problem 1

3. A strain rosette (Type A, Figure 9-21) is used to evaluate stress at a critical point in a design. Strain readings of

   $\varepsilon_a = 1000\mu$

   $\varepsilon_b = 500\mu$

   $\varepsilon_c = -750\mu$

   Determine the principal strains.

4. If torque is to be measured using a strain rosette, what type (2 gauges or 3 gauges) is best and define its orientation with the axis of the cylinder. Discuss.

5. Thrust bearings on a drilling turbine support the downward force of 20,000 lbs caused by a pressure drop across the rotor section. If the

housing of the turbine vibrates axially with periodic motion having maximum acceleration of 4 g, determine the life reduction coefficient to the bearings if total weight of the rotor section is 500 lbs. Assume the thrust bearings are completely sealed.

6. Repeat problem 5 assuming the bearings are exposed to drilling mud (q = 1.5).

7. Bearings in a roller cone drill bit are arranged as shown in Figure 9-29. Assuming the bearing cavity is:
   a. Sealed (q = 3.33)
   b. Unsealed and exposed to drilling mud (q = 1.5)

   What is the life reduction factor, K, in both cases if average load is 20,000 lbs and the dynamic bit force amplitude is 20,000 lbs?

8. An electronic module is to be mounted such as to limit the mechanical response amplitude to 10% of tool (housing) motion. Determine the spring constant of the module support. The module weighs ¼ lbs and the frequency of excitation is 10 cps.

9.11 LIST OF REFERENCES

1. US Department of Energy; "Operations and Maintenance Best Practices: A Guide to Achieving Operational Efficiency", from December 2002, hht://eere.pnnl.gov/femp/publications/O&MBestPractices.pdf.
2. Shigley, J. E. and C. R. Mischke; *Mechanical Engineering Design*, 5th Ed, McGraw-Hill, Inc. 1989.
3. Fred B. Seely and James O. Smith; *Advanced Mechanics of Materials*, 2nd Edition, James Wiley & Sons, Inc. 1955; This solution was first obtained by Prof. G. Kirsch, V.D.I., vol. 42, 1898.
4. Shigley, J. E. and L. D. Michell, *Mechanical Engineering Design*, 4th Ed., McGraw-Hill Inc. 1983.
5. Griffin, A. A.; "The Phenomena of Rupture and Flow in Solids", *Philos. Trans. Royal Soc.*, (1920) 221A, 163.
6. Personal discussions with Albert C. Martin, Director of Launch Operations for Stage 2, Saturn/Apollo Project, North American-Rockwell (2009).
7. R. C. Hibbler, *Mechanics of Materials*, 8th Ed, Prentice-Hall, 2011.
8. C. C. Perry and H. R. Lissner; *The Strain Gage Primer*, McGraw-Hill Book Co, 1955.
9. Deily, F. H., Dareing, D. W., Paff, B. H. Ortloff, J. E., and Lynn, R. D.; "Downhole Measurements of Drill String Forces and Motions," *Transactions of ASME Journal of Engineering for Industry*, Vol. 90, Series B, No. 2, May 1968, 217–225.
10. Towers, B: First Report on Fiction Experiments, Proc. Instit. Of Mech. Eng. (London), 34, 1883.
11. O. Reynolds; "On the Theory of Lubrication and Its Application to Mr. Beauchamp Tower's Experimental Determination of the Viscosity of Olive Oil", *Phil. Trans. Roy. Soc.*, London Vol 177, Pt. 1, 1886, pp. 157–234.

12. Karelitz, G. B. and J. N. Kenyon: "Oil-Film Thickness at Transition from Semifluid to Viscous Lubrication", Trans. ASME, Vol 59, 1937, pp. 239–246.
13. S. A. McKee; "The Effect of Running-IN on Journal Bearing Performance", *Mech. Eng.*, Vol. 50, 1928, pp. 528–533.
14. Stribeck, Richard, "Die Wesentlichen Eigenschften der Gleit-und Rollenlager" [the main characteristics of the sliding and roller bearings], Zeitschrift des Vereins Deutscher Ingenieure [*Journal of the Association of German Engineers*] 36, 1902, 1341–48, 1432–38, and 1463–70.
15. Dudley D. Fuller: *Theory and Practice of Lubrication for Engineers*, 2nd Ed, John Wiley & Sons, 1984.
16. Radzimovsky, Eugene I. and Donald W. Dareing; "Influence of Load Variation Upon Life Duration of Roller-Contact Bearings", Ukrainian Technical-Economical Institute, Scientific Notes, No. V, 1964.
17. Dareing, Don W. and E. I. Radzimovsky; "Effect of Dynamic Bit Forces on Bit Bearing Life", Trans AIME, *SPE Journal*, Vol. 5, No. 4, 1965, pp. 272–276.
18. Hughes Two-Cone Drill Bit; Designated a Historical Mechanical Engineering Landmark, ASME — Baker Hughes Special Publication, The Woodlands, Texas; August 10, 2009.
19. King, G. R.; Effect of Fluid Environment on Rock-Bit Bearing Performance", AAODC Paper (1959).
20. Roy R. Craig, Jr: *Structural Dynamics*, John Wiley & Sons, 1981.

Chapter 10

CASING DESIGN

Lining or casing well bores goes back to the early days of hand dug water wells when rock and dry masonry were used to line water wells to keep unconsolidated surface soil and rock layers from caving in over the exposed water bearing formation. Small diameter holes were drilled in China prior to 6000 BC to recover brine for salt extraction. These small holes were cased with bamboo. Hollowed-out logs have also been used to case shallow bore holes. During the evolution toward modern day oil field casing, tin tubes with soldered seams and joints have been used as well as iron pipe held together with rivets.

Deeper oil and gas wells of the 1900s, required casing that can withstand much greater loads generated by formation fluid pressures, hydrostatic drilling mud pressure, and buoyed weight of the casing. Design specifications include the most severe combination of these loads. Casing design is a matter of selecting standard casing having adequate dimensions and material strength to withstand these loads during operation. The goal is to choose casing having lowest grade and lightest weight. Overall casing assembly is typically a combination of different sizes and grades for lowest casing cost. The selection can be made from tables of casing properties, sizes and grades [1].

The functions of casing are

- Support unconsolidated or weak formation and thus protect the hole opening made by the drill bit.
- Protect ground water from contamination by oil production from deeper formation.
- Protect open formations, especially at casing shoe, from fracturing due to high mud weights required down hole.
- Contain fluid pressure brought into the well bore by producing reservoirs.

Several casing sizes may be nested together, each reaching to different depths depending on formation fracture gradients.

10.1 OPERATIONAL REQUIREMENTS FOR CASING

A casing program is planned from the bottom up, i.e., the smallest or inside casing depends on size of production tubing or rate of production expected from the reservoir. The size and number of casing tubulars depend on formation pressure and fracture strength of the layered formations.

Cementing casing is an important and costly part of setting casing in a well bore. Cement around casing provides support and prevents drilling mud and gas from moving upward from outside of the casing. It is a vital aspect of well control.

Basic casing strings typical of most oil and gas wells are (Figure 10-1):

- *Conductor* is a piece of large pipe (~30 in. OD) installed at the surface to keep the upper part of the hole from caving in. The installation of conductor pipe is referred to as "spudding" in a well. It can be installed prior to setting up the drill rig on land wells and may or may not be cemented in place.
    - Holds back unconsolidated surface formations
    - Contains flow line for mud returns
    - Contains diverter above flow line when shallow gas is probable

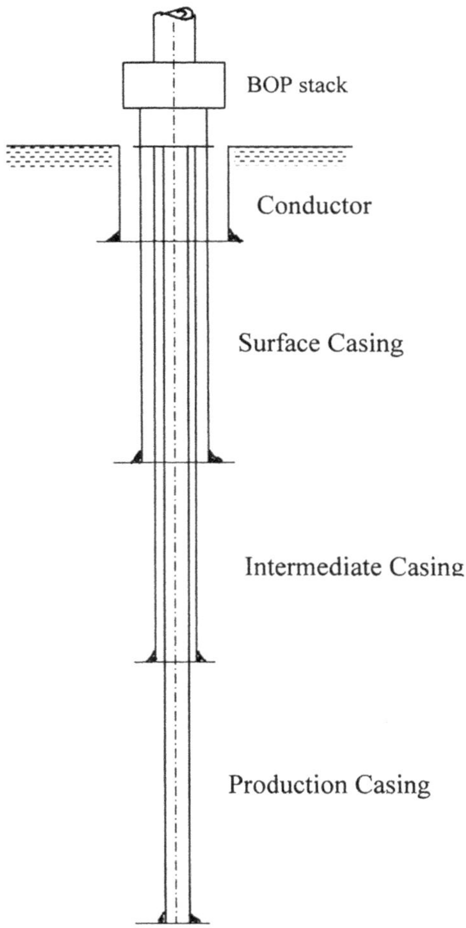

Figure 10-1. Types of casing strings.

  - Up to 30" in diameter
  - Set between 300 to 1000 ft
- *Surface Casing* is set to protect fresh water sands. It is required by law. It contains a well head flange that supports the Blow-Out Prevention (BOP) stack and later, production "Christmas Tree" equipment. Also, surface casing must support the other casing strings. The annular space must be cemented back to the surface to prevent possible gas returns outside of the BOP stack.
  - Holds back unconsolidated shallow formations
  - Protects shallow zones from drilling mud and deeper formation contamination
  - Serves as a base for the blow out preventer (BOP) stack
  - Casing seat must hold deeper formation pressure
  - Supports the suspended weight of other casing strings
  - Up to 13⅜ in. diameter
  - Set at depths from 1000 to 5000 ft (~2000 ft is typical)
  - Cemented back to the surface
- *Intermediate Casing* (or liner) is set for the same reasons.
  - Controls open hole conditions that may prevent a well from being drilled safely to the reservoir
  - Prevents lost circulation
  - Isolates troublesome shale, shallow gas, or water flows
  - Reduces amount of uncased hole, especially in deep wells
- *Production Casing* is set through production zone (except in open-hole conditions). It is cemented from the casing shoe back to the surface.
  - Designed to hold maximum shut-in pressure.

The objective in casing design is to select the lowest grade and weight of casing, to withstand specified burst and collapse pressure loads, in order to minimize overall casing cost.

10.2 GRADES OF CASING

Early grades of casing had minimum yield strengths of 25,000 psi, 40,000 psi, and 55,000 psi. In 1945, the API listed new grade symbols and standardized F25, H40, J55, and N80. Except for F25 these grades are still in use today. Since higher grades of casing have been developed to meet the load demands imposed by deeper wells (see Tables 10-1 and 10-2), API grades of N80 and higher are considered high strength steel, while grades of C75 and lower are low strengths steels.

High strength steels can withstand greater bending, tension and burst pressures, but have lower fracture toughness properties and are susceptible to stress corrosion cracking. Steel hardness increases with strength and hard metal is vulnerable to crack instability and stress corrosion. The main advantage to high strength casing is a lighter casing weight. Grade V150, for example, has minimum yield strength of 150,000 psi, but has zero notch toughness at 32°F (0°C). Its Charpy transition temperature is about 150°F to 200°F.

API BUL 5C2 [1] gives the pressure limits for each casing grade, size and weight. For example, the collapse and burst resistance of 9⅝ in. casing (40 lb/ft) Grade J55

Table 10-1. API grades.

Grade	Minimum Strength, psi (yield)	Minimum Strength, psi (ultimate)
H40	40,000	60,000
J55	55,000	75,000
K55	55,000	95,000
C75	75,000	95,000
N80	80,000	100,000
C95	95,000	105,000
P110	110,000	125,000

Table 10-2. Non API grades.

Grade	Minimum Strength, psi (yield)	Minimum Strength, psi (ultimate)
S80	55,000	95,000
S0090	90,000	105,000
SS95	80,000	100,000
S95	95,000	110,000
S105	95,000	110,000
S00125	125,000	135,000
S00140	140,000	150,000
V150	150,000	160,000
S00155	155,000	165,000

are 2570 psi and 3950 psi, while the collapse and burst resistance of casing having the same size and weight but of Grade N80 are 3090 psi and 5750 psi. Casing performance tables will be used later to show how casing is designed.

The best casing design is one which uses the lightest weight of casing and the lowest casing grade. This usually produces the lowest casing cost as cost depends directly on total weight. Environmental considerations such as corrosion or temperature may alter this approach. Casing string designs typically contain different casing weight (lb/ft) and grades in order to achieve lowest total casing cost.

10.2.1 Casing Connections and Couplings

Standard API casing thread types include;

- Short round thread (ST & C)
- Long round thread (LT & C)

- Buttress thread and coupling
- Extreme-line thread and coupling

The first three connection types are held together by couplings (Figure 10-2). Both pipe threads and coupling threads are machined on a taper and are made up hand tight when pipe threads and coupling threads are fully engaged. The threads on extreme-line couplings are machined directly into up-sets forged at the ends of casing tubes.

The taper in both the Short Thread and Coupling (ST&C) and the Long Thread and Coupling (LT & C) is ¾ in/ft on the diameter. The taper on the buttress threaded connection is also ¾ in/ft except for the 16-in. OD pipe and larger, in which case the taper is 1 in/ft.

ST & C and LT & C connections have 60 deg. V threads rounded at the tip and root. When made up past the hand tight position, the threads are tightened forming a pressure tight seal. A pressure sealing lubricant is also used between the mating threads to ensure no leakage. The thread pitch on these connections is eight (8) threads per inch.

The tensile strength of API connections when installed on plain end casing is not as strong as the casing body. The cross-sectional area of the casing body is greater than the cross-sectional area of the casing under the last engaged thread. In order to produce a 100 per cent efficient connection, the ends of casing may be upset so that the cross-sectional area under the last engaged thread is equal to or greater than the casing tubular.

One mode of failure of ST & C or LT & C connections is threads jumping over one another at high tensile loads. As pipe wall yields, fewer threads are engaged and the connection jumps apart without metal separation or fracture.

Buttress threaded connections also use a coupling, which threads over the pipe ends. The Buttress threads reduce the chance of pipe threads jumping out of the coupling at high tensile loads. In addition, the threads run out at the outer diameter of the tubular making the connection nearly 100 per cent efficient. Pressure sealing is accomplished by making up the threads beyond hand tight position and using a pressure sealing lubricant.

Extreme-line threaded connections have no coupling. The box and pin are an integral part of casing and are machined on up-sets forged at the ends of the casing tubular. Pressure is sealed by metal contact in a tapered wedged at tip end of the pin and throat of the box. An advantage of this connection is a smaller overall outside diameter at the connection.

Another metal to metal sealing type connection is the Hydril 2-step thread. The overall thread has two separate pitch diameters giving the appearance of a step. There are three possible sealing surfaces: one at the tip of the pin, one at the step and another at the shoulder off the pin. This connection has no coupling and the threads are machined on the upsets. Thread forms are square.

10.3 INSTALLATION OF CASING

Prior to the installation of casing, the entire drillstring is removed and the well bore filled with drilling mud. Mud weight is high enough to control formation pressure.

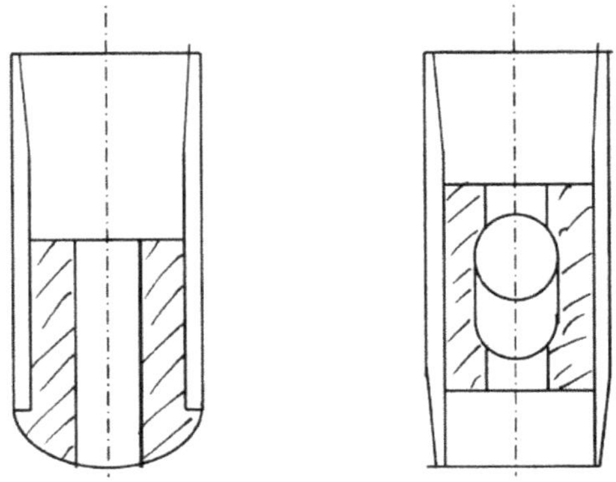

Guide shoe
(a)

Float collar
(b)

Figure 10-2. Filling annular space with cement.

A special attachment, called a guide shoe (Figure 10-2a), is placed at the end of the first casing joint to guide the casing down the well bore. A float collar (Figure 10-2b), which contains a one way value, is placed about two joints above the guide shoe. Flow is allowed downward through the valve but prohibits flow back up the casing. This valve allows casing to be floated into the hole thus taking some of the load of the rig. Another purpose is to keep the cement slurry from coming back into the casing once the slurry is in place.

The space between the float collar and guide shoe contains the last volume of cement slurry pumped into the casing. To minimize the possibility of drill mud contamination, this last portion is intentionally trapped between the float collar and guide shoe. It is drill out once the cement is set, which usually takes about 24 hours. The guide shoe may also contain a one-way valve, in which case it is called a float shoe. Once the casing is in place, the cementing operation can start.

At the beginning of a cement operation, a special rubber plug (bottom plug) is inserted inside the at the top of the drilling mud column to separate the cement slurring from drilling mud. The weight of the cement slurry (13 to 15 ppg) helps push the drilling mud up the annular space. The bottom plug is pumped down until the right amount of cement has been placed inside the casing. A second plug (top plug) is then placed on top of the cement to isolate the cement volume from drilling mud. The entrapped cement slurry volume between the two plugs is pumped down with drill mud. Once the bottom plug reaches the float collar, the diaphragm in the plug is ruptured by increased pump pressure allowing the cement slurry to continue through the float shoe and upward filling the annular space. Once the second plug reaches the float collar, all of the cement volume has filled the annular space.

10.3.1 Leak-Off Test

Leak-off tests are performed to determine the fracture strength of the formation at the casing shoe. This is conducted after every casing is set. This test establishes the maximum mud weight that the casing shoe and formation can support in anticipation of higher mud weights that might be needed further down hole. If higher mud weights are needed down hole that create pressure at the casing shoe exceeding the fracture strength of the open formation, then another casing has to be set to protect the formation at the last casing shoe.

This test is made after about 20 ft of hole is drilled past the casing shoe. The annular preventer is closed and standpipe pressure is increased stepwise by stroking the pump and displacing approximately two barrels of drilling mud at a time into the cased hole. Hydrostatic pressure beneath the casing shoe is equal to standpipe pressure plus pressure created by the mud column (Figure 10-3).

A leak off test after cementing in a new casing will define $(\gamma_m)_{fract}$ at the casing shoe. A leak off test is conducted as follows:

Step #1 — shut the annular preventer.
Step #2 — stroke the mud pump at a low rate while continually monitoring standpipe pressure. Pump pressure should be increased in small time increments.
Step #3 — monitor standpipe pressure continuously. At some point, the stand pipe pressure curve drops indicating a breakdown in formation at the casing shoe.
Step #4 — Determine hydrostatic pressure at the casing shoe corresponding to formation breakdown. A leakoff test is conducted by stroking the pump slowly and periodically stopping to see if the formation at the casing shoe is holding. The test is stopped when casing shoe pressure drops sharply as indicated in Figure 10-4.
Step #5 — Use the fracturing pressure along with the current mud weight to determine equivalent mud weight and fracturing gradient at the casing shoe.

Total bottom hole pressure at any time in the test is

$$p_{cs} = p_{sp} + 0.052\gamma_m L \tag{10-1}$$

where

p_{cs} = pressure at casing shoe, psi
p_{sp} = standpipe (or shut-in) pressure, psi
γ_m = mud density, ppg
L = distance to casing shoe, ft

When the formation breaks down, $p_{cs} = p_F$, where p_F is the formation fracture pressure.

Mud weight that would cause the formation breakdown at the casing shoe is determined by

$$p_F = 0.052\gamma_F L \tag{10-2}$$

Formation fracture gradient is derived from fracture pressure.

$$p_F = 0.052\gamma_F x \quad \text{(psi)} \tag{10-3}$$

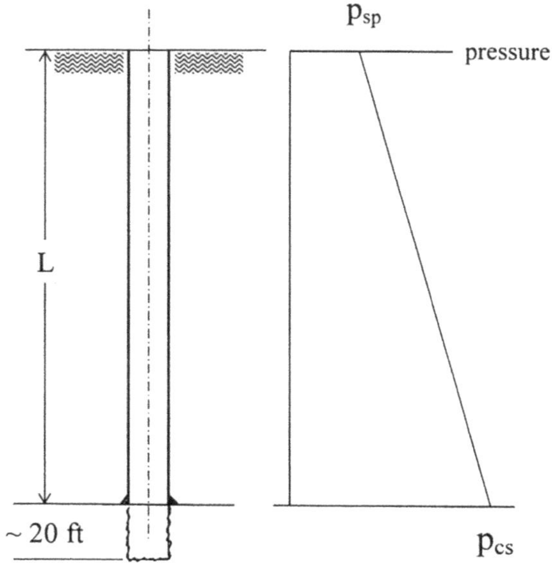

Figure 10-3. Leak-off test pressure.

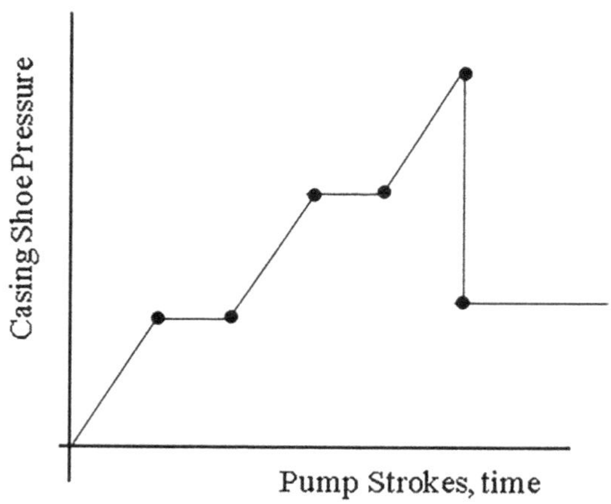

Figure 10-4. Casing shoe pressure versus time.

where

p_F = pressure required to fracture a given formation, psi
γ_F = drilling mud density causing formation breakdown at the last casing shoe, ppg
x = depth, ft

Formation fracture gradient is defined as

$$\left(\frac{dp}{dx}\right)_F = 0.052\gamma_F \quad \text{(psi/ft)} \tag{10-4}$$

This expression simply defines the slope of the pressure line from the surface corresponding to formation fracture at a given location. It defines mud weight required to fracture the formation at the casing shoe.

Local hydrostatic pressures, due to mud weight and formation pressure, can be defined in the same way.

$$\left(\frac{dp}{dx}\right)_{mud} = 0.052\gamma_m \tag{10-5}$$

$$\left(\frac{dp}{dx}\right)_{formation} = 0.052\gamma_f \tag{10-6}$$

where

γ_f = drilling mud weight required to equalize formation pressure

The three gradients (mud, formation and fracture) are useful to determine whether a given drilling mud weight is high enough to contain formation pressure down hole without fracturing a formation as the last casing shoe (Figure 10-5). For ideal conditions,

$$\gamma_F \succ \gamma_m \succ \gamma_f \tag{10-7}$$

If this condition is marginal, then an intermediate casing must be set.

10.4 STRESSES IN CASING AND MODES OF FAILURE

Casing is designed to withstand three types of stress: burst, collapse and tension. Casing couplings must also withstand these loads. Knowledge of the physical strength capability of casing is important in designing a casing string for minimum weight and cost. This section reviews basic formulas used to describe hoop stresses, burst stresses and buckling of pipe or pressure vessels.

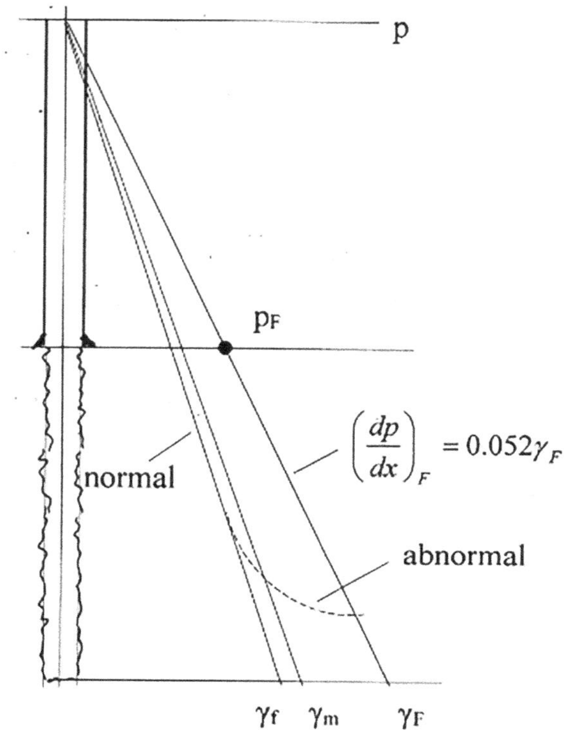

Figure 10-5. Abnormal pressure.

10.4.1 Thick Walled Cylinder Stresses

The theory of hoop and radial stresses in thick wall cylinders is derived from the theory of elasticity and is well documented in the literature [2]. These stresses are defined in terms of internal and external pressures by Lames' equations.

$$\sigma_\theta = \frac{p_i a^2 - p_o b^2 - a^2 b^2 (p_o - p_i)/r^2}{b^2 - a^2} \tag{10-8}$$

$$\sigma_r = \frac{p_i a^2 - p_o b^2 + a^2 b^2 (p_o - p_i)/r^2}{b^2 - a^2} \tag{10-9}$$

Stress distributions for the special case of internal pressure (p_i) with $p_o = 0$ are shown in Figure 10-6.

The stress state within the cylinders cross section is biaxial if axial stresses are not included. The maximum stress levels for both hoop and radial are located on the inside surface.

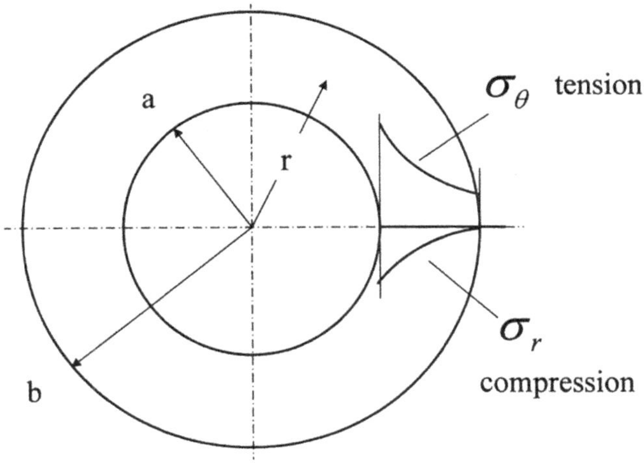

Figure 10-6. Hoop and radial stress distributions (p_o = 0).

10.4.2 Thin Wall Stress Equations

The Lame' equations show that for $\frac{D}{t} \geq 20$, the maximum hoop stress is only 5% higher than the average hoop stress across the cylinder wall [3]. Many casing sizes fall outside this ratio. The thin wall model gives a simple view of how hoop stresses are developed (see the freebody diagram in Figure 10-7).

Summing forces in the radial direction gives

$$\sigma_\theta = \frac{R}{t}p = \frac{D}{2t}p \tag{10-10}$$

This simple formula gives reasonable engineering results in many cases.

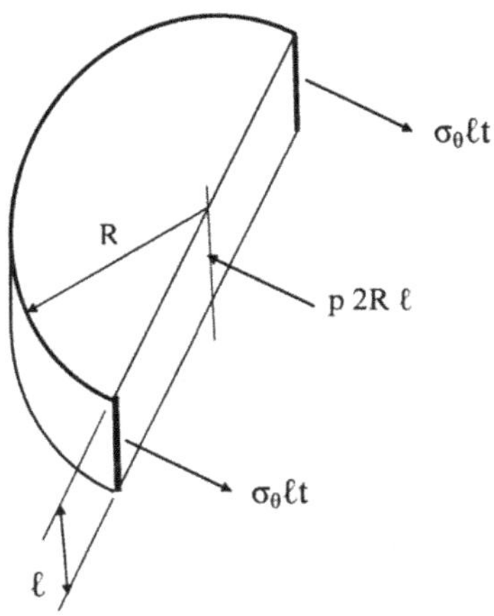

Figure 10-7. Hoop stresses in thin wall pipe.

10.4.3 Barlow Formula for Burst

Burst pressure ratings given in casing performance tables are based on the Barlow formula [1], which carries a 0.875 factor to account for the specified manufacturing tolerance of the pipe wall.

$$p_{yld} = 0.875\frac{2\sigma_{yld}}{\left(\frac{D}{t}\right)} \tag{10-11}$$

For example, consider 9.625 (47 lb/ft, Grade C75).

OD = 9.625 in
t = 0.472 in

The burst strength from Table 10-4 is given as 6440 psi. From the above data, D/t = 20.39 and the predicted burst pressure [using Eq. (10-11)] is predicted at 6438 psi.

10.4.4 Elastic Buckling of Thin Wall Pipe

One way to view the buckling of thin walled cylinders is to consider the circumference as a straight beam with end loads. This model is suggested by the shape of the primary mode as depicted in Figure 10-8. Assuming Euler's equation applies,

$$F_{cr} = \left(\frac{4\pi}{L}\right)^2 EI = \sigma_\theta A = \left(\frac{d}{2t}p\right)t\ell \quad \text{(four quarter-lobes)}$$

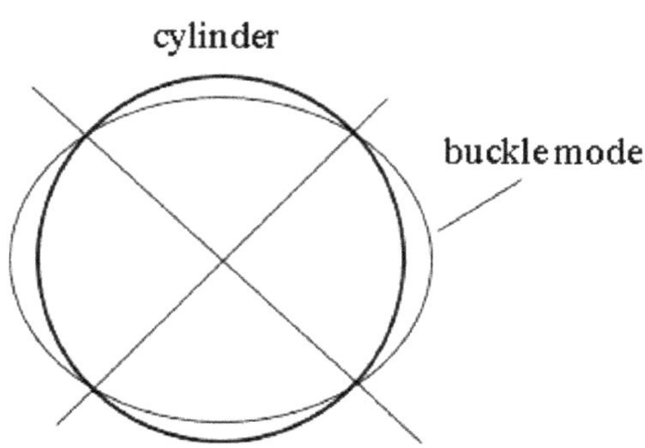

Figure 10-8. Loaded circular ring.

$$\left(\frac{4\pi}{\pi d}\right)^2 E\,\frac{t^3\ell}{12} = \left(\frac{d}{2t}\,p_{cr}\right) t\ell$$

After simplification

$$p_{cr} = \frac{8}{3} E \frac{1}{(d/t)^3} \qquad (10\text{-}12)$$

This solution treats the circumference of the cylinder as a straight beam.

10.4.4.1 Bresse's Formulation

Bresse [4] gives a more realistic model and solution to the buckling of circular rings. His formulation was developed to predict buckling of straps of unit length. Figure 10-9 shows one fourth of the ring of radius, r. The radial deflection, w(s), is expressed in polar coordinates. Curvature of the deflection arc relative to the initial curvature (1/r) is

$$\frac{1}{\rho} = \frac{d^2w}{ds^2} + \frac{w}{r^2}$$

Assuming Euler bending

$$EI\left(\frac{d^2w}{ds^2} + \frac{w}{r^2}\right) = -M$$

The bending moment at location s or θ is M = rwp. Therefore,

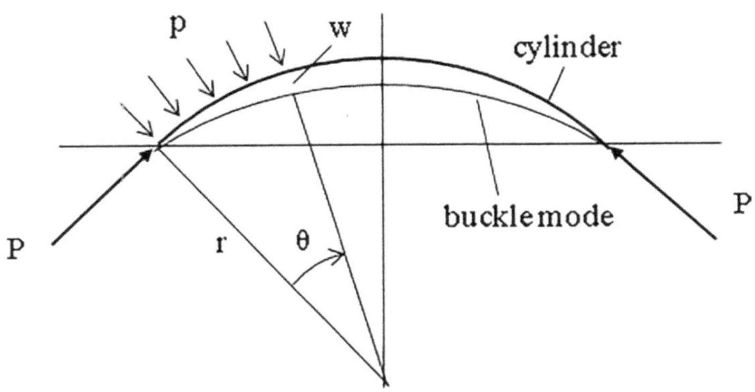

Figure 10-9. Freebody diagram of segment of cylinder.

$$EI\left(\frac{d^2w}{ds^2}+\frac{w}{r^2}\right)=-rwp \quad \text{(p is force per arc length of strap)}$$

Collecting terms

$$\frac{d^2w}{ds^2}+\left(\frac{1}{r^2}+\frac{rp}{EI}\right)w=0 \tag{10-13}$$

$$\frac{d^2w}{d\theta^2}+k^2w=0$$

where

$$k^2=\left(1+\frac{r^3p}{EI}\right) \tag{10-14}$$

The solution to Eq. (10-13) is

$$w(\theta)=C_1\sin k\theta+C_2\cos k\theta \tag{10-15}$$

Applying boundary conditions, w(0) = 0 and w(π/2) = 0 leads to

$C_2 = 0$ and
$0=C_1\sin k\frac{\pi}{2}$

Leading to a nontrival solution of

$$k\frac{\pi}{2}=\pi$$

So k = 2 produces instability or buckling of a cylinder.

$$1+\frac{pr^3}{EI}=4 \tag{10-16}$$

$$p_{cr}=\frac{3EI}{r^3} \qquad \text{with } I=\frac{t^3}{12}$$

$$p_{cr}=\frac{E}{4}\left(\frac{t}{R}\right)^3 \quad \text{(viewed as pressure since strap as unit width)}$$

$$p_{cr}=2E\frac{1}{(D/t)^3} \quad \text{(Bresse's equation)} \tag{10-17}$$

10.4.4.2 Application to Long Cylinders

When applying Bresse's equation to long cylinders E is replaced as follows

$$p_{cr} = \frac{2E}{1-\nu^2}\frac{1}{(d/t)^3}$$

In the bending of plates and shells, longitudinal strain at a point is zero, leading to

$$\sigma_z = \nu\sigma_\theta$$

$$\varepsilon_\theta = \left(\sigma_\theta - \nu^2\sigma_\theta\right)$$

$$\sigma_\theta = \frac{E}{1-\nu^2}\varepsilon_\theta$$

And since $p = \frac{2\sigma t}{d}$

$$\sigma_{cr} = \frac{E}{1-\nu^2}\left(\frac{t}{d}\right)^2$$

Consider a long, thin walled pipe with outside diameter of 8 in. and a wall thickness of 0.25 in.. If the yield strength of pipe material is 75,000 psi, determine the critical buckling pressure. Will the pipe fail by buckling or material yielding. Assume E = 30 × 10⁶ and ν = 0.25.

$$p_{cr} = \frac{2\times 30\times 10^6}{1-0.25^2}\frac{1}{(32)^3} = 1{,}953 \text{ psi}$$

Circumferential stress in pipe wall is

$$\sigma_{cr} = \frac{30\times 10^6}{1-0.25^2}\left(\frac{0.25}{8}\right)^2 = 27{,}465$$

The pipe will buckle before reaching the yield stress level.

10.4.5 API Collapse Pressure Guidelines

The API standard on casing design [1] defines three regions of casing properties leading to collapse.

1. Plastic Collapse Pressure — The equation defining the minimum collapse pressure in this case is

$$p_P = f_{ymn}\left[\frac{A_c}{D/t} - B_c\right] - C_c \tag{10-18}$$

p_P = pressure for plastic collapse
f_{ymn} = specified minimum yield strength
$A_c B_c C_c$ = empirical constants depending on casing grade and D/t ratio

2. Transition Collapse Pressure — The equation defining the minimum collapse pressure in this case is:

$$p_T = f_{ymn}\left[\frac{F_c}{D/t} - G_c\right] \tag{10-19}$$

p_T = pressure for transition collapse
$F_c G_c$ = empirical constants depending on casing grade and D/t ratio

3. Elastic Collapse Pressure — The equation defining elastic collapse pressure is:

$$p_E = 46.95\times 10^6\left\{\frac{1}{\left[D/t\left(D/t-1\right)^2\right]}\right\} \tag{10-20}$$

10.4.6 Plastic Yielding and Collapse with Tension

Tension also effects plastic yielding and collapse in casing. Tension and compression combine to create a biaxial state of stress (Figure 10-10), which is evaluated for yielding using the von Mises criteria of failure [3].

$$\sigma' \geq \sigma_{yld} \tag{10-21}$$

where σ_{yld} is material yield strength as determined from uniaxial testing and

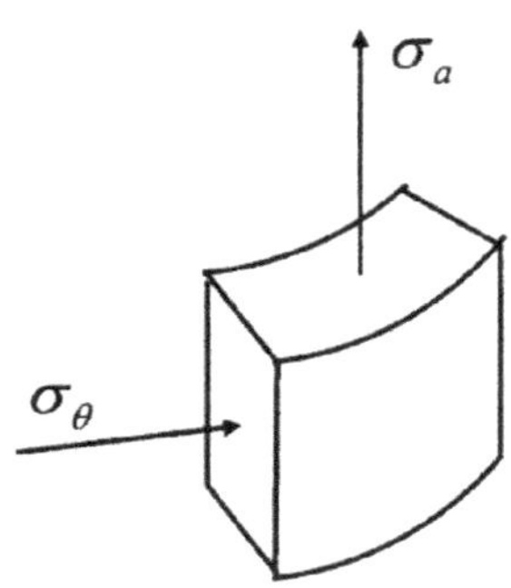

Figure 10-10. Biaxial state of stress.

$$\left(\sigma'\right)^2 = \sigma_a^2 - \sigma_a\sigma_\theta + \sigma_\theta^2 \tag{10-22}$$

According to Eq. (10-21), yielding occurs when von Mises stress, $\sigma' \geq \sigma_{yld}$.

$$\sigma_a^2 - \sigma_a\sigma_\theta + \sigma_\theta^2 \geq \left(\sigma_{yld}\right)^2 \tag{10-23}$$

In dimensionless units, yielding occurs when

$$\left(\frac{\sigma_a}{\sigma_{yld}}\right)^2 - \frac{\sigma_a}{\sigma_{yld}}\frac{\sigma_\theta}{\sigma_{yld}} + \left(\frac{\sigma_\theta}{\sigma_{yld}}\right)^2 \geq 1 \tag{10-24}$$

For example, if

$$\sigma_\theta = -40{,}000 \text{ psi}$$

$$\sigma_a = 20{,}000 \text{ psi}$$

$$\sigma_{yld} = 50{,}000 \text{ psi}$$

then

$$\frac{\sigma_\theta}{\sigma_{yld}} = -0.8 \quad (80\%)$$

$$\frac{\sigma_a}{\sigma_{yld}} = 0.4 \quad (40\%)$$

By substituting these numbers into Eq. (10-24), the left side is equal to 1.12, therefore, this biaxial stress state will cause plastic yielding.

10.4.7 Tabulation of Casing Performance Properties

Strength properties of casing sizes are conveniently display in tables (see Table 10-4). Collapse and burst strength capabilities are based on API equations outlined earlier. Casing size is listed in terms of their outside diameter. Weight changes reflect changes in inside diameter. Grade or material strength may vary for each casing size and weight. Grade of material effects collapse, burst and tension limits. A complete set of casing performance data can be found in ref [1].

10.5 ELEMENTS OF CASING DESIGN

The goal in casing design is to select the lowest grade and weight depending on burst and collapse loading. In general casing strings are not uniform in weight and grade. Casing design begins with operational requirements and design specifications. Operational requirements define the production expectations from the well. Design specifications set the expected performance and constraints on a casing design program. Information from nearby off-set wells is helpful in establishing design specifications. Examples of casing design specifications are:

- Casing diameter
- Formation fracture gradient
- Setting depth
- Drilling fluid density when casing is run
- Expected maximum bottom hole shut in pressure
- Well temperature
- Well curvature
- Load distributions to be used in design
- Factor of safety for each type of loading

Typically, three modes of failure (collapse, burst, tension) are considered. Each type of loading varies with vertical distance into the well. Pressure loads will be considered next.

10.5.1 Collapse Pressure Loading (Production Casing)

Collapse pressure is the difference between the outside and inside pressures according to

$$\Delta p = \left[p_0(0) + \gamma_0 x \right] - \left[p_i(0) - \gamma_i x \right] \tag{10-25}$$

It is common practice to assume that the most severe collapse pressure loading condition occurs when the casing is completely evacuated and external

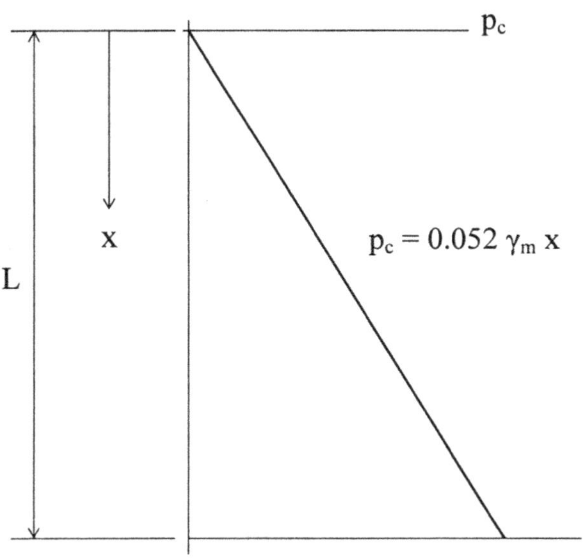

Figure 10-11. Collapse pressure loading line.

pressure is the result of hydrostatic mud pressure. External pressure for this case increases linearly with depth and is based on the weight of drilling mud when casing is run. This condition is usually assumed in casing design because it gives a conservative design against collapse. The collapse load equation then becomes

$$p(x) = 0.052\gamma_o x \tag{10-26}$$

where

p = pressure, psi
γ_o = mud density (γ_m), ppg (when casing was run-in)
x = vertical distance, ft (downward from surface)

The collapse pressure loading line is illustrated in Figure 10-11.

10.5.2 Burst Pressure Loading (Production Casing)

Burst loading is based on reservoir pressure (p_f) with gas trapped inside the casing from bottom to top. Assuming the formation at the casing shoe can support bottom-hole pressure then the *internal pressure* distribution is defined by

$$p_i(x) = p_f - 0.052\gamma_g (L - x) \tag{10-27}$$

Here, p_f is formation or reservoir pressure. At the surface, the shut-in pressure is formation pressure minus the effect of the gas pressure.

In terms of pressure gradients, Eq. (10-27) becomes

$$p_i(x) = p_f - 0.052\gamma_m (L - x) \tag{10-28}$$

The pressure gradient term has units of psi/ft.

The *external pressure* is usually assumed to follow drilling mud hydrostatic pressure.

$$p_o = 0.052\gamma_m x \tag{10-29}$$

The resulting burst pressure, p_b, then becomes the difference between the inside pressure and the outside pressure in the casing.

$$p_b = p_i - p_o \tag{10-30}$$

$$p_b = p_f - 0.052\gamma_g (L - x) - 0.052\gamma_m x \tag{10-31}$$

For the sake of simplicity, the gas pressure gradient is sometimes ignored (a conservative assumption) leaving

$$p_b = p_f - 0.052\gamma_m x \tag{10-32}$$

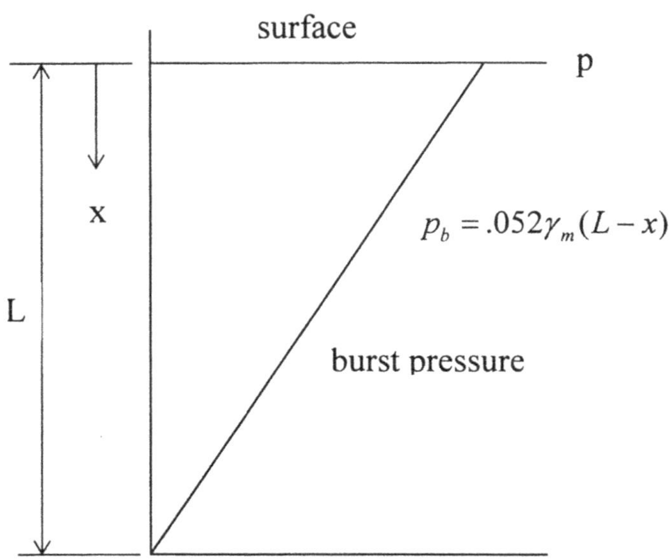

Figure 10-12. Burst pressure loading line.

Taking formation pressure as bottom hole mud pressure

$$p_f = 0.052\gamma_m L \tag{10-33}$$

Then the burst loading line becomes (Figure 10-12)

$$p_b = 0.052\gamma_m (L - x) \tag{10-34}$$

10.5.3 Summary of Pressure Loading (Production Casing)

Collapse pressure loading is illustrated in Figure 10-13 showing the pressure distribution increasing directly with depth according to

$$p_c = 0.052\gamma_m x \text{ psi} \tag{10-35}$$

The maximum collapse pressure loading is at the bottom end.

Burst pressure load is the difference in internal pressure and the external pressure. An extreme case would exist if formation gas pressure is exerted internally from bottom to top and internal pressure being partially resisted by hydrostatic mud pressure outside the casing. The resulting burst pressure for this assumption is

$$p_b = p_f - 0.052\gamma_m x \tag{10-36}$$

$$p_b = 0.052\gamma_m(L - x)$$

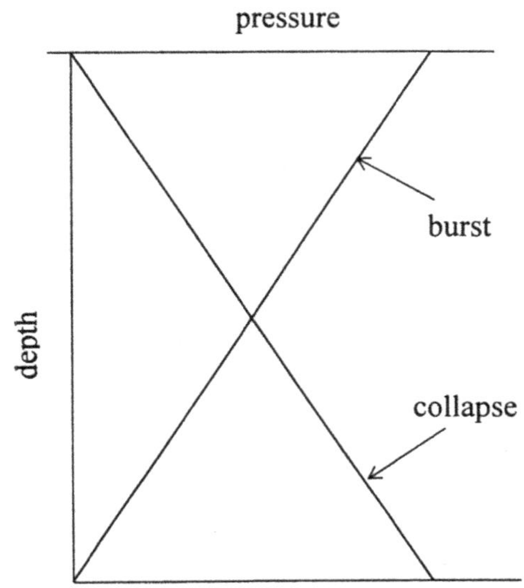

Figure 10-13. Typical loading for casing design.

The maximum burst pressure occurs at the well head.

The collapse and burst pressure loads are illustrated in Figure 10-13 by the "X" loading diagram.

A casing string designed strictly for collapse would be strongest at the bottom and weakest at the top. While this unusual casing string would not collapse, it would be susceptible to burst type failure in the event the well is shut in during a blowout. A casing string designed for burst alone would most likely collapse if the casing were evacuated.

A proper casing design is one, which can resist both collapse pressure load and burst pressure load. In addition, it is tapered and has the strongest section near the top and bottom with the weakest casing section near the middle.

10.5.4 Collapse Pressure Loading (Intermediate Casing)

The worst case of collapse loading is when the intermediate casing is empty and. This condition is usually assumed in casing design because it gives a conservative design against collapse. The collapse load equation then becomes

$$p(x) = 0.052\gamma_o x \tag{10-37}$$

where

p = pressure, psi
γ_o = mud density (γ_m), ppg
x = vertical distance, ft (downward from surface)

10.5.5 Burst Pressure Loading (Intermediate Casing)

In this situation, casing is set to protect predrilled formations. The formation fracture pressure (p_F) becomes the maximum allowable pressure at the casing shoe. This pressure limitation is predicted from drilling records in the regions as determined from a leak-off test following casing installation. In this case, the burst pressure profile is defined by

$$p_b = p_F - 0.052\gamma_g (L_{cs} - x) - 0.052\gamma_m x \tag{10-38}$$

If the formation fracture pressure gradient is unknown it is often taken as 1 psi/ft, i.e., PG_F = 1 psi/ft. If gas pressure gradient is ignored

$$p_b = p_F - 0.052\gamma_m x$$

$$p_b = 0.052(\gamma_F) L_{CS} - 0.052\gamma_m x \quad (p_F \prec p_{reservoir}) \tag{10-39}$$

After the casing is set, a leak off test is conducted to determine the true fracture gradient of the formation for use in gauging the effect of mud weight further down the hole. The last term in Eq. (10-39) represents support of drilling mud in the annular space.

If the fracture strength (p_F) of the formation at the casing shoe is greater than reservoir pressure, (p_f), then Eq. (10-38) is modified as follows

$$p_b(x) = p_f - 0.052\gamma_g (L - x) - 0.052\gamma_m x \quad (p_F \succ p_{reservoir}) \tag{10-40}$$

If gas pressure gradient is ignored,

$$p_b = p_f - 0.052\gamma_m x \tag{10-41}$$

$$p_b = 0.052\gamma_m (L - x); \quad x \le L_{cs} \tag{10-42}$$

10.5.6 Factors of Safety

In mechanical design, factors of safety are applied to both material strength and applied loads to compensate for uncertainty. Perhaps the greatest uncertainty relates to the applied load assumed in the design. Factor of Safety is defined as

$$FS = \frac{\sigma_{yld}}{\sigma_{allow}} = \frac{p_{limit}}{p_{allow}} \tag{10-43}$$

where

p_{limit} = pressure limit (table values)
p_{allow} = allowable pressure (design pressure)

Table 10-3. Factors of safety for casing loads.

Loading	High	Low
Tension (n_3)	2.25	1.40
Burst (n_2)	1.50	1.00
Collapse (n_1)	1.25	1.00

The numbers given in the casing performance tables give p_{limit}. In order to account for a given Factor of Safety, each column of collapse, burst and tension is changed according to

$$p_{allow} = \frac{p_{limit}}{FS} \tag{10-44}$$

This has to be done for each casing grade and size under consideration. The working pressures, p_c and p_b, defined by the "X" loading diagram (Figure 10-12), have to be less than or equal to the allowable pressures in each case.

$$p_c(x) \leq \left(p_{allow}\right)_c \tag{10-45}$$

$$p_b(x) \leq \left(p_{allow}\right)_b \tag{10-46}$$

Table 10-3 gives a range of safety factors that have been applied to each of the three types of casing loads.

The selection of a factor of safety may vary with the level of experience in a given area and the confidence level for each casing load prediction. Small factors of safety give lighter and, therefore, cheaper casing strings but also carry a higher risk element.

10.5.7 Effect of Tension on Casing Collapse

Tension along the axis of casing also affects the *collapse* resisting capability of casing. The collapse pressures given in the casing performance tables are based on a uniaxial state of stress assuming casing is subjected to external pressure only. This pressure creates circumferential or hoop stresses.

Casing in general can be loaded with a direct pull creating longitudinal stresses simultaneously with circumferential stress producing a biaxial state of stress. According to the von Mises criteria of failure, local yielding in casing will develop when the maximum energy of distortion created by biaxial stress state is equal to the maximum energy of distortion at yielding under a uniaxial state of stress. Equating these two energies gives

$$\sigma_{yld}^2 = \sigma_T^2 - \sigma_T\sigma_C + \sigma_C^2 \tag{10-47}$$

or

$$\left(\frac{\sigma_T}{\sigma_{yld}}\right)^2 - \left(\frac{\sigma_T}{\sigma_{yld}}\right)\left(\frac{\sigma_C}{\sigma_{yld}}\right) + \left(\frac{\sigma_C}{\sigma_{yld}}\right)^2 = 1 \tag{10-48}$$

where

σ_T = tension stress, psi
σ_C = collapse stress, psi
σ_{yld} = yield strength, psi

Equation (10-48) can also be written

$$\left(\frac{T}{T_{yld}}\right)^2 - \left(\frac{T}{T_{yld}}\right)\left(\frac{p_C}{P_{c,yld}}\right) + \left(\frac{p_C}{p_{c,yld}}\right)^2 = 1 \tag{10-49}$$

This equation establishes the relation between collapse pressure and tension, which produce yielding. The plot of this equation (Figure 10-14) shows how much the collapse pressure has to be derated for a given amount of direct pull on casing [5]. The portion of this diagram that is applicable to casing design is the portion in the fourth quadrant. These coordinates can be tabulated for accuracy and convenience.

For example, if the direct pull on 9⅝, 40 lb/ft, N80 casing at a given location is 110,000 lbs, then

$$\frac{T}{T_{yld}} = \frac{110,000}{916,000} = 0.12$$

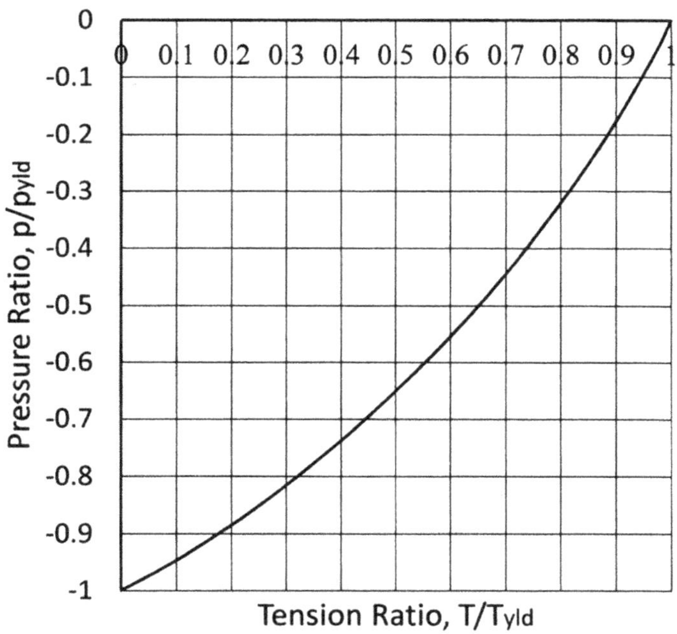

Figure 10-14. Yielding condition for biaxial state of stress.

According to Figure 10-13, the casing's resistance to collapse pressure has to be derated by a factor of 0.93. At this location, collapse strength is derated from 3090 psi to 2874 psi.

10.5.7.1 Tension Forces in Casing

The neutral point here is defined as the point in the casing where axial force or stress is zero. This point is different from the neutral point in drill collars, which is based on buckling. Tension calculations are usually based on casings being completely surrounded by drilling mud (inside and out). In reality this is not true but the calculation is conservative.

Based on this assumption, the internal force distribution along casing is defined by (see Figure 10-15).

$$F(x) = \gamma A(L - x) - L\gamma_m A \tag{10-50}$$

where x is measure from the top and $w = \gamma A$

$$F = w(L - x) - w_m L \tag{10-51}$$

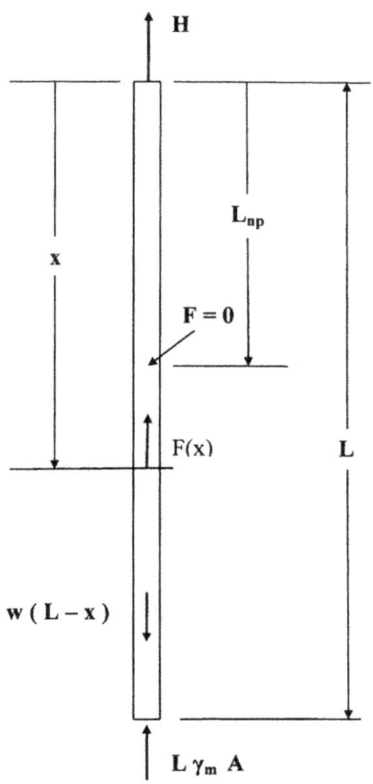

Figure 10-15. Internal force in vertical pipe.

This equation is used to define the magnitude of the true axial force in casing. Accordingly, at x = 0

$$F = W(BF) \tag{10-52}$$

and at x = L

$$F = -w_m L \tag{10-53}$$

The location of the point where F = 0 ($\sigma_a = 0$)

$$0 = \gamma AL - \gamma AL_{np} - \gamma_m AL \tag{10-54}$$

$$0 = (w - w_m)L - wL_{np}$$

$$L_{np} = \left(1 - \frac{w_m}{w}\right)L \tag{10-55}$$

$$L_{np} = (BF)L \tag{10-56}$$

If casing is empty and plugged at the bottom the force distribution would be completely different from what is predicted by Eq. (10-51). Equation (10-56) defines the location of the neutral point; the point of zero stress. The internal tension above the neutral point is determined from

$$F(x) = w\left(L_{np} - x\right) \tag{10-57}$$

Casing collapse pressure strength has to be derated to account for the effect of tension. The strength ellipse in Figure 10-14 relates the pressure ratio, $\frac{p}{p_{yld}}$, to the tension ratio, $\frac{T}{T_{yld}}$. The collapse strength is derated from p_{yld} to p, which becomes the allowable collapse pressure, p_a before FS is applied.

10.6 DESIGN OF 9⅝-IN. PRODUCTION CASING

As an example, consider the design of a production casing having the following specifications:

- Depth of well = 9500 ft
- Casing size = 9⅝ in.
- Mud weight is 12 ppg

The objective of the design is to select a distribution of casing having lowest weight and grade. The design starts at the bottom by considering collapse loading (Figure 10-16). The maximum collapse pressure load is shown as 5928 psi ($p = 0.052(12)9500 = 5928$ psi).

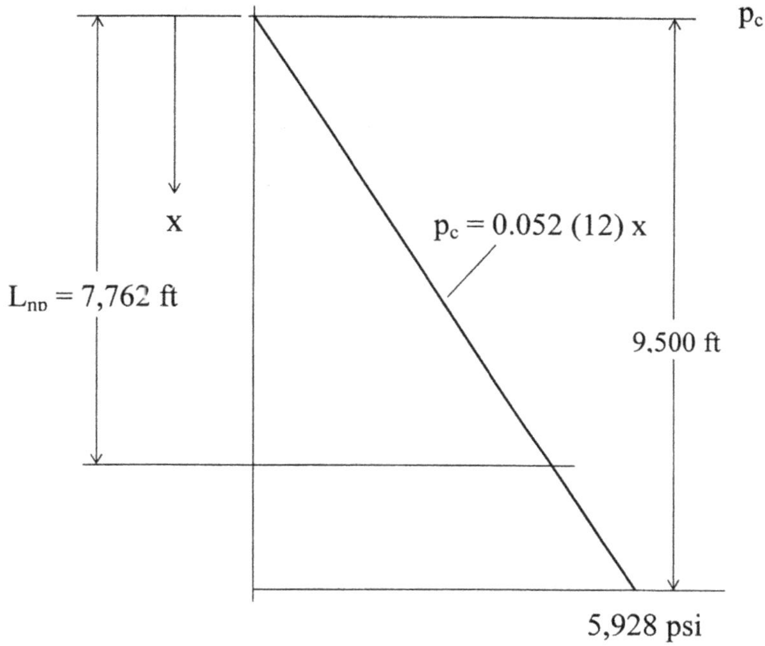

Figure 10-16. Collapse loading (9⅝ casing, 12 ppg mud).

10.6.1 Design without Factors of Safety

For the sake of simplicity, FS are not considered at this point. They will be included later. From Table 10-4, the lowest weight and grade combination having a collapse strength equal to or greater than 5928 psi is 47 lb/ft (S-95), which has a collapse strength of p_c = 7100 psi (burst strength is p_b = 8150 psi).

Each casing section will be tabulated as the design moves upward.

Section #1   9500 - x   47 lb/ft, (S 95)

At what point can casing weight and grade be reduced? Casing 43.5 lb/ft (S 95) has a collapse strength of p_c = 5600 psi (burst strength of p_b = 7510 psi).

The location at which the applied collapse pressure is 5600 psi is determined from the pressure load function.

$$p_c(x) = 0.052\ (12)\ x = 5600 \text{ psi}$$

$$x = \frac{5600}{0.052(12)} = 8{,}974 \text{ ft}$$

Table 10-4. Strength capability of 9⅝ in. casing (non API grades).
(Condensed from Lone Star Steel Casing Tables, ref 6)

Nominal Weight (lbs/ft)	Grade	Collapse Pressure (psi)	Burst Pressure (psi)	Body Tension (lb × 1000)	Coupling Tension (lb × 1000)
36.00	H40	1740	2540	410	
	K53	2020	3520	564	489
	S80	2980	3520	564	605
40.00	K55	2570	3950	630	561
	S80	4230	3950	630	694
	C75	2980	5390	859	694
	N80	3090	5750	916	737
	SS95	4230	5750	916	837
	S95	4230	6820	1088	858
	C95	3330	6820	1088	847
43.50	C75	3750	5930	942	776
	N80	3810	6330	1005	825
	SS95	5600	6330	1005	936
	S95	5600	7510	1193	959
	C95	4130	7510	1293	948
	P110	4430	8700	1381	1106
47.00	C75	4630	6440	1018	852
	N80	4750	6870	1086	905
	SS95	7100	6870	1086	1027
	S95	7100	8150	1289	1053
	C95	5080	8150	1289	1040
	P110	5310	9440	1493	1213

This locates the beginning of Section #2.

Section #1	9500 – 8974	47 (S 95)
Section #2	8974 – x	43.5 (S 95)

The next lowest weight is 40 (S 95) having collapse strength of 4230 psi (burst strength is 6820 psi). The beginning of the third section then is

$$x = \frac{4{,}230}{0.052(12)} = 6{,}778$$

The status of the casing design is now

Section #1	9500 – 8974	47 (S 95)
Section #2	8974 – 6778	43.5 (S 95)
Section #3	6778 – x	40 (S 95)

The next lightest casing weight is 36 (S 80) having a collapse strength of 2980 psi. This casing weight and grade is selected for Section #4, which starts at

$$x = \frac{2{,}980}{0.052(12)} = 4{,}776 \text{ ft}$$

The status of the casing design is now

Section #1	9500 – 8974	47 (S 95)
Section #2	8974 – 6778	43.5 (S 95)
Section #3	6778 – 4776	40 (S 95)
Section #4	4776 – x	36 (S 80)

Each of these sections is shown in Figure 10-17. This process could continue along the collapse load line but at this point burst loading becomes the important consideration. The two dots in Section #4 indicate that 36 (S 80) can

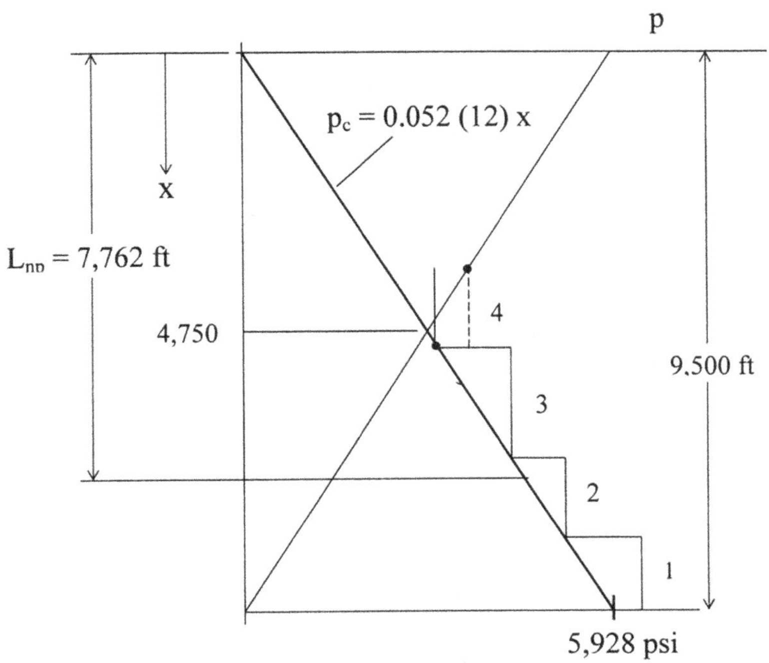

Figure 10-17. Collapse and burst loading.

support collapse loading from 4776 ft and upward. At the same time this casing can withstand the burst loading up to the upper dot.

The fourth casing weight and grade has a burst strength of p_b = 3520 psi. The length of this section is controlled by the burst loading line. The location where the burst load is equal to 3520 psi is determined as [using Eq. (10-34)]

$$x = L - \frac{p_b}{0.052(12)} = 9500 - \frac{3520}{0.052(12)} = 3{,}859$$

which locates the upper dot. From this point upward, burst pressure controls the design.

The casing string has four different casing weights (Figure 10-17):

Section #1	9500 – 8974 ft	47 (S 95)	
Section #2	8974 – 6778 ft	43.5 (S 95)	
Section #3	6778 – 4776 ft	40 (S 95)	
Section #4	4776 – 3859 ft	36 (S 80)	(burst controls)

Continuing upward on the *burst pressure* line (Figure 10-18), casing weight of 40 lb/ft gives several grade options. We choose grade S 95 because its burst strength is 6820 psi, which is greater than the maximum burst load (5928 psi) at the surface. Adding Section #5 completes the casing design.

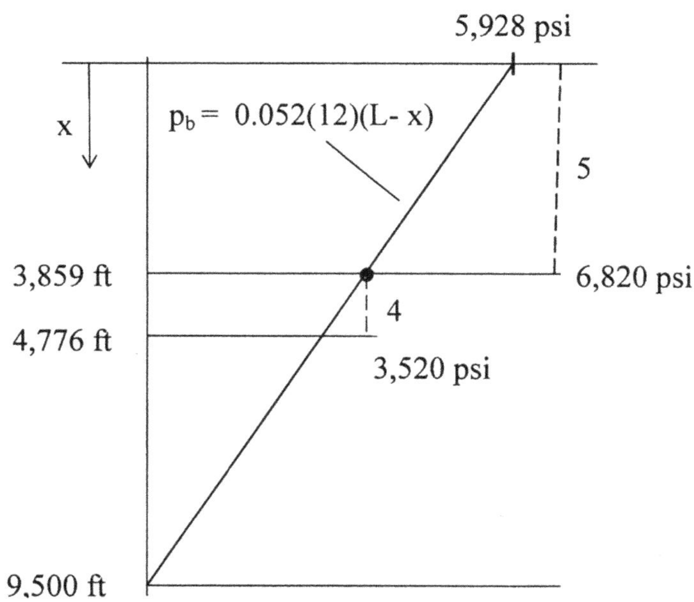

Figure 10-18. Burst load line and casing sections.

	Interval (ft)	Wt (Grade) (lb/ft)	Weight (lbs)	
Section #1	9500 – 8974 ft	47 (S 95)	24,722	
Section #2	8974 – 6778 ft	43.5 (S 95)	95,526	
Section #3	6778 – 4776 ft	40 (S 95)	80,080	
Section #4	4776 – 3859 ft	36 (S 80)	33,012	(burst controls)
Section #5	3859 – 0	40 (S 95)	154,360	(burst controls)
			Total 387,700 lbs	

10.6.2 Effect of Tension

The neutral point in this example is determined [Eq. (10-56)] to be L_{np} = 0.817(9500) = 7762 ft from the top. From the diagram in Figure 10-17, we note that the starting point (x = 6778 ft) for Section 3 is under tension. The tensile force [Eq. 10-57)] at the bottom of Section #3 is

$$T = (7762 - 6778)\,43.5 = 42{,}804 \text{ lbs}$$

With T_{yld} = 1,088,000 lb for 9⅝" 40 (S 95) casing

$$\frac{T}{T_{yld}} = \frac{42{,}804}{1{,}088{,}000} = 0.0393$$

The collapse strength of 40 (S 95) at this point must be derated (Figure 10-14) by a factor of ~0.98. The derated collapse strength becomes 4145 psi (from 4230 psi). This has the effect of moving the starting point for Section #3 up to

$$x = \frac{4{,}145}{0.052(12)} = 6{,}643 \text{ ft}$$

instead of at 6778 ft.

The starting point for section #4 is also affected by tension. Earlier calculations show Section # 4 starts at 4776 ft. The magnitude of the tension at 4776 ft is

$$T = (7{,}762 - 6{,}643)43.5 + (6{,}643 - 4{,}776)40 = 123{,}357 \text{ lbs}$$

The tension ratio for the beginning of section 4 is

$$T = \frac{123{,}357}{564{,}000} = 0.2187$$

From Figure 10-13

$$\frac{p_c}{p_{c,yield}} = 0.86$$

which means that yield strength of 36 (S 80) in collapse must be derated from 2980 psi to 2563 psi because of tension.

This effect elevates the beginning of Section #4 from 4776 to

$$x = \frac{2{,}563}{12(0.052)} = 4{,}107 \text{ ft}$$

Burst pressure strength is not affected by tension so the remaining sections are unchanged. The updated design with the effects of tension is

	Interval (ft)	Wt (Grade) (lb/ft)	Weight (lbs)	
Section #1	9500 – 8974 ft	47 (S 95)	24,722	
Section #2	8974 – 6643 ft	43.5 (S 95)	101,398	
Section #3	6643 – 4107 ft	40 (S 95)	101,440	
Section #4	4107 – 3859 ft	36 (S 80)	8928	(burst controls)
Section #5	3859 – 0	40 (S 95)	154,360	(burst controls)
			Total 390,848 lbs	

10.6.3 Check Tension Load in Casing Body and Couplings

Strength for both cases will be based on applied tension at the surface for the sake of simplicity. In reality, tension strength should be checked for each casing weight and grade used in the design. The tension at the surface is [see Eq. (10-52)] is

$$T = BF \sum w_i L_i$$

$$T = 0.817\ (390{,}848) = 319{,}323 \text{ lb}$$

The tensile strength of 9⅝ 40 (S95) is (Table 10-3) 1,088,000 lbs. The factor of safety in tension is, therefore,

$$FS = \frac{1{,}088{,}000}{319{,}323} = 3.41$$

which is well above the required FS of 1.8.

The tension strength of LT & C for 40 (S95) is 858,000 lb. The factor of safety of this coupling is

$$F.S. = \frac{858{,}000}{319{,}323} = 2.69$$

This is well above the required factor of safety of 1.8 for tension.

10.6.4 Design with Factors of Safety

Factors of Safety will now be applied to the design assuming

$n_1 = 1.125$ (collapse)
$n_2 = 1.125$ (burst)
$n_3 = 1.8$ (tension)

Starting at the bottom with 47 (S95) and applying the FS of 1.125, the allowable collapse strength becomes

$$p_a = \frac{7{,}100}{1.125} = 6{,}311 \text{ psi}$$

This number is greater that the applied collapse pressure (5928 psi) at x = L = 9500 ft so the design starts with 47 (S 95).

Beginning of Section #2
Using 43.5 (S95) having a yield collapse pressure of 5600 psi

$$p_a = \frac{5{,}600}{1.125} = 4{,}978 \text{ psi}$$

The lower end of Section #2 is determined as

$$x = \frac{4{,}978}{0.052(12)} = 7{,}978 \text{ ft}$$

Beginning of Section #3
Using 40 (S95) having a yield collapse pressure of 4230 psi

$$p_a = \frac{4{,}230}{1.125} = 3{,}802 \text{ psi}$$

The lower end of Section #3 is determined as

$$x = \frac{3{,}802}{0.052(12)} = 6{,}093 \text{ ft}$$

Beginning of Section #4
Using 36 (S80) having a yield collapse pressure of 2980 psi

$$p_a = \frac{2{,}980}{1.125} = 2{,}649 \text{ psi}$$

The lower end of Section #4 is determined as

$$x = \frac{2{,}649}{0.052(12)} = 4{,}545 \text{ ft}$$

These four casing sections are displayed in Figure 10-19. Note the change in casing design between Figures 10-17 and 10-19 as a result of applying the factors of safety. The casing intervals are

Section #1	9500 – 7978 ft	47 (S95)	
Section #2	7978 – 6093 ft	43.5 (S95)	
Section #3	6093 – 4545 ft	40 (S95)	
Section #4	4,545 –	36 (S80)	(burst controls)

The burst strength of 36 (S80) is 3520 psi. After applying the FS in burst, the allowable burst pressure becomes

$$p_a = \frac{3{,}520}{1.125} = 3{,}129 \text{ psi}$$

The applied burst pressure at 4545 ft is

$$p = 0.052(12)[9.500 - 4545] = 3092 \text{ psi}$$

The allowable burst pressure is greater than the applied (3129 > 3092) burst pressure. The *top* of Section #4 is located at

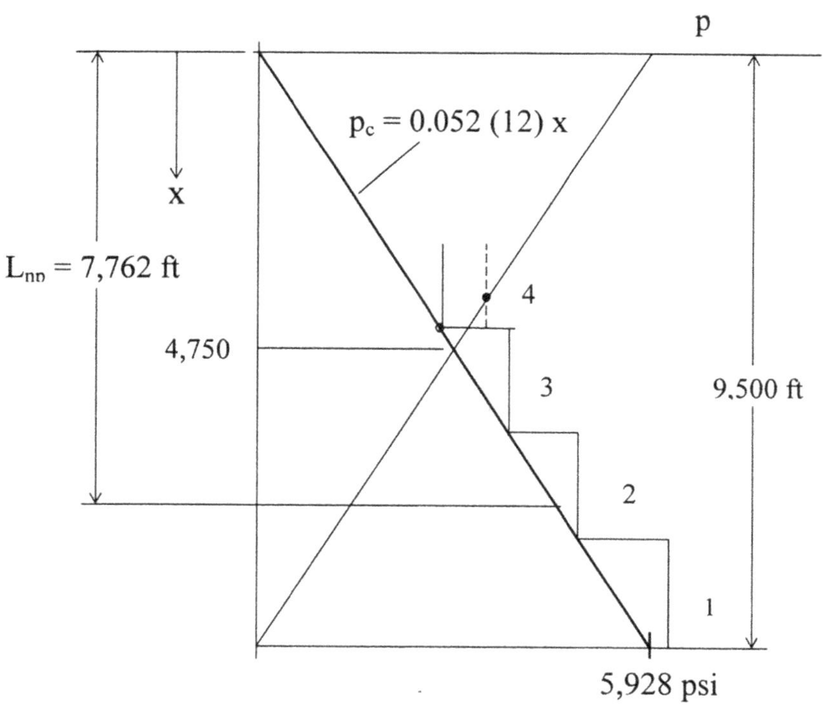

Figure 10-19. Casing design adjusted for factors of safety.

$$x = 9{,}500 - \frac{3{,}129}{0.052(12)} = 4{,}486 \text{ ft}$$

The length of Section #4 is 4545 – 4486 = 59 ft. Because of the logistic of handling only 59 ft of a different casing weight and grade, we elect to extend Section #3 and begin Section #4 based on the burst strength of 40 (S95).

Extension of Section #3
Using 40 (S95) having a yield burst pressure rating of 6820 psi

$$p_a = \frac{6{,}820}{1.125} = 6{,}062 \text{ psi}$$

The upper end of Section #4 is located at

$$x = 9{,}500 - \frac{6{,}062}{0.052(12)} = 215 \text{ ft}$$

Section # 4
Casing 43.5 (S95) has a burst strength of 7510 psi. Applying the FS gives an allowable pressure of 6676 psi, which is greater than the maximum expected burst pressure (5928 psi) at the surface.

The casing design now becomes (see Figure 10-19)

	Interval	wt (grade)	weight	
Section #1	9,500 – 7,978 ft	47 (S95)	71,534	
Section #2	7,978 – 6,093 ft	43.5 (S95)	81,997	
Section #3	6,093 – 215 ft	40 (S95)	235,120	
Section #4	215 – 0 ft	43.5 (S95)	9,353	(burst controls)
		Total	398,004 lbs	

Effect of Tension
The bottom of Section #3 is under a tension of

$$T = (7762 - 6093)43.5 = 72{,}602 \text{ lbs}$$

The tension and pressure ratios are

$$\frac{T}{T_{yld}} = \frac{72{,}602}{1{,}088{,}000} = 0.067$$

$$\frac{p}{p_{yld}} = 0.97 \text{ (from Figure 10-14)}$$

giving an allowable collapse pressure of p_a =0.97(4230) = 4103 psi. Applying the FS (1.125) gives p_a = 3647 psi.

The depth at which the allowable collapse pressure and applied pressures are the same is

$$x = \frac{3{,}647}{0.052(12)} = 5{,}845 \text{ ft}$$

The effect of tension is to raise the beginning of Section #3 from 6093 ft to 5845 ft. Burst strength is not affected by tension so the rest of the design stays the same. The final design is

	Interval	wt (grade)	weight	
Section #1	9500 – 7978 ft	47 (S95)	71,534	
Section #2	7978 – 5845 ft	43.5 (S95)	81,997	
Section #3	5845 – 215 ft	40 (S95)	235,120	
Section #4	215 – 0 ft	43.5 (S95)	9,353	(burst controls)
			Total 398,004 lbs	(air weight)

Check Tension and Coupling Strengths
The tension at the surface is [see Eq. (5-16)] is

$$T = BF\sum w_i L_i$$

$$T = 0.817\ (398{,}004) = 325{,}169\ \text{lb}$$

The factor of safety in tension for top section is

$$FS = \frac{1{,}005{,}000}{325{,}169} = 3.09$$

The tension strength of LT & C for 40 (S95) is 936,000 lb. The factor of safety of this coupling is

$$F.S. = \frac{936{,}000}{325{,}169} = 2.88$$

Both are well above the required factor of safety for tension (1.8).

10.7 DESIGN OF 5½ IN PRODUCTION CASING

Consider the following casing program.

- 13⅜ in. surface casing string
- 9⅝ in. intermediate casing string; set at 2000 ft
- 5½ in. production casing string; to be set at 10,000 ft.

The *collapse pressure* loadline (Figure 10-20) is based on an external pressure gradient of 12 ppg mud and an empty casing,

$$p_c = 0.052(12)x$$

The collapse pressure line passes through the zero pressure point at the surface (x = 0) and

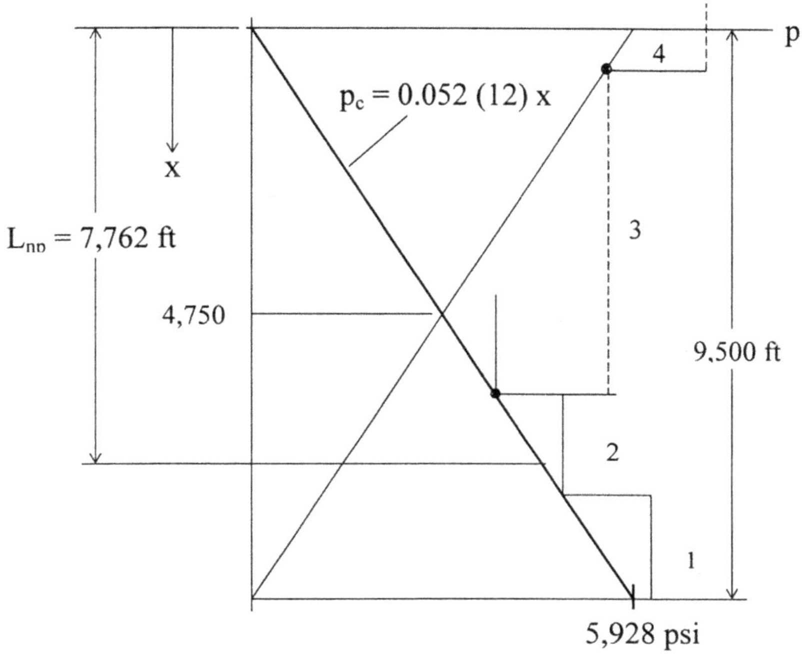

Figure 10-20 Modified casing design with new section #3.

$$p_c = 0.052(12)10{,}000 = 6{,}240 \text{ psi}$$

at x = 10,000 ft.

The *burst pressure* load line is based on formation gas pressure *equivalent* to 11 ppg mud inside the casing counter balanced by an external pressure equivalent to 12 ppg mud on the outside of the casing.

$$p_b = 0.052(11)10{,}000 - 0.052(12)x$$

The burst pressure line passes through $p_b = 5{,}720$ psi at the surface (x = 0) and say

$$p_b = 0.052(11)10{,}000 - 0.052(12)8{,}000 = 728 \text{ psi}$$

at x = 8000 ft.

These two lines establish the X pressure diagram or collapse-burst pressure loads on the casing.

Factors of safety for the design are

$n_1 = 1.125$ (collapse)
$n_2 = 1.125$ (burst)
$n_3 = 1.8$ (tension)

The factors of safety reduce the allowable pressure for each pressure mode to an allowable pressure according to

$$p_a = \frac{P_{yld}}{n}$$

This has the effect of reducing the numbers in each performance column by the appropriate factor of safety.

Casing grade and weight selection starts at the bottom where collapse pressure loading is the greatest. From Table 10-5, 17.00 (SS95) has collapse strength of 8580 psi. Applying the FS = 1.125 gives an allowable pressure of 8580/1.125 = 7627 psi. This collapse design strength is shown in Figure 10-21 by the solid vertical line intersecting the pressure axis at p = 7627 psi; 17.0 (C90) is not strong enough when the F.S. is applied.

Section #1	10,000 – x	17.00 (SS95)

The next lightest 5½ in casing is 15.5 lb/ft, (K55). Its collapse strength is 4040 psi and after applying the factor of safety (1.125) its design collapse strength is 3591 psi. This collapse pressure is developed at a depth of

$$x = \frac{3{,}591}{0.052(12)} = 5{,}755 \text{ ft}$$

which is the beginning of Section #2. The casing weight and grade can be reduced from 17.0 lb/ft (SS95) to 15.5 lb/ft (K55) at this point.

Section #1	10,000 – 5,755 ft	17.00 (SS95)
Section #2	5,755 – x	15.5 (K55)

Table 10-5. Strength capability of 5.5 in. casing (non API grades).
(Condensed from Lone Star Steel Casing Properties Tables, ref 6)

Nominal Weight (lbs/ft)	Grade	Collapse Pressure (psi)	Burst Pressure (psi)	Body Tension (lb × 1000)	Coupling Tension (lb × 1000)
15.5	K55	4040	4810	248	239
17.0	K55	4910	5320	273	272
	C75	6040	7250	372	327
	N80	6280	7740	397	348
	C90	6730	8710	447	356
	SS95	8580	7740	397	356
	S95	6940	9190	471	392
	S105	8980	9190	471	392
	P110	7480	10640	546	445

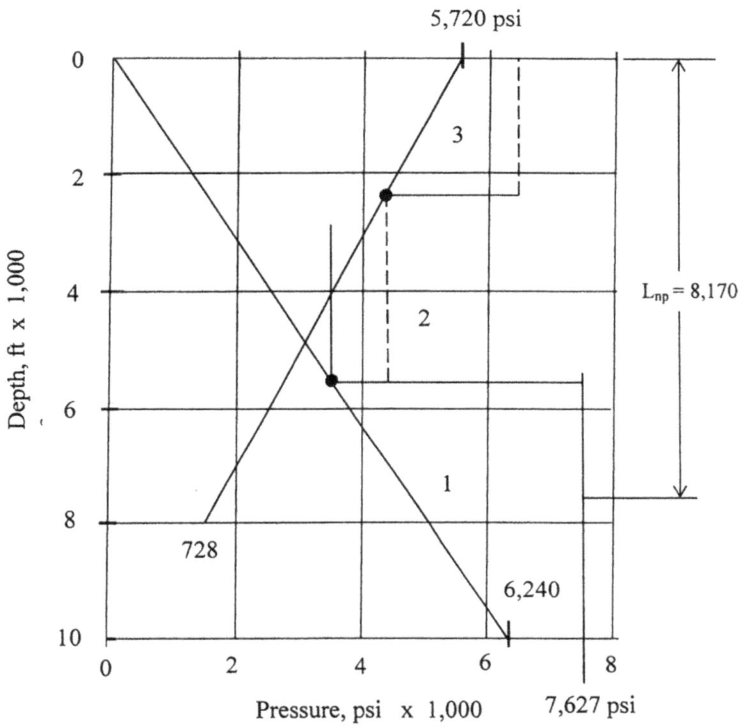

Figure 10-21. Casing design.

The burst strength of the 15.5 (K55) casing is 4810 psi and after applying the factor of safety (1.125), gives allowable burst strength of 4275 psi. This burst strength line is shown in Figure 10-20 as a vertical dashed line with pressure coordinate of 4275 psi. This pressure line intersects the burst pressure load line at

$$4275 = 0.052(11)10{,}000 - 0.052\ (12)\ x$$
$$x = 2316 \text{ ft}$$

identified by the upper dot. At this depth, burst loading dictates the design and casing weight and grade have to be increased.

Section #1	10,000 – 5,755 ft	17.0 (S95)
Section #2	5,755 – 2,316 ft	15.5 (K55)

According to casing properties tables, 17.0 lb/ft (C75) has a burst strength of 7250 psi and an allowable design burst pressure of 7250/1.125 = 6444 psi, which is greater than the maximum burst pressure load of 5720 psi at the surface.

The casing design contains three sections.

Section #1	10,000 – 5755 ft	17.0 (S95)
Section #2	5755 – 2316 ft	15.5 (K55)
Section #3	2316 – 0	17.0 (C75)

The casing design must now be checked for the effects of tension at the bottom of Section #2.

Effect of Tension

The neutral point in the casing string determined by

$$x = L\ BF$$

or

$$x = 10{,}000\ (0.817) = 8170\ \text{ft}$$

assuming uniform pipe size.

The beginning of Section #2 is located above the neutral point. Its tension is

$$T = (8170 - 5755)17 = 41{,}055\ \text{lbs}$$

The tension ratio at this point is

$$\frac{T}{T_{yld}} = \frac{41{,}055}{248{,}000} = 0.166$$

The collapse pressure strength has been derated to

$$\frac{p_c}{p_{c,yld}} = 0.91$$

$$p_c = 0.91(4{,}040) = 3{,}676\ \text{psi}$$

Applying a factor of safety of 1.125 to the collapse strength of 15.5 (K55), reduces the allowable collapse pressure from 3676 psi to 3268 psi.

$$x = \frac{3{,}268}{0.052(12)} = 5{,}237\ \text{ft}$$

The start of Section #2 moves up from 5755 ft to 5237 ft. The final design becomes

Section #3	0 – 2,316 ft	17.0 (C75)	39,372
Section #2	2,316 – 5,237 ft	15.5 (K55)	45,275
Section #1	5,237 – 10,000 ft	17.0 (S95)	80,971
		Total	165,618 lbs

Tension has no effect on the burst strength, so the remaining portion of the design is unchanged Figure 10-22.

Check Body Strength and Coupling Strength
The total buoyed weight of the casing string is

$$T = 0.817\ (165{,}618) = 135{,}310 \text{ lbs}$$

The yield strength of the top casing (17 C75) is 372,000 lbs.

$$FS = \frac{372{,}000}{135{,}310} = 2.75 \text{ (casing body)}$$

The yield strength of the top coupling is 327,000 lbs.

$$FS = \frac{327{,}000}{135{,}310} = 2.42$$

Both FS are greater than the required 1.8 factor of safety in tension.

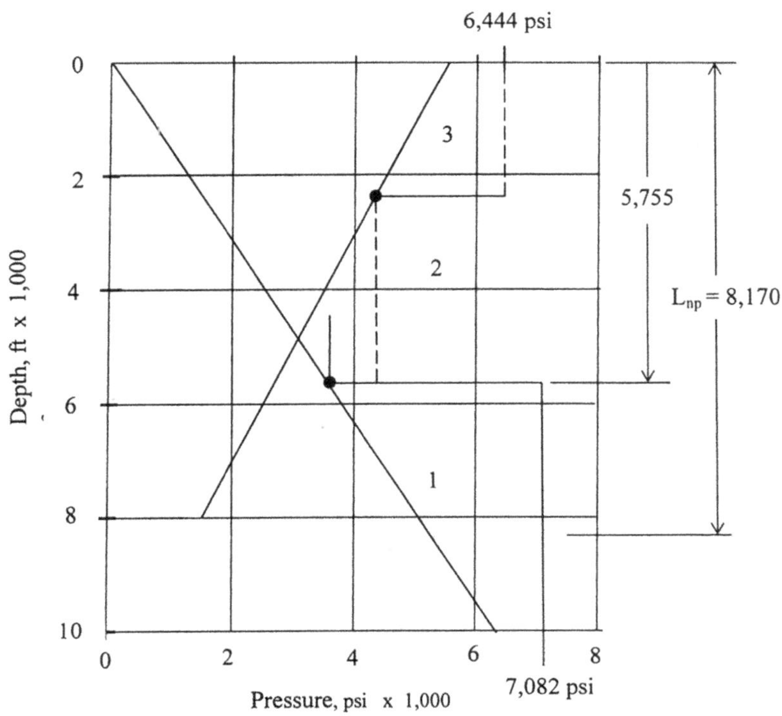

Figure 10-22. Adjusting casing design to account for effect of tension on collapse.

10.7.1 Summary of Equations

$$p_{cs} = p_{sp} + 0.052\gamma_m L \quad \text{(used in leak off test)}$$

$$\left(\frac{dp}{dx}\right)_F = 0.052\gamma_F \quad \text{(psi/ft)} \quad \text{(fracture gradient)}$$

$$\left(\frac{dp}{dx}\right)_{mud} = 0.052\gamma_m \quad \text{(drilling fluid pressure gradient)}$$

$$\left(\frac{dp}{dx}\right)_{formation} = 0.052\gamma_f \quad \text{(formation pressure gradient)}$$

$$\left(\frac{\sigma_a}{\sigma_{yld}}\right)^2 - \frac{\sigma_a}{\sigma_{yld}}\frac{\sigma_\theta}{\sigma_{yld}} + \left(\frac{\sigma_\theta}{\sigma_{yld}}\right)^2 \geq 1 \quad \text{(von Mises criteria of failure)}$$

Design Loads on Casing

$$p(x) = 0.052\gamma_o x \quad \text{(collapse pressure loading)}$$

$$p_b = 0.052\gamma_m(L - x) \quad \text{(burst pressure loading)}$$

$$\left(\frac{T}{T_{yld}}\right)^2 - \left(\frac{T}{T_{yld}}\right)\left(\frac{p_C}{P_{c,yld}}\right) + \left(\frac{p_C}{p_{c,yld}}\right)^2 = 1 \quad \text{(von Mises criteria in terms of loads)}$$

$$L_{np} = (BF)L \quad \text{(location of neutral point in casing)}$$

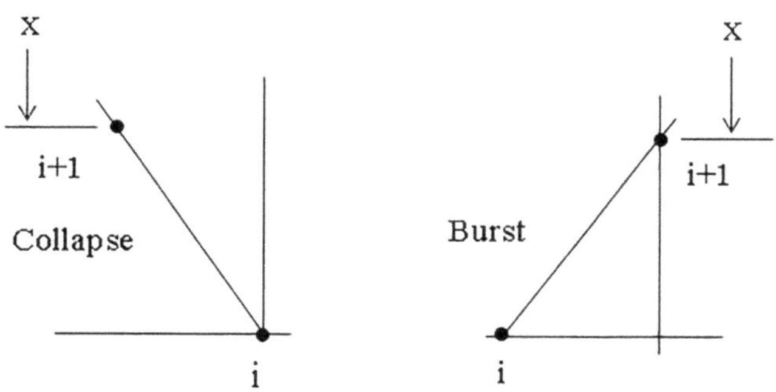

Beginning of next (i + 1) casing (Collapse mode)

$$p_c = 0.052\gamma_m x \qquad x = \frac{p_c}{0.052\gamma_m}$$

Beginning of next (i + 1) casing (Burst mode)

$$p_b = 0.052\gamma_m(L - x) \qquad x = L - \frac{p_b}{0.052\gamma_m}$$

10.8 PROBLEMS

1. A leak off test is conducted after a casing is set at 10,400 ft. The test showed that the open formation began to fracture when shut-in standpipe pressure reached 1800 psi and mud weight is 13.0 ppg. Determine the maximum allowable mud weight for drilling beyond the casing shoe.

2. A leak off test is conducted at 8000 ft with 12 ppg in the hole. Formation breakdown occurred at 1575 psi standpipe pressure. What is the formation fracture gradient at the casing shoe? What is the maximum allowable mud weight downhole before setting another casing string?

3. Surface casing was set to 2800 ft with drilling mud weight of 10 ppg. A leak off test showed that formation fracture pressure is equivalent to 11.8 ppg mud. Determine the standpipe pressure when the formation fracture was initiated in the leak off test.

4. The fracture pressure gradient at a casing shoe is $(PG)_F = 0.5709$ psi/ft. What is the maximum allowable formation pressure gradient downhole?

5. Using Lame's equation, calculate and plot σ_r and σ_θ across the thickness of 9.625" (47.00 lb/ft) casing when subjected to an *internal* pressure of 5000 psi. Note: ID = 8.681" and t = 0.472". Take 10 incremental steps across the thickness. Plot the results.

6. Calculate the hoop stress assuming thin wall theory for problem 5. Over-lay the resulting hoop stress on the σ_θ plot in problem 5. Discuss the difference.

7. Using Bresse's equation, determine the critical buckling pressure on 9 5/8 (40 lb/ft) K55 Casing with wall thickness of 0.395 inches. Assume E = 30 × 106 and ν = 0.25. Which will happen first, elastic collapse or material yielding?

8. Design a 7 in. production casing string for the following conditions:
   - 12,000 ft well depth
   - Reservoir pressure equivalent to 14 ppg mud
   - Mud weight is 14.3 ppg.

   Do not consider safety factors or the effect of tension on collapse strength.

9. Repeat problem 7 assuming the following safety factors.
   - Collapse FS = 1.125
   - Burst FS = 1.125

   Also include the effect of tension on derating casing in collapse.

10.9 LIST OF REFERENCES

1. Technical Report on Equations and Calculations for Casing, Tubing, and Line Pipe Used as Casing or Tubing; and Performance Properties Tables for Casing and Tubing; ANSI/API Technical Report 5C3; First Edition, December 2008; TN879.53 A64.
2. Lame, "Lecons sur la theorie --- de l'elasticite", Paris 1852; also see S. Timoshenko and J. N. Goodier, *Theory of Elasticity*, 2nd Ed. McGraw-Hill, 1951.
3. William Riley, Leroy Sturges and Don Morris; *Mechanics of Materials*, 6 Ed., John Wiley & Sons, Inc. 2007.
4. Buckling of rings was solved by M. Bresse, *Cours de mecanique appliqué*, Paris, Part I, p. 334, 1866 (also see Fred B. Seely and James O. Smith; *Advanced Mechanics of Materials*, James Wiley & Sons, (1955), p. 614).
5. S. Timoshenko and S. Woinowsky-Krieger; *Theory of Plates and Shells*, McGraw-Hill Book Co. (1959).
6. Holmquist, J. L. and Nadai, A: "A theoretical and Experimental Approach to the Problem of Collapse of Deep Well Casing", *API Drilling and Production Practice*, 1939.
7. Lone Star Steel Company; Casing and Tubing Technical Data, 1984, Dallas, TX.

Chapter 11

WELL CONTROL

Well control refers to drilling techniques that keep formation fluids from erupting at the surface. There are three levels of well control: 1) drilling mud, 2) annular preventer and 3) mechanical rams within a blowout preventer stack (BOP). Maintaining mud pressure slightly higher than formation pressures is non-disruptive and is the first line of defense against a blow-out. The annular preventer is used to shut-in a well and to remove a kick. Mechanical rams (pipe, shear, blind) are used as a last resort. This chapter covers the application of these well control techniques.

Figure 11-1. The Lucas gusher at Spindle Top, Texas.

On October 27, 1900, after several tries, Captain Anthony F. Lucus, an experienced engineer and salt mine driller in Louisiana, selected a spot around the dome (or spindle) and began drilling. On Christmas, 1900, the drilling crew had reached a depth of 880 ft. On New Years Day, 1901, the drilling crew drilled down to 1020 ft within a week. On January 10, 1901, the well began to flow and erupted into a gusher (Figure 11-1). Once under control, the oil well produced nearly 100,000 barrels per day, more than the total national oil production at that time. The well, which became known as "Spindle Top", set the stage for modern day drilling and production.

It is apparent that the consequences of a well blow out are not good. There is loss of the well, equipment, time and possibly loss of life and personnel injury. There is also damage to the environment and perceived losses in reputation and good will. Since the early days of rotary drilling much has been learned about the early detection and control of formation pressures while drilling.

11.1 DRILLING MUD, FORMATION AND FRACTURE PRESSURES

Formation pressure refers to the pressure of oil, gas, or other fluids that naturally occur in the pores of geologic formations. Formation pressure can be normal or abnormal.

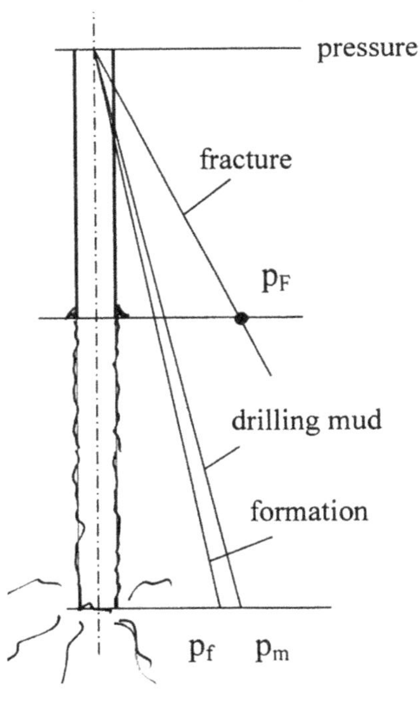

Figure 11-2. Normal pressure gradients.

Pressure gradient is a useful term to indicate the rate at which hydrostatic pressure increases with depth. It is expressed in terms of psi per ft of depth and is equal to

$$\frac{dp}{dx} = 0.052\gamma$$

For example, fresh water has a density of $\gamma = 8.34$ ppg. Its pressure gradient is 0.443 psi/ft.

Pressure gradients may vary with depth, especially formation pressure. Fracture pressure gradient depends on rock type and drilling location; Gulf coast, Rocky Mountain, etc. In an ideal situation, each of these pressure lines can be represented as shown in Figure 11-2. The slope of each pressure line is its pressure gradient.

11.1.1 Mud Pressure

Mud weight reduces rate of penetration exponentially. The practical selection of mud weight, therefore, depends on formation pressure, fracture gradient of the formation and drilling rate. In normal pressure areas, mud weight (mud pressure gradient) can be kept constant.

Hydrostatic pressure is the result of gravity and is always in the drilling mud whether or not the pumps are running. Total pressure is the algebraic sum of hydrostatic and dynamic pressures. This means that both hydrostatic and dynamic pressures can be analyzed separately and then added together to determine pressure at any point in the circulating system.

Hydrostatic mud pressure in the well bore is determined from

$$p(x) = 0.052\gamma_m x \tag{11-1}$$

where

p = local hydrostatic pressure, psi
γ_m = mud density, ppg (note: 1 ft^3 = 7.48 gal)
x = vertical depth, ft

For example, the hydrostatic pressure at a true vertical depth of 10,000 ft, assuming 12 ppg mud is

$$p = 0.052(12)(10{,}000) = 6240 \text{ psi}$$

The pressure gradient is the rate at which pressure increases per foot hole depth.

$$\frac{dp}{dx} = 0.052\gamma_m \quad \text{(psi per foot of hole depth)} \tag{11-2}$$

For 12 ppg mud

$$\frac{dp}{dx} = 0.624 \text{ psi per ft}$$

11.1.2 Normal Formation Pressure

Normal formation pressure varies hydrostatically from the surface, depending on the density of ground water. In the Gulf Coast area, *normal* formation pressure is based on the density of sea water (9.0 ppg; pressure gradient, PG = 0.468 psi/ft). In the Rocky Mountains, *normal* formation pressure is based on fresh water (8.34 ppg; pressure gradient, PG = 0.443 psi/ft). Formation pressure poses no threat to drilling as long as drilling mud density is equal to or greater the density of the formation fluids. However, if drilling mud density is too high, the formation can fracture or breakdown causing lost circulation and possible loss of well pressure control.

11.1.3 Abnormal Formation Pressure

Abnormal formation pressure is formation pressure that is not normal. Under this condition, mud weight has to be adjusted as drilling progresses to control the higher than normal formation pressures (Figure 11-3). High pressure zones develop when oil and gas are trapped by impermeable layers of rock and usually occur in faulted, folded or under rock strata that has zero permeability. Abnormal pressure zones can be encountered prior to reaching the reservoir.

Formations may contain pockets of abnormal formation pressure. In many cases these are petroleum bearing formations, which prompted drilling in the area. The location of these formations is usually known in production drilling. In wild cat drilling, they are not. Close monitoring of drilling mud returns is important in either case because mud returns is an indication of an influx of high pressure fluids coming into the well bore.

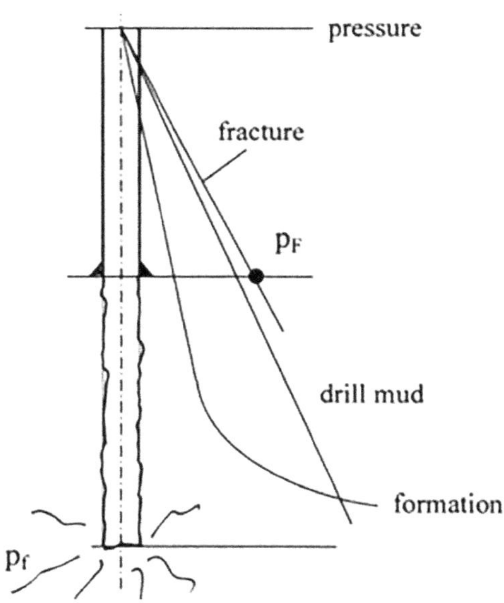

Figure 11-3. Abnormal pressure gradient.

11.1.4 Fracture Pressure

Fracture pressure is the pressure level in the well bore that causes formations to breakdown or fracture. Fracture pressure at a casing shoe establishes the fracture gradient line. It is important that the drilling mud pressure line always stay within the fracture gradient line to prevent breakdown in the formation at the casing shoe. A formation breakdown can lead to gas migration up the outside of the last casing or can cause lost circulation, both are highly undesirable while drilling.

11.2 INTENSITY OF A WELL KICK

When a well is invaded by formation fluids (gas, oil, water) formation pressure is greater than the pressure of the drilling mud. The influx of these fluids reduces the weight of the mud column, creating a greater pressure imbalance and allowing more and more formation fluids to enter the well bore. This fluid invasion is called a "well kick". The intensity of a well kick depends on three factors:

1. Difference between formation pressure and well bore pressure.
2. Length of the producing formation exposed to the well bore
3. Permeability of the producing formation.

11.2.1 Differential Pressure

The difference between formation pressure and well bore pressure, called pressure differential, is the key to well control. This was apparently recognized at the Spindletop well and history indicates that water was weighted up by having cattle muddy a nearby pond for use in the well, giving rise to the term "drilling mud". Today, drilling muds are very sophisticated containing barite to weight the drilling mud solution and bentonite to add viscosity for supporting bit cuttings.

While mud weight keeps formation fluids under control, it affects the rate of penetration of the drill bit. It is well known that high pressure differentials can reduce ROP by a factor of 10. High pressure differentials can also lead to stuck pipe.

11.2.2 Length of the Exposed Producing Formation

Oil and gas reservoirs are incased by an impermeable cap rock, which stopped movement of the hydrocarbons over millions of years. On the top side of the cap rock formation pressures can be normal. On the bottom side of the cap rock, pressure and temperature of the trapped hydrocarbons are much higher. This pressure differential across the cap rock causes a huge increase in drilling rate once the cap rock has been penetrated. The sudden increase in drilling rate can be detected immediately at the rig floor. The well bore is usually drilled 2 ft to 5 ft beyond a drilling break in order to check the occurrence of a well kick. Obviously, the length of the formation exposure affects the magnitude of the kick.

11.2.3 Permeability of the Producing Formation

Permeability is a property of rock that defines how fast fluid flows through a given rock when subjected to a pressure differential. Standard tests are conducted in a laboratory to determine rock permeability. A high permeability means high production rates. Low permeability means small rates of flow. A producing formation having high permeability produces a greater kick than a smaller permeability. While it is difficult to know permeability ahead of time, extra care should be taken in approaching production formations where high permeability is suspected.

11.3 DETECTING A WELL KICK WHILE DRILLING

There are a few indicators that a well is taking a kick during drilling operations [1]. Each give an immediate indication that the well bore is taking in formation fluids. Current technology and equipment is adequate to control the kick and avoid the serious consequences of a well out of control. The indicators are:

1. Drilling break, increase in ROP
2. Increase in rate of flow coming out of the well
3. Increase in mud volume in the mud pit
4. Decrease in pump pressure

A well kick can occur at any time, so it is important to be vigilant in monitoring these indicators. Early detection of a kick is important as large kicks are more difficult to control.

11.3.1 Drilling Break

Drilling rate is affected considerably by the pressure balance condition between drilling mud in the well bore and fluids in the pores of the formation. Drilling plans usually call for a slightly overbalance pressure condition equivalent to about ½ ppg (260 psi at 10,000 ft). This pressure is ample to keep formation fluids out of the well bore. When formation pressures are known precisely, overbalance pressure differences can be adjusted downward for maximum rate of penetration.

The well bore will take on formation fluids or gases when formation pressure is greater that well bore pressure. Even though rate of penetration is highest under this under balance condition, the invading formation fluids can dilute the density of the drilling mud and produce an unstable flow condition driven by formation fluids. If this unstable condition goes unchecked, the well will blow out.

11.3.2 Increase in Flow Rate

Drilling into a reservoir rock produces an immediate increase in rate of penetration. This increase in ROP can be as high as a factor of 2 or 3. It is a good idea to check all drilling breaks for flow, by picking up the drillstring and shutting off the pumps. If drilling mud continues to flow out of the flow line the well is taking a kick and the well should be shut in.

This is a simple matter of the continuity of flow. If formation fluids are invading the well bore, the output will be greater than flow produced by the surface pumps. This can be detected by flow meters in the output flow lines. If the invading fluids are hydrocarbons, the effect is to lighten the mud column, which causes more and more hydrocarbon to flow into the wellbore allowing the kick to become progressively larger.

11.3.3 Increase in Mud Pit Volume

An increase in the mud tank volume indicates that more and more mud is being displaced out of the well bore, as the invading fluid makes its way into the well bore. This is another sign that formation pressure is greater than the mud column pressure. A measure of the effect of gas cutting on the stability of a mud column, is the increase in mud pit volume. If the increase stabilizes, gas cutting is not a concern.

The amount of pressure reduction is determined by the increase in mud pit volume (see Figure 11-4) as follows:

$$p_1 = 0.052\gamma L_1 \tag{11-3}$$

$$p_2 = 0.052\gamma L_2 \tag{11-4}$$

$$p_1 - p_2 = 0.052\gamma (L_1 - L_2) \tag{11-5}$$

$$\Delta p = 0.052\gamma L_1 \frac{L_1 - L_2}{L_1} \tag{11-6}$$

$$\Delta p = 0.052\gamma L_1 \frac{\Delta Vol}{Vol} \tag{11-7}$$

where

(Vol) = total annulus volume
ΔVol = increase in mud pit volume

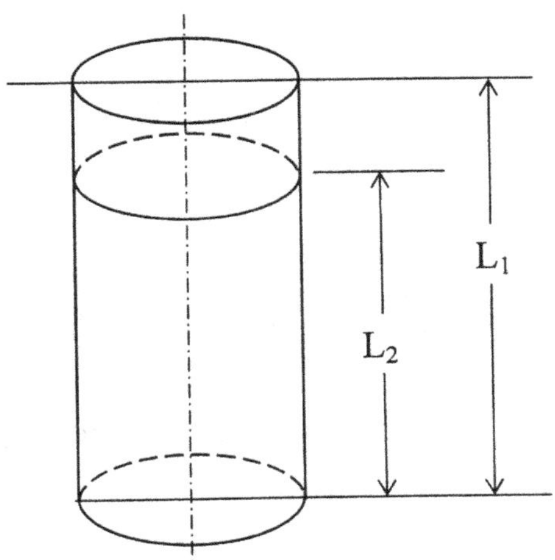

Figure 11-4. Volume of drilling mud in the well.

If the mud column becomes unstable, the annular preventer is closed around the outside of the drill pipe. This creates a closed volume back to the mud pumps and traps the invading gas within the drillstring and well bore annulus.

11.3.4 Decrease in Pump Pressure

Pump is working to overcome fluid friction inside the drillstring and provide pressure for bit cleaning. The annulus parasitic loss is small by comparison. So pump pressure will not change much until the kick fills a large portion of the annular space. It takes a significant kick to reduce pump pressure.

11.4 TAKING A KICK WHILE MAKING A TRIP

Wells are vulnerable to a well kick while; 1) pulling the drillstring out of the well bore, 2) Standing empty and 3) Going back into the well bore.

11.4.1 Coming out of the Well Bore

This is one of the most hazardous conditions in the drilling operation. During this time, the Kelly is disconnected from the drillstring, there is no circulation of drilling fluid, and the upward movement of the drillstring can pull in formation fluids into the well bore through the swabbing action of the drill collars.

As drill pipe is being removed, top mud level will drop to fill the void left by the removed pipe. It is important to keep the well bore filled with additional mud in order to maintain bottom hole pressure for well control purposes. This is especially true when drill collars are removed as the volume of drill collars can be five to ten times the volume of drill pipe. This is done by use of a fill-tank. Failure to keep the well bore full can cause the well to kick at a time when it is difficult to control.

The magnitude of "swabbing" the well bore while removing the drillstring depends on annular space between drill collars and well bore, mud weight and speed of pipe removal. Normally, the drop in bottom hole pressure due to swabbing is about 25 to 100 psi, which is less that the parasitic loss in annulus or reduction in pressure when the mud pumps are shut off.

11.4.2 Empty Hole — Drillstring Completely Removed

A well kick during the time of complete removal of a drillstring probably started during the trip out of the well bore. A likely cause is lack of drilling mud in the hole especially during removal of the drill collars. The best indicator is continual mud flow from the well. If this happens it is good practice to close the blind rams to prevent a continuation of flow. Before the blind rams are opened, check casing pressure to see if gas is making its way to the surface.

11.4.3 Going into the Well Bore

Well kicks while going back into the well bore are usually caused by formation fluids drawn into the well bore while removing pipe. Formation gases left in the well bore may be displaced farther up the hole by the incoming pipe. In this case, the gases expand rapidly as they move up the well bore, dropping bottom hole pressure enough to unload the well bore.

Lost circulation can also occur as a result of pressure under drill bits. This action is similar to pistons moving into a partially trapped fluid.

11.5 EFFECT OF GAS CUTTING ON WELL BORE PRESSURE

When formation gas invades a well bore, it travels to the surface and expands within the drilling mud. Gas invasion is due to an underbalanced pressure condition at the bottom of the well bore. The expansion of gas traveling up the well bore follows Boyles's Law (pV = mRT) and increases inversely with local fluid pressure. The gas reduces the average density of the drilling mud and this effect is called *gas cutting*.

11.5.1 Gas Cut Mud

The amount of gas cutting is the percent by volume of mud occupied by gas and is measured at the surface under atmospheric conditions. The percentage of gas cut mud is

$$n_s = \frac{Vol_{gas}}{Vol}$$

A ratio is also expressed in % of gas to total volume. Gas cut mud is measured at the surface.

$$\gamma_s = \frac{w}{Vol} \quad \text{(Vol is total volume at surface)}$$

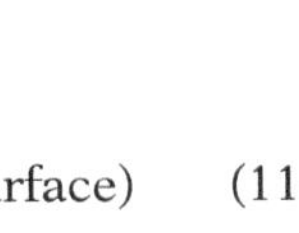

$$w = \gamma_m \left(Vol - Vol_{gas}\right)$$

Let n_s represent *gas cut ratio*, $n_s = \frac{Vol_{gas}}{Vol}$, then

$$w = \gamma_m Vol\left(1 - n_s\right)$$

$$\gamma_s = \gamma_m \left(1 - n_s\right) \quad \text{(average density of mud/gas at the surface)} \tag{11-8}$$

where

γ_s = density of gas cut mud at the surface
γ_m = density of drill mud

This last expression also defines the average density of mud/gas downhole.

$\gamma = \gamma_m \left(1 - n\right)$ (note, γ and n are without subscripts indicating values at any x downhole)

For example, if the drilling mud is cut by 50%, half of a given volume of gas-mud at the surface (under atmospheric conditions) is gas and half is drilling mud. The density of the gas cut mud is

$$\gamma_s = \gamma_m \left(1 - 0.5\right) = 0.5\gamma_s \tag{11-9}$$

The density of gas is negligible compared to the density of drilling mud. If the drilling mud is cut by 10%, the gas cut mud density at the surface is

$$\gamma_s = \gamma_m \left(1 - 0.1\right) = 0.9\gamma_m \tag{11-10}$$

The practical significance of gas cutting is reduction in bottom hole pressure and possible increases in rates of gas influx. If left unchecked, drilling mud density is steadily reduced. In some cases, gas cutting produces a steady state density distribution along the well bore.

11.5.2 Relation between n_s and n

Consider a volume of gas cut mud (Figure 11-5) at the surface containing a volume of gas ΔV_s brought about by an underbalanced pressure condition at the bit [2, 3]. The volume of drilling mud ΔV_m is unchanged. The volume of gas at the surface and downhole locations are related by Boyles' Law (pV = mRT) giving (assuming constant temperature).

$$\left(p_s + p\right) \Delta V = p_s \Delta V_s \tag{11-11}$$

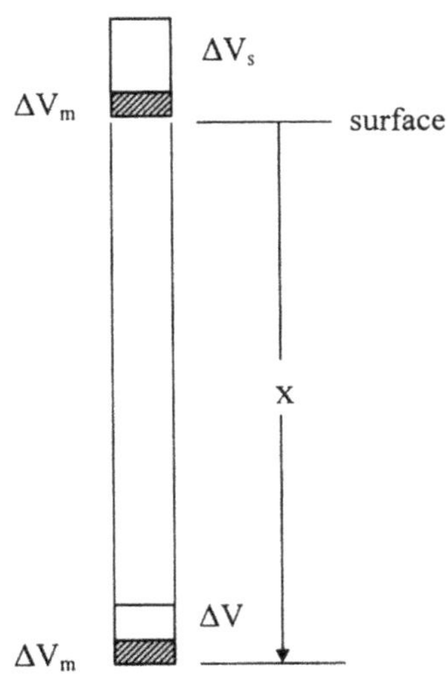

Figure 11-5. Gas expansion during migration to the surface.

where

p_s = atmospheric (absolute) pressure, atms
p = downhole *gauge* pressure, atms
ΔV = volume of gas at depth, x
ΔV_s = volume of gas at the surface
ΔV_m = volume of drilling mud (same at x and surface)

Dividing both sides by ΔV_m gives

$$\left(p_s + p\right)\frac{\Delta V}{\Delta V_m} = p_s\frac{\Delta V_s}{\Delta V_m} \tag{11-12}$$

$$\left(p_s + p\right)\frac{\Delta V}{V_T - \Delta V} = p_s\frac{\Delta V_s}{V_{Ts} - \Delta V_s} \tag{11-13}$$

Dividing the left side by V_T and the right side by V_{Ts}

$$\left(p_s + p\right)\frac{n}{1-n} = p_s\frac{n_s}{1-n_s} \tag{11-14}$$

where

V_T = total volume of gas and mud at depth, x
V_{Ts} = total volume of gas and mud at the surface

n and n_s are fractions of the total volume occupied by gas at location x down hole and surface locations, respectively.

11.5.3 Pressure Gradient and Pressure Distribution

Local hydrostatic pressure due to variable fluid density caused by gas cutting (Figure 11-6) is developed as follows.

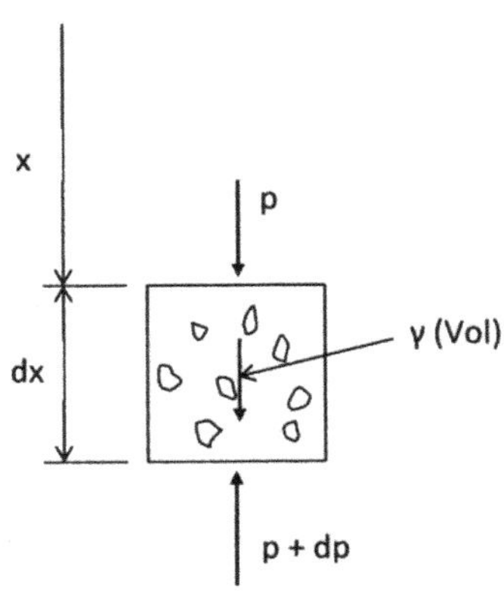

Figure 11-6. Hydrostatic pressure in gas cut mud.

Summing forces in the x direction gives

$$\left(p + dp\right) A = Adx\gamma + pA \tag{11-15}$$

$dp = \gamma dx$ (same expression as constant mud weight throughout well bore)

But *average* density, considering part gas, is

$$\gamma = \frac{\left(V_T - \Delta V\right) \gamma_m}{V_T} \tag{11-16}$$

$$\gamma(x) = (1 - n)\gamma_m \tag{11-17}$$

By substituting into Eq. (11-15)

$$dp = (1 - n)\gamma_m dx \tag{11-18}$$

In terms of surface conditions [from Eq. (11-14)]

$$n = \frac{p_s n_s}{\left(1 - n_s\right)\left(p + p_s\right) + p_s n_s} \tag{11-19}$$

and

$$1 - n = \frac{1}{1 + \dfrac{p_s n_s}{\left(1 - n_s\right)\left(p + p_s\right)}} \tag{11-20}$$

11.5.4 Determination of Pressure Distribution Downhole

Substituting Eq. (11-20) into Eq. (11-18) gives

$$\gamma_m dx = dp + \frac{p_s\, n_s}{\left(1 - n_s\right)} \frac{dp}{\left(p + p_s\right)} \tag{11-21}$$

Integrating and noting that p = 0 at x = 0 gives

$$\gamma_m x = p + \frac{p_s\, n_s}{1 - n_s} \ln \frac{p_s + p}{p_s} \tag{11-22}$$

Since surface pressure is atmospheric ($p_s = 1$ atm) Eq. (11-22) becomes

$$\gamma_m x = p + \frac{n_s}{1 - n_s} \ln(1 + p) \tag{11-23}$$

The pressure distribution function, p(x), is implicit in Eq. (11-23) and is best determined numerically. A plot showing a typical pressure distribution due to gas cutting is shown in Figure 11-7. Note that pressure in Eq. (11-23) is expressed in units of atms. The left side of this equation must also be expressed in atms as illustrated in the example below.

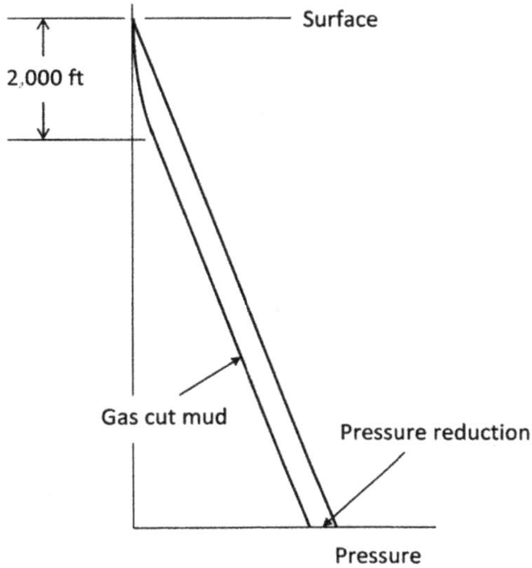

Figure 11-7. Pressure distribution caused by gas expansion.

For example, the reduction in bottom hole pressure for

γ_m = 10 ppg
x = 6000 ft
n_s = 0.1 (10% gas cut at the surface)

is 8.4 psi and is determined as follows.

$$\frac{(0.052)10(6{,}000)}{14.7} = p + \frac{1}{9}\ln(1+p) \qquad \text{(p has units of atms)}$$

$$212.24 = p + \frac{1}{9}\ln(1+p)$$

After several tries for p, a value of p = 211.67 atms gives, for the right side of the above equation

$$211.67 + \frac{1}{9}\ln(212.67) = 212.26$$

which is close to 212.24. Therefore, p = 211.67 atms is the solution. The reduction in bottom hole pressure is (212.24 – 211.67)14.7 = 8.4 psi.

Even gas cutting of 50% causes only a bottom hole pressure reduction of 80 psi. Gas cutting, therefore, does not cause a large change in bottom hole pressure provided overall mud circulation is steady state. On the other hand, an underbalanced pressure condition of 80 psi can lead to gas cutting of 50% at the surface, if gas invasion of the well bore is allowed to reach steady state.

These results check closely with the data presented by Obrien and Goins [4] and Bartlett and Edmiston [5].

Gas cutting at shallow depths is more critical than gas cutting at deep depths. For example, if 10 ppg mud is cut to 5 ppg at a drilling depth of 2000 ft, bottom hole pressure is reduced from 0.052 (10) 2000 = 1040 psi to 1040 – 70 = 970 psi.

$$\frac{(0.052)10(2,000)}{14.7} = 70.748 = p + \frac{1}{9}\ln(1+p)$$

By trial and error, p is determined to be 66.5 atm. This means that underbalanced drilling of (70.748 – 66.5) 14.7 = 62.45 psi, will produce 50% gas cutting mud at the surface.

This causes the formation to produce more gas, which further cuts the drilling mud leading to an unstable flow condition. A kick can quickly turn into a blow-out in this case.

Shallow gas is potentially more hazardous than deep gas because formation permeability and porosity is normally high causing a rapid influx of gas into the well. Reaction time is short, and the well cannot be shut-in sometimes because formation fracture gradient is low and an underground blow out could occur. Sometimes the BOP stack is not even installed. Shallow gas is diverted away from the rig by a diverter system.

11.6 DETERMINING THE MAGNITUDE OF A KICK

It is necessary to determine the magnitude of a *kick* in order to know how much to increase the mud weight to equalize well bore pressure with formation pressure. Once this is determined, the weight of the new mud can be established. The shut-in situation is illustrated in Figure 11-8.

The annular preventer (Figure 11-9) is located at the top of the blowout preventer (BOP) stack. A key component of an annular preventer is an elastomer (rubber) ring, which deforms around drill pipe sealing off the annular space. The elastomer ring is activated mechanically. The pressure limit of annular preventers is around 1500 psi.

Because drilling mud is not heavy enough to equalize formation pressure down hole, back pressure is generated in the standpipe when the well is shut-in. In this case, standpipe pressure increases until bottom hole pressure is equal to formation pressure. Bottom hole formation pressure is determined from shut-in pressure by

$$p_f = p_{si} + 0.052\gamma_m L \qquad (11\text{-}24)$$

where

p_{si} = shut-in pressure (standpipe pressure), psi

The magnitude of the well kick is equal to the shut-in standpipe pressure, p_{si}. The new mud weight required to balance the formation pressure is

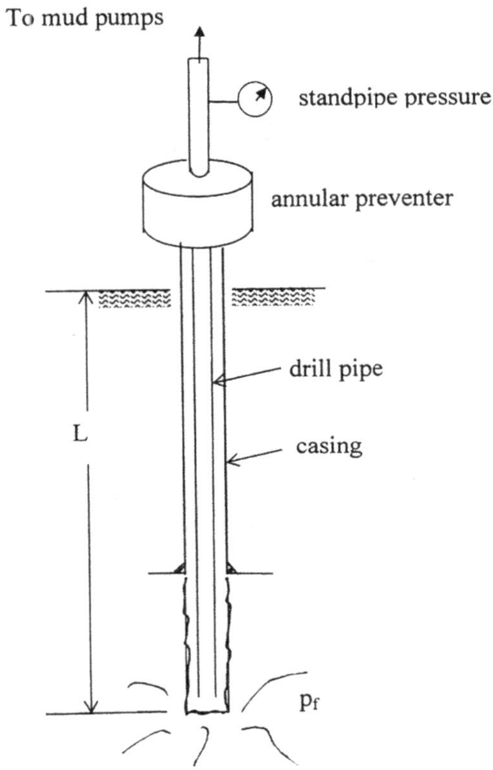

Figure 11-8. Shut-in standpipe pressure.

$$\gamma_{new} = \gamma_m + \Delta\gamma_m \tag{11-25}$$

where

$$\Delta\gamma_m = 19.23\frac{p_{si}}{L} \tag{11-26}$$

Assuming

- standpipe pressure is 300 psi (magnitude of kick)
- mud weight inside drill pipe is 12 ppg
- hole depth is 8000 ft

then formation pressure is

$$p_f = 0.052(12)8{,}000 + 300 = 5{,}292 \text{ psi}$$

The mud weight required to balance this formation pressure is

$$5{,}292 = 0.052(\gamma_m + \Delta\gamma_m)8{,}000 \text{ psi}$$

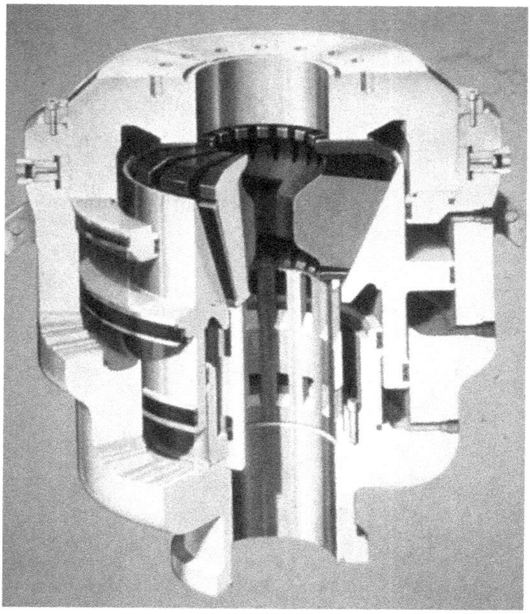

Figure 11-9. Annular blowout preventer (Used by Permission from Baker Hughes).

$$\gamma_m + \Delta\gamma_m = 12.72 \text{ ppg}$$

The well took a 0.72 ppg kick and the 12 ppg mud has to be brought up to 12.72 ppg to establish a balanced pressure condition. Final mud weight would probably be 13.22 ppg in order to have a 0.5 ppg overbalance while drilling. Alternatively, the magnitude of the kick can be determined by using Eq. (11-26) directly.

In cases where the magnitude of the well kick is very high (10,000 psi), the well may be shut in with ram type closing devices (Figure 11-10).

Blowout preventer stacks contain various types of rams, including the capability to shear pipe in two.

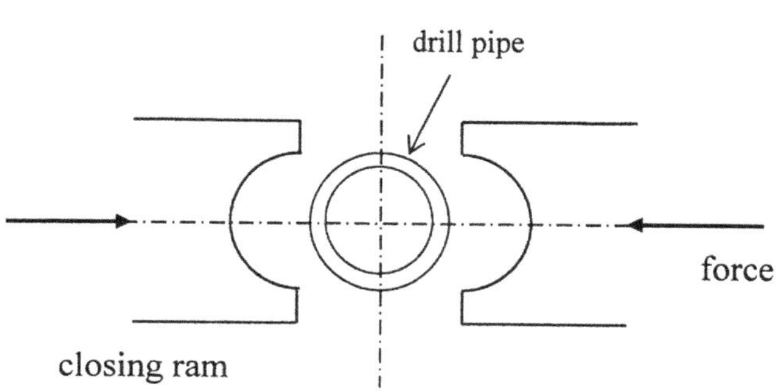

Figure 11-10. Pipe ram blowout preventer.

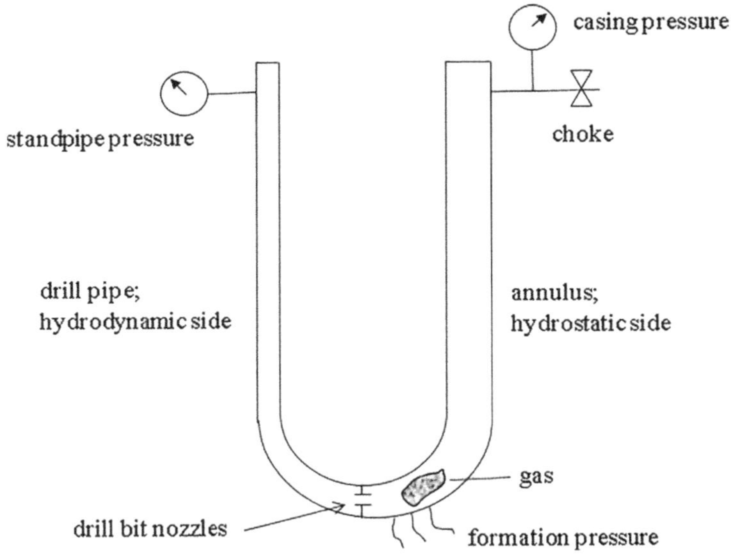

Figure 11-11. U-Tube representation of a shut-in well.

11.7 U-TUBE REPRESENTATION OF A SHUT-IN WELL

It is convenient to visualize the drillstring and annular space as a U-tube (Figure 11-11) when tracking standpipe pressures and kick position. The left side represents the inside of the drillstring with standpipe pressure gauge at the top. The right side represents the annulus with drilling choke and casing pressure gauge at the top. Bit nozzles separate the two sides while the invading formation fluid is on the annulus side. The drilling choke is adjusted during the circulation to maintain equilibrium between bottom hole pressure and formation pressure.

11.8 CALCULATIONS USED IN REMOVING A KICK

After a well has been shut-in and the magnitude of the kick established, the next step is to remove or circulate out the invading gas, which is now trapped in the annulus. These objectives are accomplished by controlling bottom hole formation pressure with the choke at the annular preventer. If bottom hole pressure is too low, additional formation fluids invade the well bore and nothing is accomplished. If bottomhole pressure is too high, the formation could break down leading to lost circulation or loss of well control.

The kick should be circulated out early to prevent gas from migrating to the surface carrying with it the bottom hole formation pressure. This pressure migration comes about because the total downhole drilling mud volume is fixed and as a result the invading gas does not have a chance to expand; gas pressure, therefore, stays constant during the migration.

11.8.1 Kill Rate-Kill Rate Pressure

An important step, which is taken prior to a well kick, is to determine pump pressure required to circulate at a specified pumping rate. Pump stroking rate for this test is usually about ½ the stroke rate under normal drilling conditions. For a duplex pump the kill rate is about 30 strokes/minute and for triplex pumps about 45 spm to 50 spm.

Since pumping pressure varies approximately to the square (~1.8) of the flow rate, the lower stroke rate and flow rate (Q) reduces standpipe pressure to about ¼ the standpipe pressure associated with normal drilling pressure. The lower stroke rate is called Kill Rate (spm). The corresponding pressure caused by parasitic (system losses) and bit pressure drop is called Kill Rate Pressure.

$$KRP = p_S + \Delta p_{bit} \tag{11-27}$$

This leaves ample pump pressure to circulate against the formation pressure or kick. The difference between the two curves (in Figure 11-12) is the available pump pressure reserved to pump against a kick. Kill rate pressure is measured daily.

The sum of KRP and formation pressure (kick) is the Initial Circulating Pressure (ICP).

$$ICP = KRP + kick \tag{11-28}$$

The bit pressure curve corresponds to a given set of bit nozzles.

11.8.2 Volume of Drilling Mud inside Drillstring

The volume inside the drillstring is the sum of volumes within the drill pipe and drill collars. Often the total volume inside a drillstring is approximated

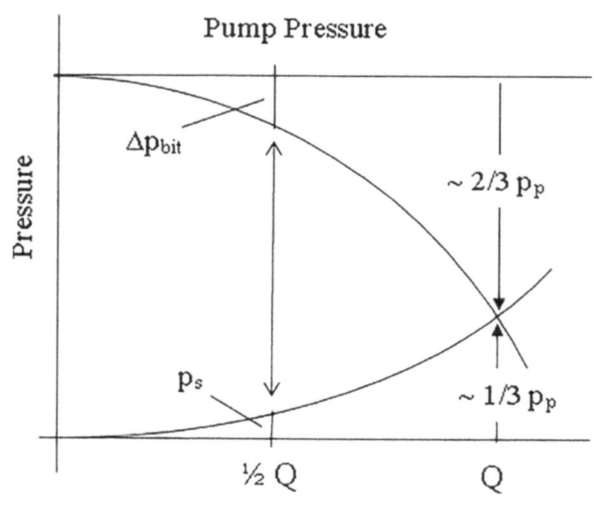

Figure 11-12. Kill rate pressure.

considering only the capacity of the drill pipe. The volume capacity of drill pipe (inside volume) is determined by

$$V_{dp} = C_{dp} L \text{ bbl} \tag{11-29}$$

where C_{dp} is the volume per foot having units of bbl/ft of drill pipe. This number is tabulated for various drill pipe and drill collar sizes in Tables 11-3 and 11-4 in the Appendix.

Pump strokes required to displace drilling mud inside drill pipe is

$$strokes = \frac{V_{dp}}{volume/stroke} \tag{11-30}$$

Time to displace drilling mud inside drill pipe is

$$time = \frac{strokes}{spm} \text{ (minutes)} \tag{11-31}$$

11.8.3 Volume of Drilling Mud in Annulus

The volume in the annular space around a drillstring is the sum of annular volumes between the well bore and drillstring (drill pipe and drill collars). Often the total annular volume is approximated considering only the annular volume around the drill pipe. In this case the annular volume over the drill pipe length is determined by

$$V_a = C_a L \text{ bbl} \tag{11-32}$$

where C_a is the annular capacity also given in units of bbl/ft of pipe. Annular capacities for different combinations of well bore and pipe sizes are given in the Appendix (Table 11-5).

Pump strokes required to displace drilling mud in annulus is

$$strokes = \frac{V_a}{volume/stroke} \tag{11-33}$$

Volume per stroke for both duplex and triplex pumps are given in the Appendix (Tables 11-6 and 11-7).

Time to displace drilling mud in annulus is

$$time = \frac{strokes}{spm} \text{ (minutes)} \tag{11-34}$$

11.8.4 Effect of Mud Weight on Circulation Pressure

From the hydraulic equations in Ch. 4, note that parasitic pressure losses are proportional to drilling mud density.

$$p_2 = p_1 \frac{\gamma_2}{\gamma_1} \tag{11-35}$$

11.9 CIRCULATING OUT A KICK

Once a kick has been detected and trapped, it must be removed before drilling can continue. The objective during the removal process is to protect the formation while controlling and removing the kick. If circulating pressures are too high, the formation could be damaged causing lost circulation and even lose control of the well. Two methods commonly used to circulate out a kick are 1) drillers method and 2) wait and weight method.

11.9.1 Drillers Method

The driller method relies on two separate mud circulations. The kick is first circulated out *with* the original mud. Afterwards, the new (heavier) mud is

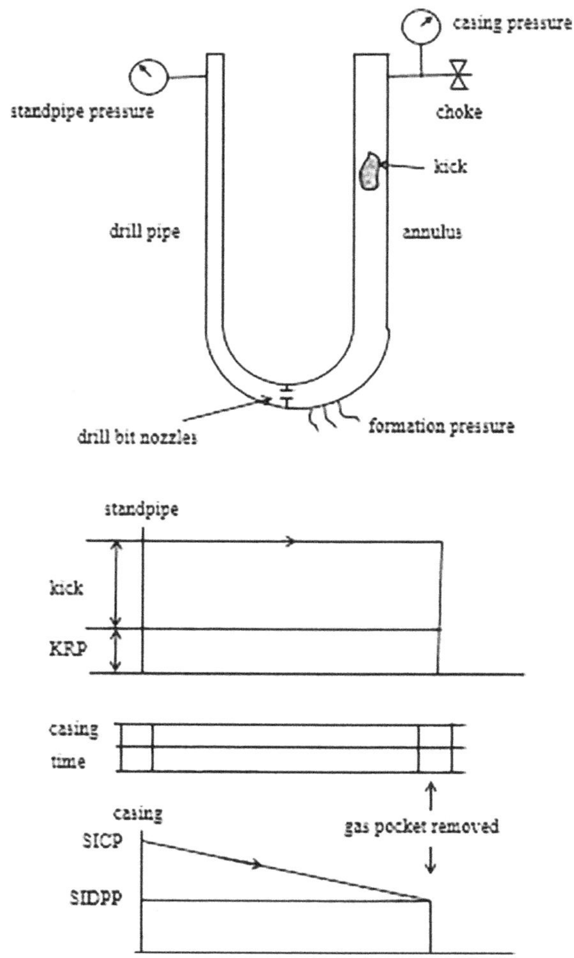

Figure 11-13. Removing kick with original mud-drillers method.

circulated into the enclosed volume to kill the well. These two steps will now be discussed using Figures 11-13 and 11-14.

Figure 11-13 shows standpipe pressure and casing pressure while the kick is being circulated out. The standpipe pressure remains constant because the hydrodynamics of fluid flow inside the drillstring is unchanged as the kick is being pushed up and out of the annulus. Casing pressure, however, continually reduces from the Shut-in Casing Pressure (SICP) as the choke is continually adjusted to maintain constant standpipe pressure.

The next step is to circulate in the new mud. Figure 11-14 shows how standpipe pressure and casing pressure vary during this step. Casing pressure is held constant until the inside of the drillstring is displaced. At this point the inside of the drillstring is filled with the new heavier mud while the annulus still contains the initial lighter mud. Once the heavier mud completely fills the inside of the drillstring, the standpipe pressure is held at the final circulating pressure

$$FCP = \frac{\gamma_n}{\gamma_o} KRP$$

with the casing choke. If circulation is stopped at any time the standpipe pressure should be zero because the new mud should balance the formation pressure. Casing pressure, however, would not read zero because part of the annulus still contains the initial mud. When the inside of the drillstring and the annulus have been completely filled with new mud, both standpipe pressure and casing pressure should be zero when the pumps are turned off.

A brief outline of this method is given below.

1. Open choke and bring pump strokes up to kill rate (spm).
2. Adjust casing choke (annular preventer) so that standpipe pressure equals shut-in pressure plus predetermined kill rate pressure.
3. Maintain this standpipe pressure while completely displacing the annulus by adjusting the choke. Note that initial casing shut-in pressure is greater than shut-in drill pipe pressure because the kick is less dense than the drilling mud. Also monitor pump strokes to keep track of the kick location.
4. Record the final casing pressure after the kick is out and the pump is running at the kill rate. This pressure should be close to the initial shut-in stand pipe pressure.
5. Close the well in and note the shut-in standpipe and casing pressures. Both should be the same.
6. Begin to circulate the new mud into the drillstring by opening the choke and maintaining back pressure equal to the final casing pressure reading as noted in step 4. Maintain this casing pressure until drillstring has been displaced with new mud. Track this displacement by monitoring the number of pump strokes. Also, at this point, the standpipe pressure should be equal to the final circulating pressure (FCP).
7. Maintain the final circulating pressure in the standpipe until the annulus has been displaced by the new mud. The final circulating pressure should be equal to $FCP = \frac{\gamma_n}{\gamma_o} KRP$. Final circulating pressure is

maintained by the choke. Casing pressure should gradually reduce to zero as the new mud fills the annulus.

8. Shut the pump off and check standpipe and casing pressures. They both should be equal to zero.

11.9.2 Wait and Weight Method

Following this method, the kick is removed, and the new mud inserted with only one circulation. The name implies that time is required to bring the mud up to its new weight. Once the desired mud is in the mud pits, circulation begins. Figure 11-15 shows pressure changes in both standpipe and casing during this process.

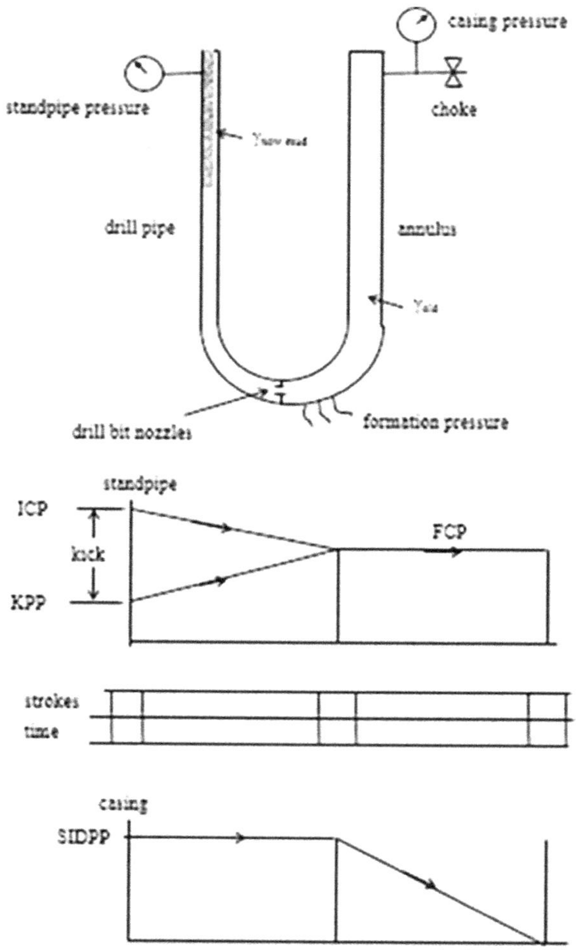

Figure 11-14. Circulating in new mud-drillers method.

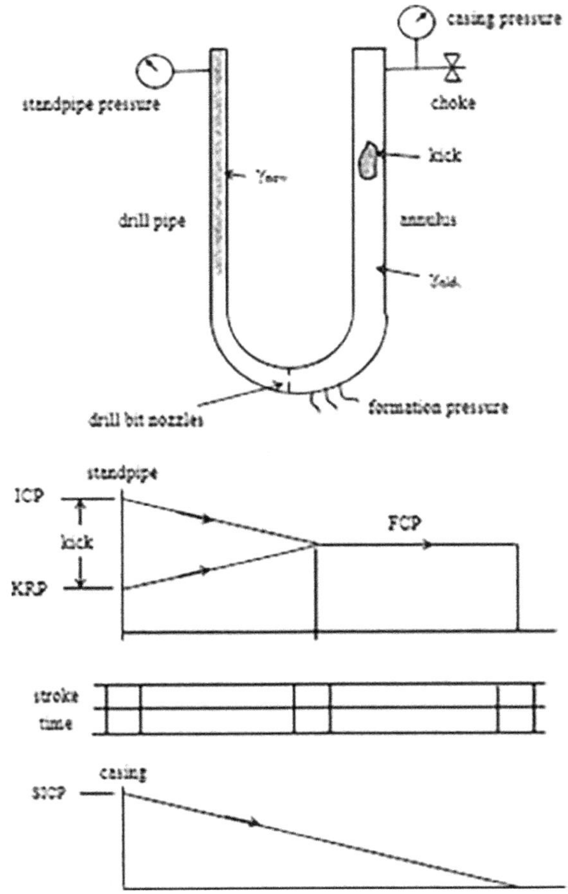

Figure 11-15. Wait and weight method for removing a kick.

After the drilling mud has been completely displaced inside the drillstring, standpipe pressure is held constant at the final circulating pressure (FCP) until the annulus has been displaced. This is done by the choke in the annular preventer.

Casing pressure changes continuously from the initial casing shut-in pressure (SICP) down to zero, when the annulus is completely filled with the new drilling mud. As the heavy mud follows the kick up the annulus it raises the mud column pressure and reduces casing pressure at the choke.

An advantage of the wait and weight method is its ability to reduce well head pressure in the shortest amount of time. It does, however, require a period of shut-in time while the heavier mud is being mixed so this should be done as quickly as possible.

A brief outline of this method is given below.

1. Prepare new mud while the well is shut-in with the kick.
2. Open the choke and establish pump strokes equal to the kill rate.

3. Immediately adjust the choke to establish a standpipe pressure equal to the sum of the kick plus the predetermined kill rate pressure (ICP).
4. Continually adjust the choke as the new mud is displaced into the drillstring so that standpipe pressure reduces with pump strokes according to the pressure-pump stroke line. When the drillstring is completely full of the new mud as determined by the pump stroke counter, the standpipe pressure should be equal to the final circulating pressure as determined by $FCP = \frac{\gamma_n}{\gamma_o} KRP$.
5. Continue circulating until the annulus is filled with the new mud by maintaining the FCP in the standpipe. Monitor the pump stroke counter to determine when the annulus has been completely displaced.
6. Close-in the well in and check standpipe and casing pressures. Both readings should be equal to zero.

11.10 EXAMPLE OF KILLING A WELL

Both methods will now be illustrated by use of an example based on the following information.

Pump	6 × 12 triplex	
Kill Rate	45 spm	
Kill Rate Pressure	900 psi	
Hole Size	8½ in.	
Drill Pipe	5 in. (19.5 lbs/ft)	
Hole Depth	9720 ft	
Shut-in Standpipe Pressure	400 psi	(SIDPP)
Shut-in Casing Pressure	700 psi	(SICP)
Original Mud Weight	12.8 ppg	

It is helpful to label various parts of the U-tube representation with known input. Initially, shut-in casing pressure is greater than shut-in standpipe pressure because of a lower average mud density caused by the cut mud.

A work sheet is given in the Appendix (Table 11-2) to assist in organizing the input data and preparing the data for use in the two methods. Data for the example problem is shown in the Well Control Work Sheet (Table 11-1).

11.10.1 Drillers Method

The position of the kick is monitored by the number of pump strokes (4628 strokes) required to displace the annulus fluids.

During the initial circulation, standpipe pressure is maintained at a constant level of

$$\text{ICP} = \text{kick} + \text{KRP} \quad (\text{kick} = \text{SIDPP}) \tag{11-36}$$

$$ICP = 400 + 900 = 1300 \text{ psi}$$

Table 11-1. Well control work sheet.

Length of Drillstring		
Length of Drillstring	9,700 ft	(given)
Vertical Depth of Well	9,700 ft	(given)
Pressure Information		
SIDPP	400 psi	(measured)
SICP	700 psi	(measured)
Magnitude of Kick	400 psi	(SIDPP)
Original Mud Density, γ_o	12.8 ppg	(given)
New Mud Density, γ_n	13.6 ppg	
Pump Information		
Kill Rate (KR)	45 spm	(specified)
Kill Rate Pressure (KRP)	900 psi	(measured daily)
Capacity of Drill Pipe	169.8 bbl	
Strokes to fill Drill Pipe	1,796 strokes	
Time to fill Drill Pipe	40 minutes	
Capacity of Annulus	437.4 bbl	
Strokes to fill Annulus	4,628 strokes	
Time to fill Annulus	103 minutes	
Final Circulating Pressure	956.3 psi	

Bottom-hole pressure will be equal to formation pressure (p_f) during this circulation phase.

Once the kick is removed, the heavier drilling mud is circulated into the well (see Figure 11-13). Standpipe pressure changes continually from an initial circulating pressure (ICP) to a final circulating pressure (FCP) as shown in the top left diagram. The magnitude of the FCP is determined as follows:

$$FCP = \frac{\gamma_n}{\gamma_o} KRP \tag{11-37}$$

$$FCP = \frac{13.6}{12.8} 900 = 956.25 \text{ psi}$$

The reduction of the standpipe pressure according to the pressure line in the top diagram is maintained by the choke. The magnitude of casing pressure should be equal to the initial shut in drill pipe pressure (SIDPP) when the drill pipe is filled with the new mud.

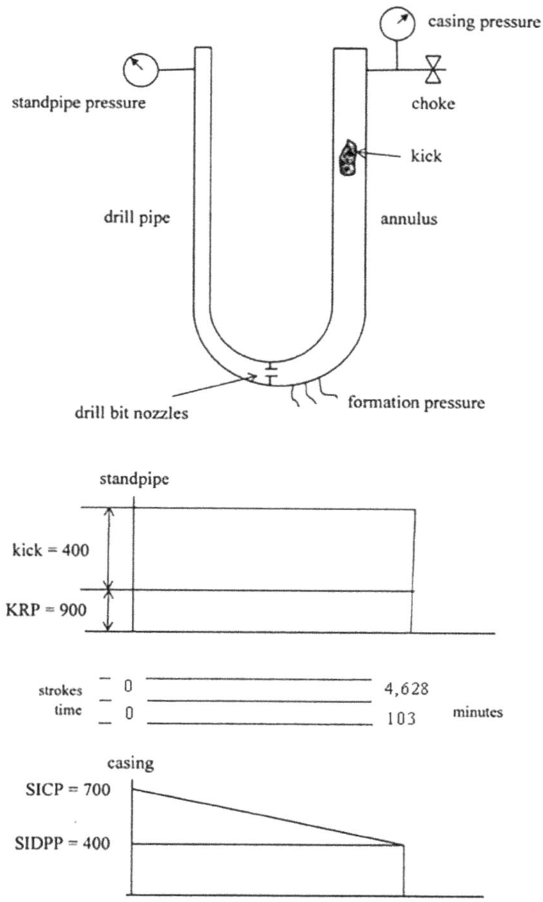

Figure 11-16. Removing kick with original mud-drillers method.

When the new drilling mud starts up the annulus, Final Circulating Pressure is maintained (956.25 psi) while casing pressure reduces steadily to zero once the annulus is completely filled with the new heavier mud.

11.10.2 Wait and Weight Method

The top left pressure drawing is the same as for the Drillers Method when the new drilling mud is being circulating into the enclosure even though the kick is in the annulus. At the beginning of circulation, standpipe pressure is 400 + 900 psi or 1300 psi because the original mud is inside the drillstring. When

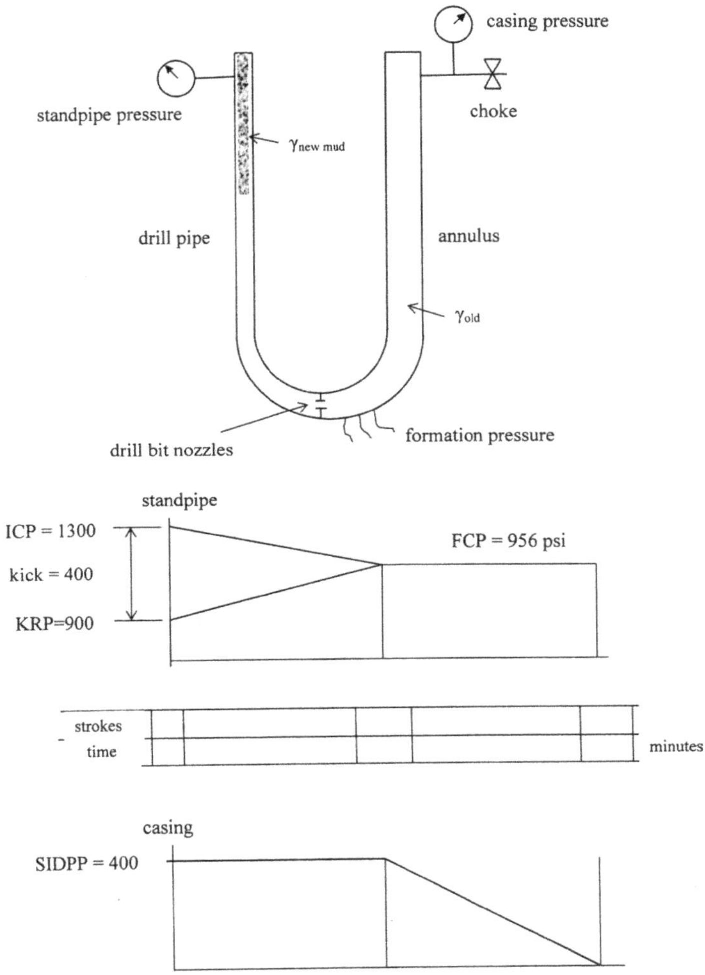

Figure 11-17. Circulating in new mud-drillers method.

the inside has been completely displaced by the new mud, standpipe pressure should be equal to the final circulating pressure (FCP). Standpipe pressure is steadily adjusted downward between these two pressure levels by adjusting the choke.

It is useful to monitor the position of the kick by the pump strokes. When the volume of strokes reaches annulus volume (4628 strokes), the kick is at or near the surface. The kick is quite noisy as it moves through the choke.

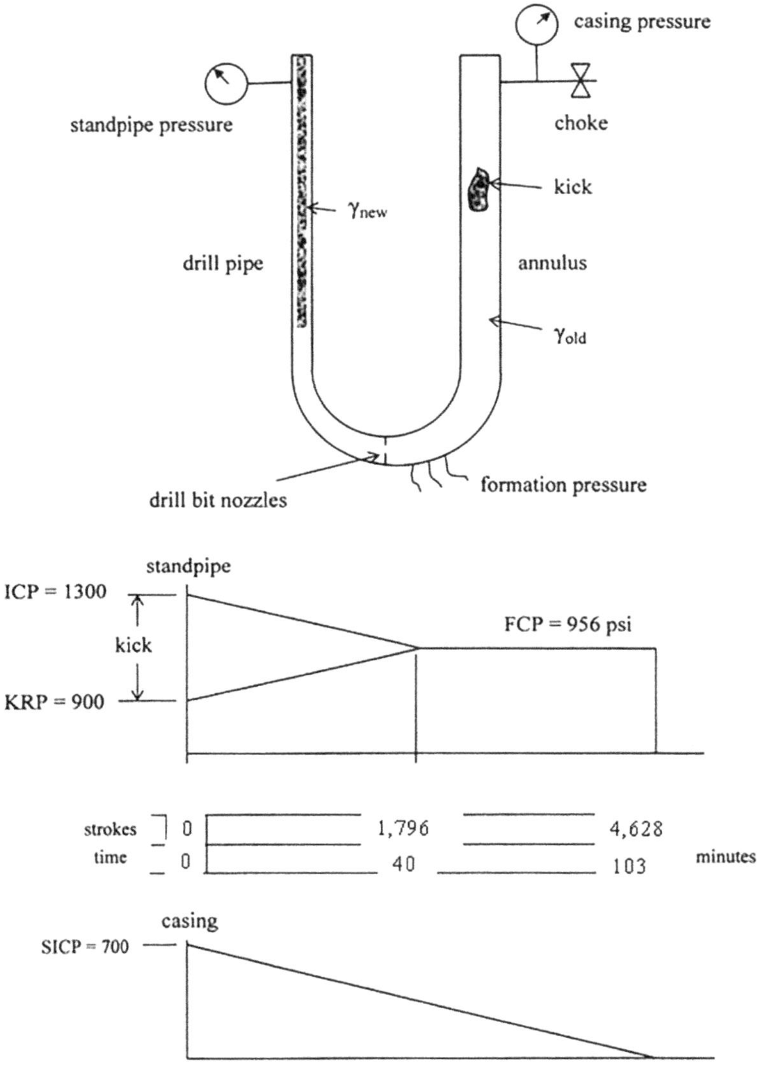

11.10.3 Summary of Equations

$p_f = p_{si} + 0.052\gamma_m L$ (formation pressure)

$\Delta\gamma_m = 19.23 \frac{p_{si}}{L}$ (required density increase to balance formation pressure)

$KRP = p_S + \Delta p_{bit}$ (kill rate pressure, determined daily)

$ICP = KRP + SIDPP$

$$FCP = \frac{\gamma_n}{\gamma_o} KRP \quad \text{(Final circulating pressure)}$$

$$strokes = \frac{V_{dp}}{volume/stroke} \quad \text{(number of strokes)}$$

$$time = \frac{strokes}{spm} \quad \text{(minutes)}$$

11.11 PROBLEMS

1. The shut-in drill pipe pressure (SIDPP) after a well kick is 350 psi. If the well depth is 8220 ft and mud weight is 13.6 ppg, determine the increase in mud weight required to balance the formation pressure. What is the new mud weight if overbalance drilling is 0.5 ppg?

2. Mud tank volume increased by 10 barrels at a time when the well bore is 9000 ft deep. If a 9½ well bore is drilled and the drillstring contains 4½ in. (16.6 lb/ft) drill pipe, estimate the reduction in bottom hole fluid pressure as a result of the kick. Mud weight is 11.5 ppg.

3. Standpipe pressure for a shut-in well is 500 psi. Hole depth (TVD) is 8600 ft and mud weight is 11.8 ppg. If the gas kick is allowed to migrate to the surface before being circulated out of the hole, determine the burst pressure applied to casing at the surface under this extreme condition.

4. Determine the reduction in well bore pressure at 10,000 ft if 12 ppg mud is cut by 25%.

5. The shut-in drill pipe pressure after a kick is 920 psi. The measured depth of the well is 12,500 ft, however, survey calculations shown that the true vertical depth is 9000 ft. Mud weight is 13.0 ppg. What is the magnitude of the formation pressure? Calculate the new mud weight required to balance the formation pressure.

6. Drilling conditions at a given depth are:

   KRP = 900 psi
   SIDPP = 600 psi
   True Vertical Depth = 10,000 ft
   Mud weight = 12 ppg
   Flow rate = 360 gpm

   A kick is detected, and the well is shut-in with SIDPP = 600 psi. Determine the initial circulating pressure (ICP). What is the final circulating pressure (FCP)?

7. Determine the number of pump strokes and time required to fill the drill pipe from the surface down to the bit for the following conditions.

   Well depth = 7850 ft
   Drill pipe size = 4½ in. (16.6 lb/ft)
   Mud pump = 6½ × 15 (Duplex Pump with 15 in. stroke)
   Kill rate = 30 spm

8. Determine the number of pump strokes and time required to displace drilling mud from the annulus space between a 9 in. well bore and 5 in. drill pipe.

   Measured depth = 10,100 ft
   True vertical depth = 8700 ft
   Drill Pipe size = 5 in. (19.5 lb/ft)
   Mud pump (Triplex) = 7 in. liner and 12 in. stroke
   Kill rate = 40 spm

9. Determine the final circulating pressure (FCP) for the following conditions.

   Kill rate pressure (KRP) = 850 psi
   Well depth = 7500 ft
   Shut-in drill pipe (SIDPP) = 320 psi
   Mud weight (initial) = 12.6 ppg

10. Fill out the well control worksheet for the following shut-in conditions. Show how standpipe pressure and casing pressure will vary with number of pump strokes and pumping time for the Drillers Method.

    Hole depth = 9,850 ft
    Hole size = 7 7/8 in.
    Drill Pipe = 4½ in. (ID = 3.826 in.), 16.6 lb/ft
    Pump = 6 × 15 (Duplex)
    Mud weight = 13.0 ppg
    Kill Rate Pressure = 450 psi at 30 spm
    SIDPP = 750 psi
    SICP = 1150 psi

11. Fill out the well control worksheet for the following shut-in conditions. Show how standpipe pressure and casing pressure vary with number of pump strokes and pumping time as applied to the Wait and Weight Method.

    Measured depth = 14,200 ft
    TVD = 12,050 ft
    Hole size = 9 7/8 in.
    Drill pipe = 5 in. (19.5 lb/ft)
    Casing = 10¾ in. at 10,000 ft; TVD is 9120 ft

Pump = 6 × 12 (Triplex)
Mud weight = 12.8 ppg
Kill Rate Pressure = 800 psi at 45 spm
SIDPP = 400 psi
SICP = 550 psi

11.12 LIST OF REFERENCE

1. Blowout Prevention Manual, TR81-26, Maurer Engineering, Inc, Well-Control School, October 1981.
2. Strong, M. W., "Mud Technique in Iran", Trans AIME (1939), 132, 11(67).
3. Robert J. White, Bottom-Hole Pressure Reduction Due to Gas Cut Mud, Society of Petroleum Engineers (AIME), July 1957.
4. Goins, W. C., Jr and Riley Sheffield, *Blowout Prevention*, 2nd Ed., Gulf Publishing Company (1983), Houston, TX.
5. Bartlett, L. E. and Edmiston, D. L.; "Kick Control", Atlantic Richfield Co., Drilling Manual (1970).

11.13 APPENDIX

Table 11-2 is a well control work sheet for student use in organizing key operational data needed to solve well control problems.

Table 11-2. Well control work sheet.

Length of Drillstring		
Length of Drillstring	____ ft	(given)
Vertical Depth of Well	____ ft	(given)
Pressure Information		
SIDPP	____ psi	(measured)
SICP	____ psi	(measured)
Magnitude of Kick	____ psi	(SIDPP)
Original Mud Density, γ_o	____ ppg	(given)
New Mud Density, γ_n	____ ppg	
Pump Information		
Kill Rate (KR)	____ spm	(specified)
Kill Rate Pressure (KRP)	____ psi	(measured daily)
Capacity of Drill Pipe	____ bbl	
Strokes to fill Drill Pipe	____ strokes	
Time to fill Drill Pipe	____ minutes	
Capacity of Annulus	____ bbl	
Strokes to fill Annulus	____ strokes	
Time to fill Annulus	____ minutes	
Final Circulating Pressure	____ psi	

Table 11-3. Capacity of drill pipe (bbl/ft).

$$C_{dp} = \frac{ID^2}{1029}$$

Pipe Size OD, in.	Weight lbs/ft	Capacity bbl/ft
4½	15.5	0.0066
	16.6	0.0142
5	19.5	0.0175
	25.5	0.0153
5½	21.9	0.0217
	24.7	0.0208

Table 11-4. Capacity of drill collars (bbl/ft).

$$C_{dc} = \frac{ID^2}{1029}$$

Drill Collar Size (in.) OD	ID	Capacity bbl/ft
any size	2	0.0039
	2.25	0.0049
	2.5	0.0607
	2.75	0.0073
	3	0.0087

Table 11-5. Capacity of annulus or hole (bbl/ft).

$$C_H = \frac{DH^2 - DP^2}{1029}$$

Hole Size or Casing ID	Pipe OD, in.				
	2 7/8	3½	4	4½	5
4 ¾	0.013				
5 5/8	0.022	0.018			
6	0.026	0.022			
6¼	0.029	0.025			
6 ½	0.032	0.028	0.024		
6 ¾		0.032	0.028	0.024	
7 3/8		0.04	0.036	0.032	
7 5/8		0.044	0.04	0.036	0.031
7 7/8		0.048	0.044	0.04	0.035
8 3/8			0.051	0.048	0.043
8 ½			0.054	0.05	0.045
8 ¾			0.058	0.054	0.049
9			0.061	0.057	0.052
9 5/8			0.073	0.069	0.065
9 7/8			0.078	0.074	0.069
11				0.097	0.092
12 1/4				0.125	0.12

Table 11-6. Capacity of duplex double acting mud pumps (bbl/stk) (90% efficiency).

Liner Dia. in.	12″ stk 2″ rod	15″ stk 2¼″ rod	18″ stk 2½ rod
5	0.0804	0.0982	0.1147
5½	0.0988	0.1212	0.1423
6	0.1189	0.1461	0.1725
6½	0.1408	0.1736	0.2052
7	0.1644	0.2031	0.2406
7½		0.2348	0.2787

Table 11-7. Capacity of triplex mud pumps (bbl/stk) (90% efficiency).

Liner Dia. in.	8″ stk	9″ stk	10″ stk	12″ stk
5	0.0437	0.0491	0.0546	
5½	0.0529	0.0595	0.0662	0.0791
6	0.0628	0.0708	0.0787	0.0945
6½	0.0739	0.0831	0.0923	0.1024
7	0.0857	0.0964	0.1071	0.1283

Chapter 12

DIRECTIONAL DRILLING

Directional drilling refers to the process of drilling along a prescribed well path to a target located at a given depth and some lateral distance from a vertical axis through the drill site. To a large extent, directional drilling is like flying an airplane. A flight plan is established based on prevailing winds and weather conditions between point of departure and destination. Power is essential. The crew navigates and makes course changes when necessary. Directional drilling also requires a plan, which defines the road map or well path from surface location and target.

One of the earliest applications of directional drilling was a relief well drilled by John Eastman in January 1934. In early 1933 an oil well (Alexander No. 1) near Conroe, Texas blew out, swallowed the drilling rig into a crater more than 150 ft across and spilled oil at a rate of 7000 barrels a day creating a large surface pool. The relief well was spudded 400 ft northeast of the blow out. It was drilled vertically to 1960 ft, then deflected toward the blowing well bore with a removable whipstock. Progress of the relief well was monitored with a magnetic single shot and Eastman was able to intercept the active formation at a depth of 5135 ft.

Prior to 1934, oak logs were shaved to an angle and dropped into the well bore to deflect drill bits off of cement plugs. Large boulders have also been used for the same purpose. Since then, directional drilling has advance allowing well bores to be drilling accurately along a prescribed path intersecting underground reservoir target within a few feet.

Initially, drilling operations focused on vertically straight well bores. It became apparent that hole deviation depended on bit weight, collar size and formation. The main criteria for an acceptable wellbore path was well bore deviation; the hole should not stray outside of a 3 deg. to 5-deg. imaginary cone with apex at the rig. During the 1950s the industry felt that local well bore curvature or dog-leg severity (DLS) was a better measure of an acceptable well path because of its potential negative effect on drilling operations [1].

Drill collar stabilizers were introduced in the 1950s to help keep the well bore moving fairly straight and vertical. These tools were slightly under gauge in terms of drill bit diameter. Their purpose was to keep drill collars lined up with the direction of the hole. They were designed with open blades to allow drilling mud to move up the annular space unobstructed. At first, several stabilizers were placed near the drill bit. Later stabilizers were spaced according to beam theory as a means of pushing drill bits in a preferred direction. Several stabilized placements and directional effects are shown in the Appendix.

In 1951 MacDonald and Lubinski [2] proposed using conventional stabilizers to allow greater WOB without collar buckling with the added benefit of eliminating high curvature or dog-legs. Further work by Woods and Lubinski [3, 4] showed that it is possible to drill a slightly inclined well bore by establishing an equilibrium angle or a result bit force angle that is in line with the centerline of the well bore. This work provided the bases for the use of multiple stabilizers to build angle, hold angle or drop hole angle. Walker [5] gives an overview of the development of this technology.

12.1 APPLICATIONS OF DIRECTIONAL DRILLLING

12.1.1 Multiple Wells From Offshore Platforms

A common application of directional drilling is multiple production wells from offshore platforms. Without directional wells, development of offshore fields would not be economical. The purpose of multiple directional wells is to penetrate the reservoirs over a wide area to minimized or eliminate costly offset subsea completions, flow lines and number of platforms required to develop an offshore oil field.

Early methods of directional drilling required intervals of drilling, measurements of hole angle and inclination, hole correction, drilling, etc. Well bores were deflected off vertical by whipstocks at a kick-off point some 2000 ft below the mudline. The specified orientation of the well was established by multiple whipstock settings. After that, the straight inclined portion of the well was maintained using stabilized drill collars. When hole corrections were needed, adjustments were made to stabilizer locations. This process required numerous trips, which were time consuming and costly.

12.1.2 Fault Drilling

Another application is drilling across a fault plane for better reservoir production. This eliminates the hazard of drilling a vertical well through a steeply inclined fault plane, which tends to deflect the well bore off course, causing the wellbore to completely miss the reservoir or penetrate the reservoir in a nonproductive location.

12.1.3 Sidetracking and Straightening

Plugging back and sidetracking is sometimes necessary when fishing is unsuccessful, dogleg severity is extreme, or casing has failed. This usually involves a whipstock to deflect the well away from the primary wellbore, drilling laterally and then turning the hole back toward the target.

12.1.4 Salt Dome Drilling

Salt domes have a tendency to creep, especially when a bore hole is drilled into them. To avoid this problem, wells are drilled vertically and then deflected into the reservoir, sometimes underneath overhanging portions of the dome. Salt is impermeable and commonly provides a trap for oil.

12.1.5 Drain-Hole Drilling

Drain-holes are special directional bore holes that are drilled horizontally off of a vertical bore hole. After the 90-deg. turn has been made, the horizontal may be drilled out to a distance of two or three thousand feet. Horizontal drain-holes give a better sweep efficiency than vertical holes through production zones. The rate of build is approximately 20 deg. per 100 ft as compared to 3 deg. per 100 ft for typical directional wells.

12.1.6 Inaccessible Locations

Sometimes reservoirs are found under river beds, cities, mountains or other locations that cannot be reached economically by locating the rig directly over the target. Directional wells can reach targets within these reservoirs from drilling sites set off from inaccessible locations.

12.1.7 Relief Wells

Relief wells are drilled to intersect an out-of-control well to provide a means of killing the blowout. It is, therefore, important that the well path of the blown well be known and for this reason all wells should be tracked and well paths documented for possible relief drilling purposes. The size of the intersecting target depends on formation permeability and fracture pressure.

12.2 PLANNING THE DIRECTIONAL WELL

Planning directional wells for reservoir development starts with the reservoir analysis, which defines the best production well pattern. Petroleum reservoirs typically contain layers of gas, oil and water each layered according to density with gas on top, oil in the middle and connate water on bottom. The strategy for best production is to produce the oil, which is the middle layer and use the gas pressure to drive the oil to the surface. The gas will expand with little drop in pressure. Therefore, directional wells are designed to penetrate the cap rock at the oil layer.

Directional well paths depend on shape of reservoir and location of a drilling and production platform. As many as 40 directional wells may be drilled from one platform. A platform may be located offshore or in environmentally sensitive land locations. With multiple wells from one platform, well paths of each must be documented with respect to the others. The point of entry into the reservoir is also important.

There are three basic types of well paths (see Figure 12-1). Well paths may be more complicated depending on shape of the reservoir. Broken and faulted reservoirs may require entry from the side or even from below.

12.2.1 Formation Considerations

One of the challenges in controlled drilling is coping with the effects of formation layers with varying hardness and inclinations. These factors can

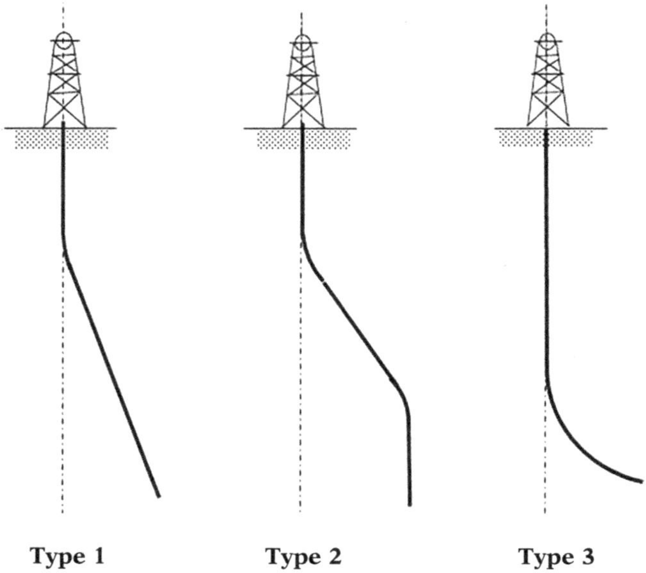

Figure 12-1. Three basic types of directional wells.

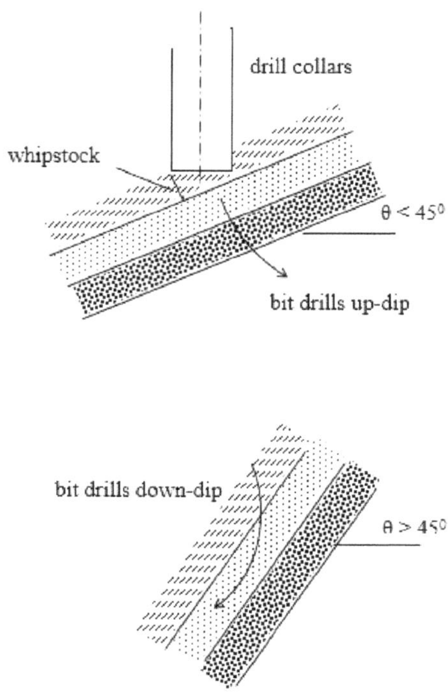

Figure 12-2. Effect of formation dip angle on hole deviation.

apply side forces causing the bit to drill off course. If geological conditions were homogeneous, side forces would be predictable and directional control would be much easier. Some of the formation conditions that influence hole path are:

- Dipping formations
- Offset Ledges
- Hard Stringers

Formation dip will cause the bit to drill up-dip or down-dip (Figure 12-2). If the formation dip angle is less than 45 deg. the bit will tend to drill up-dip. If the dip angle is greater than 45 deg. the bit will tend to drill down-dip. The explanation most commonly given for this phenomenon is the mini whipstock theory for which small whipstocks are formed in the formation directly in front of the bit: the formation cracks perpendicular to the bedding plane and not along the axis of the wellbore. These tiny whipstocks push the bit up dip. This theory is supported by micro bit tests.

12.3 WELL PATH NAVIGATION

Well path navigation is essential for tracking progress of the drill bit along a planned well path. In many respects, well path navigation is similar to flight and space navigation.

Geo-space navigation is based on incremental steps ($\Delta x, \Delta y, \Delta z$) between two successive stations, the current location (station 2) and the previous location (station 1). The objective is to predict the location of station 2 reference from station 1. The location coordinates (x,y,z), referenced from the rig site, is the sum of all incremental steps.

A significant breakthrough in marine navigation was the invention of the chronometer in the mid-1700s. The chronometer is an instrument (or clock) that keeps the time at Greenwich, England very accurately. At noon time, when the sun is directly overhead, the difference between noon and Greenwich Mean Time would allow the navigator to locate the longitude of a current location. Latitude was determined by measuring the suns angle or by measuring the angle to the North Star (in the Northern Hemisphere). Both latitude and longitude defined the coordinates of a ship. The chronometer allowed James Cooke, who was an exceptional navigator, to explore the earth during the late 1700s.

Drill bit navigation toward a reservoir target is similar, but different. Two basic measurements are required to track a well path are: well bore inclination and direction. Well bore inclination is measured by means of a pendulum relying on gravity. Well bore direction is measured by means of a compass; a magnetic compass or a gyroscopic compass may be used. Both inclination and direction are measured near the drill bit. These two measurements are used to determine the incremental steps along a well path between two survey stations. Tracking the well path establishes when and if well corrections are needed.

12.3.1 Definition of Terms

Definition of terms, commonly used in directional drilling, is given below.

Inclination; *drift angle* (α) — Angle between a vertical line and a line, which is tangent to the current well path (measured in degrees).

Azimuth angle (ε) — Direction of tangent line projection onto a horizontal plane. The direction of the projected line is measured from true North.

Course Length (CL) — Distance between two successive survey stations measured along the bore hole. In general, this is an arc distance.

Course Deviation — Length of a straight-line made by projecting a course length onto a horizontal plane.

Direction — The direction deviation or inclination of a borehole. It may be expressed in quadrant degrees (N15E) or azimuth (ε, measured from true North).

Departure (Δx) — Incremental East or West displacement of a course deviation.

Latitude (Δy) — Incremental North or South displacement of a course deviation.

Vertical Depth (Δz) — In practice, the difference between total vertical depth of two survey stations.

Measured Depth — Actual length of a borehole from its surface location to any specified survey station.

12.3.2 Wire Line Survey

Hole inclination and direction are illustrated in the single shot survey shown in Figure 12-3. The survey instrument is lowered on a wire line to a special tool

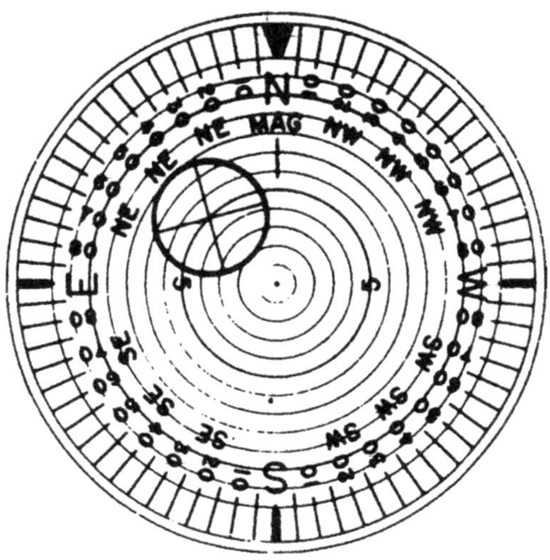

Figure 12-3. Single shot survey target.

located directly above the bit. The small circle with the cross-hairs shows the inclination at this survey station is 5 deg. and the direction of the inclination is N45E or azimuth of 45 deg. Azimuth angle (ε) is always measured from true north regardless of the quadrant. Survey instruments are also designed to measure inclinations up to 90 deg.

Following each survey, decisions are made to continue drilling with the current BHA or make hole corrections with a different set of bottom hole tools.

Current methods for tracking well path depend on these two parameters plus distance drilled between survey locations. Each method predicts the incremental steps in x, y, z directions between the location of the latest survey location (Station 2) relative to the previous survey station (Station 1) using: 1) footage drilled, 2) hole direction (ε), and 3) hole inclination (α).

12.3.3 Measurement While Drilling

Measurement While Drilling (MWD) tools measure downhole parameters and transmit this information back to the surface by means of mud pulse telemetry. This is done without disturbing drilling operations. MWD is used in directional drilling to monitor hole direction, inclination and tool face orientation. They also measure and transmit other downhole parameters, such as force, torque, pressure and temperature.

The downhole MWD tool is a complex package of sensors, data processing instruments and a binary generator. Reliability is of utmost importance. The tool can be self-monitoring.

The pressure pulsar taps out the numerical value of each parameter in binary code, which travels up the mud column to a pressure sensor mounted in the standpipe (Figure 12-4). The pressure message is then decoded by surface instruments for read-out in digital form. A fundamental difference in commercial MWD tools is the manner in which binary information is generated and transferred. Methods of transmission are:

- Positive pressure pulse
- Negative pressure pulse
- Phase Shifting of pressure signal having constant frequency

Initially survey data was obtained at 100 ft intervals. Rotary Steering Tools (RST) systems gather survey information over much shorter intervals. MWD methods have been successfully used to transmit downhole information from hole depths down to 20,000 ft. Lear and Dareing [6] show that drillstring vibrations have little or no effect on pressure pulse transmission.

Automatic computers are based on electronically manipulating binary numbers. Binary numbers contain only the numbers, one (1) and zero (0). These two numbers can be generated by electronic switches that can be turned "off and on". The position of each switch establishes the value of a binary number. For example, consider the binary number, 1011. The "off" position represents, say, the ones (1) , while the "on" position represents the zeros (0). The decimal number or real number corresponding to this particular binary number is

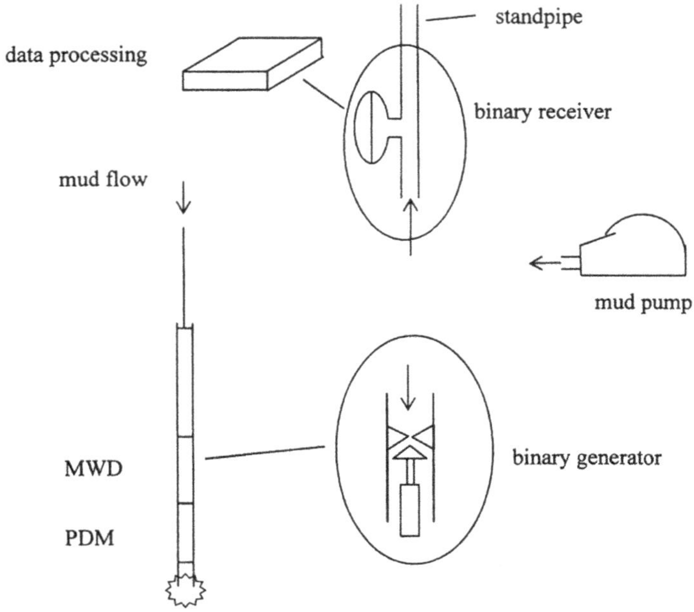

Figure 12-4. Mud pulse system.

$$\begin{array}{l} 1 \quad\quad 0 \quad\quad 1 \quad\quad 1 \quad = \\ 1 \times 2^0 + 0 \times 2^1 + 1 \times 2^2 + 1 \times 2^3 = \\ 1 \quad + \quad 0 \quad + \quad 4 \quad + \quad 8 \quad = 13 \end{array}$$

Binary numbers can be added, subtracted, multiplied and divided with electronic networks. Binary numbers are converted back to the decimal system at the surface.

Four navigation methods will now the discussed. The first method, Tangent Method, is of historical interest only. It is least accurate of the four methods but does illustrate unit vectors used in the other three. The objective of each method is to predict the incremental steps from one station to the next.

12.3.4 Tangent Method of Tracking Well Path

The tangent method provides a simple way to calculate incremental changes in xyz coordinates between Station 1 and Station 2 (see Figure 12-5). It predicts location #2 using data at #1.

$$\Delta x = CL \sin\alpha \sin\varepsilon \quad \text{(departure)} \tag{12-1a}$$

$$\Delta y = CL \sin\alpha \cos\varepsilon \quad \text{(latitude)} \tag{12-1b}$$

$$\Delta z = CL \cos\alpha \quad \text{(vertical increment)} \tag{12-1c}$$

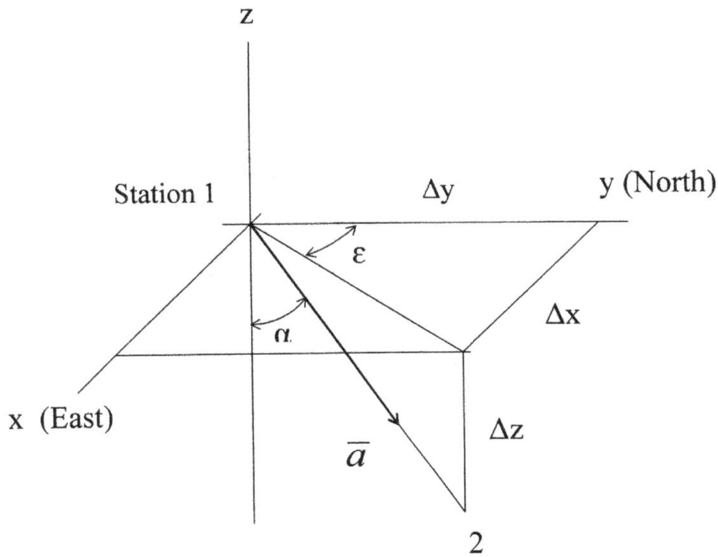

Figure 12-5. Incremental steps between two survey stations.

where

CL = course length
α = hole inclination, measured in degrees
ε = hole direction (Azimuth), measured in degrees

This method gives an approximate location of Station 2 because the well path is not necessarily a straight line between Stations 1 and 2. However, the smaller the closure distance the more accurate the projected location of station 2 by the tangent method.

This diagram also helps define the unit vector tangent to the well path at station #1.

$$\bar{a} = \sin\alpha \sin\varepsilon \bar{i} + \sin\alpha \cos\varepsilon \bar{j} - \cos\alpha \bar{k} \quad (12\text{-}2)$$

This unit vector is referenced from the right hand x,y,z axes as shown in Figure 12-5.

The straight line or tangent method of calculating well path, while relatively simple, creates error at each survey station and this error cumulates with each stepwise calculation.

12.3.5 Average Angle Method

The average angle method is mathematically similar to the tangent method. Instead of using hole inclination and direction at Station 1, these angles are replaced with the simple averages of the corresponding angles at Stations 1 and 2.

$$\alpha_a = 1/2(\alpha_1 + \alpha_2) \tag{12-3}$$

$$\varepsilon_a = 1/2(\varepsilon_1 + \varepsilon_2) \tag{12-4}$$

Incremental steps in x,y,z directions are calculated by using

$$\Delta x = CL \sin\alpha_a \sin\varepsilon_a \tag{12-5a}$$

$$\Delta y = CL \sin\alpha_a \cos\varepsilon_a \tag{12-5b}$$

$$\Delta z = -CL \cos\alpha_a \tag{12-5c}$$

These equations are strictly empirical but do give rather good results.

12.3.6 Radius of Curvature Method

This method is considered one of the more accurate well path predictors. The well path between stations 1 and 2 is assumed to follow a curved path lying on the outside surface of an imaginary cylinder (Figure 12-6a). Station 2 is the last survey station. Station 1 is the previous survey station. α_1 and α_2 are drift (inclination) angles at the two stations.

Angles ϕ_1 and ϕ_2, measured in degrees, are related to hole direction at the two stations by

$$\varepsilon_1 = \phi_1 + 90^0 \tag{12-6}$$

$$\varepsilon_2 = \phi_2 + 90^0 \tag{12-7}$$

The radius, R_H, of the imaginary cylinder containing the well path relates to arc distance, s, as shown in the plan view.

The assumed well path makes a circle on the flatted imaginary cylinder (Figure 12-6b). The radius of this circular arc is represented by R_v; the assumed well path is not a helix. R_v is determined as follows:

$$CL = R_v(\alpha_2 - \alpha_1)\frac{\pi}{180} \tag{12-8}$$

from which

$$R_v = \left(\frac{CL}{\alpha_2 - \alpha_1}\right)\frac{180}{\pi} \tag{12-9}$$

The arc distance "s" in the horizontal plane and vertical depth Δz are

$$s = R_v(\cos\alpha_1 - \cos\alpha_2) \tag{12-10}$$

$$\Delta z = R_v(\sin\alpha_1 - \sin\alpha_2) \tag{12-11}$$

From a top view (Figure 12-7c)

$$s = R_H(\phi_2 - \phi_1)\frac{\pi}{180} \tag{12-12}$$

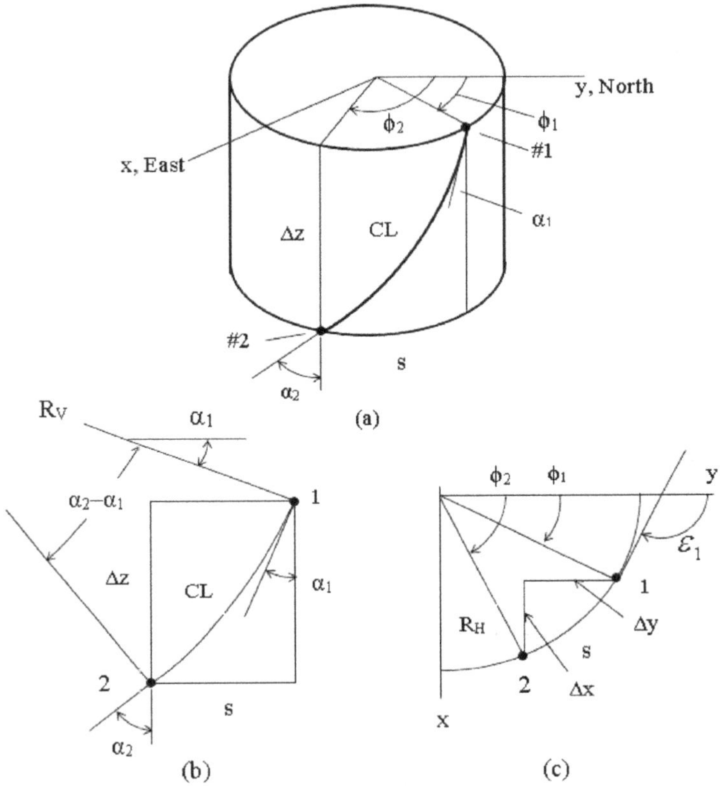

Figure 12-6. Variables used in radius of curvature method.

giving

$$R_H = \frac{s}{(\phi_2 - \phi_1)} \frac{180}{\pi} \tag{12-13}$$

where ϕ has units of degrees.

Departure and latitude can now be determined as follows:

$$\Delta x = R_H(\sin\phi_2 - \sin\phi_1) \quad \text{(departure)} \tag{12-14}$$

$$\Delta y = R_H(\cos\phi_2 - \cos\phi_1) \quad \text{(latitude)} \tag{12-15}$$

Space coordinates (x,y,z) of station 2 are determined as before. This method of calculating well paths is very accurate.

As an example, consider the following:

$\alpha_1 = 3^0$ $\alpha_2 = 5^0$
$\varepsilon_1 = 11^0$ $\varepsilon_2 = 13^0$
Course Length = 90 ft

By substitution

$$R_v = \left(\frac{90}{5-3}\right)\frac{180}{\pi} = 2,578 \text{ ft}$$

$$s = 2,578(\cos 3 - \cos 5) = 6.28 \text{ ft}$$

$$\Delta z = 2,578(\sin 3 - \sin 5) = -89.76 \text{ ft}$$

$$\phi_1 = 11 - 90 = -79^0$$

$$\phi_2 = 13 - 90 = -77^0$$

$$R_H = \frac{6.28}{(-77+79)}\frac{180}{\pi} = 179.9 \text{ ft}$$

$$\Delta x = 179.9(\sin(-77) - \sin(-79)) = +1.306 \text{ ft}$$

$$\Delta y = 179.9(\cos(-77) - \cos(-79)) = +6.189 \text{ ft}$$

12.3.7 Minimum Curvature Method

This method is also called the circular arc method because the well path between stations 1 and 2 is assumed to be an arc of a circle. The arc, in general, lies in a space plane (Figure 12-7), which is defined by unit vectors tangent to the well path at survey stations 1 and 2.

The two unit vectors are defined in terms of hole angle (α) and direction (ε).

$$\bar{a}_1 = \sin\ \alpha_1 \sin\ \varepsilon_1 i + \sin\ \alpha_1 \cos\ \varepsilon_1 j + \cos\ \alpha_1 k \qquad (12\text{-}16)$$

$$\bar{a}_2 = \sin\ \alpha_2 \sin\ \varepsilon_2 i + \sin\ \alpha_2 \cos\ \varepsilon_2 j + \cos\ \alpha_2 k \qquad (12\text{-}17)$$

Arc distance between Stations 1 and 2 is the course length, CL, and can be related to radius of curvature of the arc by

$$\cos\ \beta = \bar{a}_1 \bullet \bar{a}_2 \qquad (12\text{-}18)$$

Since

$$R\beta = CL = s \qquad (12\text{-}19)$$

$$R = \frac{CL}{\beta^0}\frac{180}{\pi} \qquad (12\text{-}20)$$

The latitude, departure and vertical distances between stations 1 and 2 are the components of vector, $\bar{B}$. This vector is

$$\bar{B} = \bar{A}_1 + \bar{A}_2 \qquad (12\text{-}21)$$

where

$$\bar{A}_1 = R \tan\ \frac{\beta}{2}\bar{a}_1 \qquad (12\text{-}22)$$

$$\bar{A}_2 = R \tan\frac{\beta}{2}\bar{a}_2 \qquad (12\text{-}23)$$

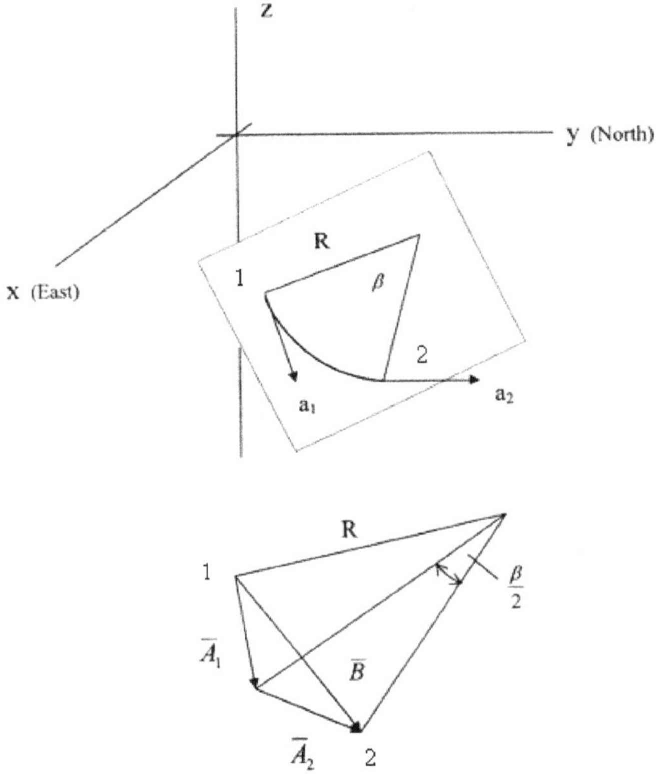

Figure 12-7. Variables used in minimum curvature method.

Using the previous example

$$\bar{a}_1 = \sin\ 3 \sin\ 11i + \sin\ 3 \cos\ 11j - \cos 3k$$

$$\bar{a}_2 = \sin\ 5 \sin\ 13i + \sin\ 5 \cos\ 13j - \cos 5k$$

$$\bar{a}_1 = 0.01i + 0.0514j - 0.9986k$$

$$\bar{a}_2 = 0.0196i + 0.0849j - 0.9962k$$

$$\cos\beta = 0.9994$$

$$\beta = 2.0417^0$$

$$R = \frac{90}{2.042}\frac{180}{\pi} = 2{,}525.7 \text{ ft}$$

$$\overline{A}_1 = 2525.7(0.0178)\bar{a}_1$$

$$\overline{A}_2 = 2525.7(0.0178)\bar{a}_2$$

$$\overline{B} = 45.0056(0.0293i + 0.1363j - 1.9948k)$$

The components of vector $\overline{B}$ define the incremental steps.

$\Delta x = 1.318$ ft
$\Delta y = 6.134$ ft
$\Delta z = -89.777$ ft

These numbers are in close agreement with the incremental steps determined by the radius of curvature method.

21.3.8 Dog Leg Severity

Dog Leg Severity (DLS) is a term used to quantify the severity of well path curvature, either intentional or unintentional. DLS affects bending in drilling tools and casing as well as drillstring friction, applied WOB and fatigue. The rate of change of hole angle with respect to distance drilled is (see Figure 12-8)

$$s = \beta R \tag{12-24}$$

$$s = \beta^0 \frac{\pi}{180} R \tag{12-25}$$

$$\frac{\beta^0}{s} = \frac{180}{\pi R} \text{ (deg./ft)} \tag{12-26}$$

Dog Leg Severity (DLS) is defined as the change in hole angle over a given drilled length. It is expressed in deg./100 ft of drilled hole.

$$DLS = \frac{\beta^0}{s} 100 \tag{12-27}$$

By substitution

$$DLS = \frac{18{,}000}{\pi R} \text{ (deg./100 ft)} \tag{12-28}$$

Dog Leg Severity defines the curvature of the well bore and affects bending of all tubulars passing through the dog leg.

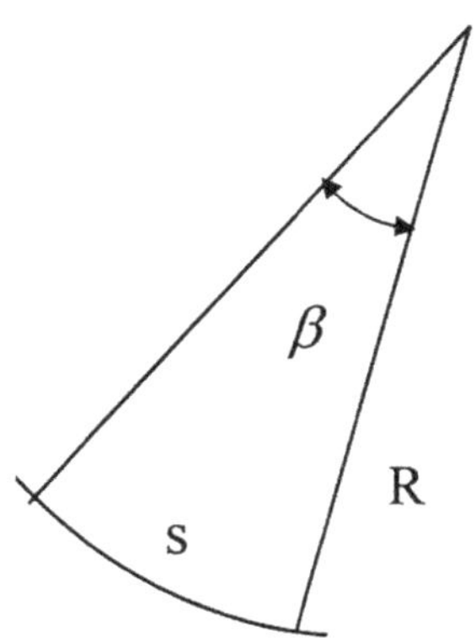

Figure 12-8. Dog leg severity.

Vectors, which define DLS are, in general, 3-D vectors. Once the angle, β, has been established between two survey stations by Eq. (12-18)

$$R = \frac{CL}{\beta(rad)} \tag{12-29}$$

and

$$\beta(rad) = \frac{(CL)DLS}{100}\left|\frac{\pi}{180}\right| \tag{12-30}$$

$$R = \frac{18{,}000}{\pi}\frac{1}{DLS}\ \text{ft} \tag{12-31}$$

For example

R = 1500 ft (DLS = $3.8^0/100$ ft)

R = 150 ft (DLS = $38^0/100$ ft)

The vector method provides a mathematical basis for calculating DLS in three-dimensional space. The DLS based on the minimum curvature method (previous example) is

$$DLS = \frac{18{,}000}{\pi}\frac{1}{2525.7} = 2.27^0/100\ \text{ft}$$

Consider, a Type I well plan having dimensions as shown in Figure 12-9.

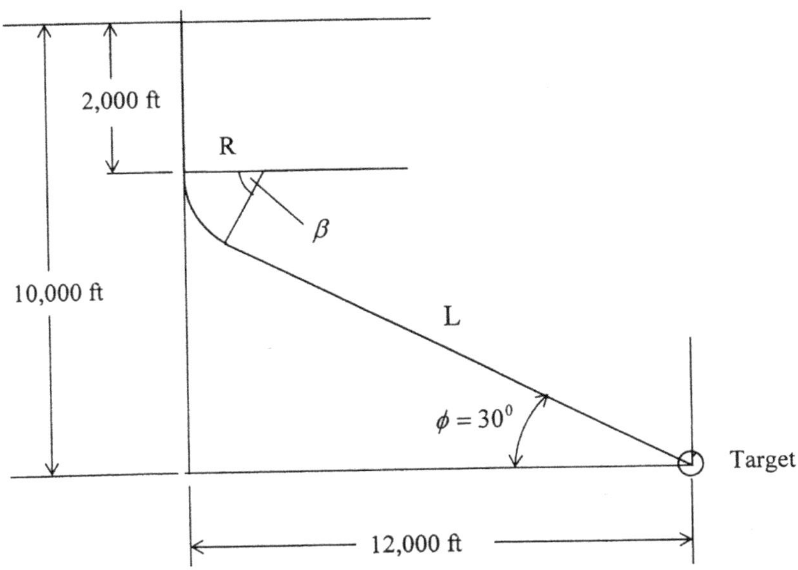

Figure 12-9. Type 1 well plan.

The dog leg severity (DLS) and total distance (s + L) to be drilled from the kick off point to the target are determined as follows.

From the geometry, $\beta = 60^0$. Radius of curvature of the build portion is determined from

$$8000 = R \cos 30 + L \sin 30$$

and

$$12{,}000 = R(1 - \sin 30) + L \cos 30$$

Simultaneous solution gives R = 1856.4 ft and L = 12,784 ft. Total footage from KOP is

$$s + L = R\frac{\beta\pi}{180} + L = 1{,}944 + 12{,}784 = 14{,}728 \text{ ft}$$

Dog-leg severity is

$$DLS = \frac{60^0}{1944}100 = 3.1^0/100\,\text{ft}$$

12.4 TOOLS FOR DIRECTIONAL CONTROL

Deflection tools are used to deflect directional wells away from the vertical at a specified kick-off point. They may also be used to correct hole direction in order to keep drilling along a given well path.

12.4.1 Whipstocks

Whipstocks are used to initiate deflection of a well, to sidetrack cement plugs and to straighten crooked holes. They can be compounded (multiple whipstocks) to direct a well bore along a desired path.

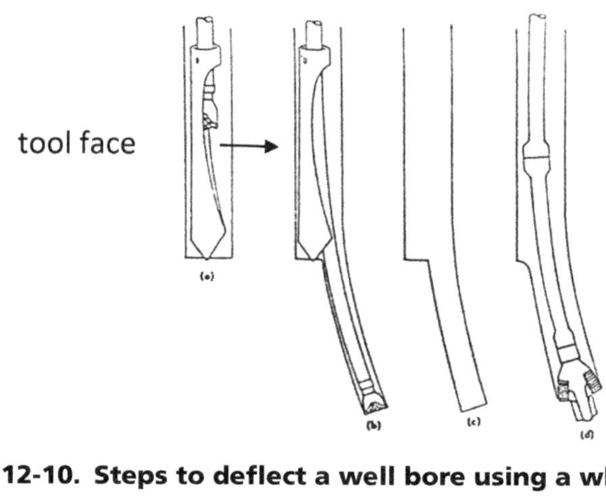

Figure 12-10. Steps to deflect a well bore using a whipstock.

The standard removable whipstock consists of a long steel wedge with a concaved surface (called the tool face) to guide the drill bit away from an established well bore (Figure 12-10). The steel wedge has a chisel point at its lower end to keep the tool face from turning once it has been set in a desired direction. After the assembly has been lower to the bottom of the hole and the tool face oriented in the desired direction, weight is applied to anchor the chisel point and to shear a pin, which allows the bit to slide along the concave wedge surface. The bit then drills a rat hole off of the whipstock, usually 12 ft or 16 ft below the anchor point.

The entire whipstock assembly is removed and this small undersized hole is opened up with a pilot bit or hole opener. A survey is then made to determine the amount of angle build. If more angle is needed, the process is repeated. This method of deflecting a hole is time consuming but produces good results.

12.4.2 Jet Bits

Jet bits are effective in soft formations to a depth of approximately 5000 ft. One nozzle is oversized, and the other two nozzles are either closed off or substantially reduced in size (see Figure 12-11).

The portion of the total flow, which travels through the oversized nozzle depends on the ratio of the area of this nozzle to total flow are (TFA) of all three nozzles.

$$Q_n = \frac{A_n}{TFA} Q \tag{12-32}$$

where

Q_n = volume flow through oversized nozzle, gpm
Q = total volume flow, gpm
A_n = area of oversized nozzle, in.^{2}
TFA = total flow area through bit nozzles, in.^{2}

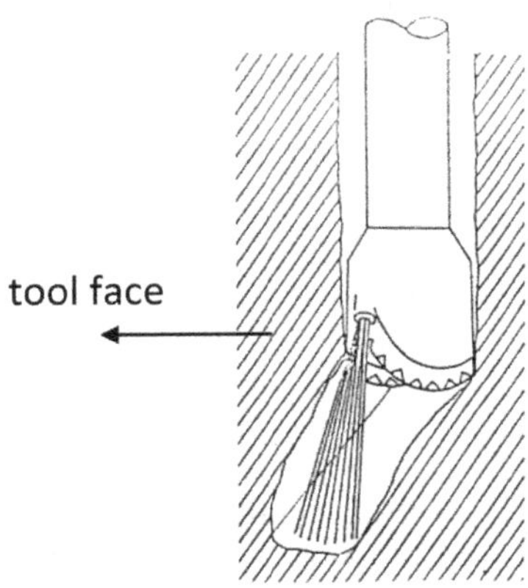

Figure 12-11. Jet bit deflection method.

The magnitude of impact force from flow through the oversized nozzle is

$$F = Q_n \left[\frac{\gamma \Delta p}{3008} \right]^{\frac{1}{2}} \tag{12-33}$$

where

F = impact force, lbs
γ = mud weight, ppg
Δp = bit pressure drop, psi

Bit pressure drop depends on total flow area (TFA) of the bit. The magnitude of the total fluid impact force depends on total volume flow through each nozzle. The fluid impact force from the oversized nozzle is, however, greater than the fluid impact forces from the other two nozzles. This greater impact force causes greater erosion, which deflects the hole in the direct of the oversized nozzle, which represents the tool face of the bit.

A jet nozzle size of ½ in. to ¾ in. is commonly used for the large nozzle with a fluid velocity of approximately 350 in. per second. Hole and bit cleaning efficiency as well as erosion efficiency will be diminished if high flow rates cannot be delivered. A jet bit assembly normally contains a near bit stabilizer to assist in the angle building when the bit is rotated after jetting.

The procedure for deflecting a well bore with a jet bit is: 1) set the tool face, 2) fix the rotary table and 3) pump through the system. The drillstring is set on bottom and spudded periodically. Jetting and spudding continue at frequent intervals until the desired curvature is established. Rotating the drillstring disorients the oversized nozzle and allows drilling to proceed normally. Well bores can be deflected by this method without pulling the jetting assembly out of the hole. Surveys are taken at regular intervals to check the path of the well bore.

12.4.3 Downhole Drilling Motors

Downhole motors were introduced [7] in the late 1960s as a means of deflecting a direction well away from vertical without using whipstocks. A bent sub was placed on top of the motor to bring about lateral drilling (Figure 12-12a). Tool face is within the bend. It was oriented in a desired plane of drilling to bring about a directional change. This arrangement began to replace whipstocks. Tool face, hole direction, and inclination were measured with a downhole sensor lowered on a wire line. Measurements were taken intermittently until the desired hole angle was achieved. Positive Displacement Motors (PDM) emerged as a preferred directional tool during this time.

The advantage of using PDMs in directional drilling was huge. It replaced the time consuming, multiple steps of setting a whipstock and surveying to establish hole direction. The bent sub and PDM did the same thing faster. At first, surveys were made by placing a survey tool in a side pocket and using a hard wire to bring hole direction and inclination information back to the surface. Later survey instruments were inserted and retrieved from within the drill collars to track progress. Once the angle had been established, a stabilized bottom hole assembly was used to maintain hole direction to the target. When the bit drilled off course, a bent sub and PDM were used again to bring the hole back on course.

As a rule, a bent sub and motor turn a well path about 75% of the bent sub angle over 30 ft of drilling. This means that a 2 deg. bent sub creates a rate of build of about 1½ deg. per 30 ft or a dog leg severity of 5 deg. per 100 ft. The maximum allowable bent sub angle is limited by the mount of off-set at the bit. If this offset is large, it is difficult to get the bottomhole assembly into the well bore. As a result, the maximum bent sub angle is limited to 2½ deg.

Bit torque is transmitted across motors and this torque creates a rotational reaction of the tool face, which depends on hole depth, friction, bit torque, and drillstring size. This reactive torque is a major consideration in setting and maintaining tool face orientation during a correction run.

Eventually, the bent sub was replaced by a bend within the motor housing (Figure 12-12b). The bent housing configuration brought about a more responsive direction change. This configuration was subsequently replaced by a tilted bearing assembly (Figure 12-12c), which located the tool face even closer to the drill bit. This configuration produced even a greater rate of directional change, which can be expressed by $\frac{\Delta\theta}{\Delta s} = \frac{\alpha}{L}$. By reducing the distance, L, the rate of directional change is increased. By moving the bend angle close to the drill bit, the angle, α could be reduced and still achieve the same rate of build.

The realigned bearing or tilted bearing assembly gave another advantage, too. The motor housing itself could be used to drill ahead as well as for making hole corrections. By rotating the motor housing, the drill bit would be continually disoriented, and drilling could proceed with motor power. Rotation of the motor housing at low speeds simply disoriented the drill bit. When hole corrections were needed, the tool face in the housing was set and the correction was made by motor power. This, along with MWD, eliminated costly tripping time.

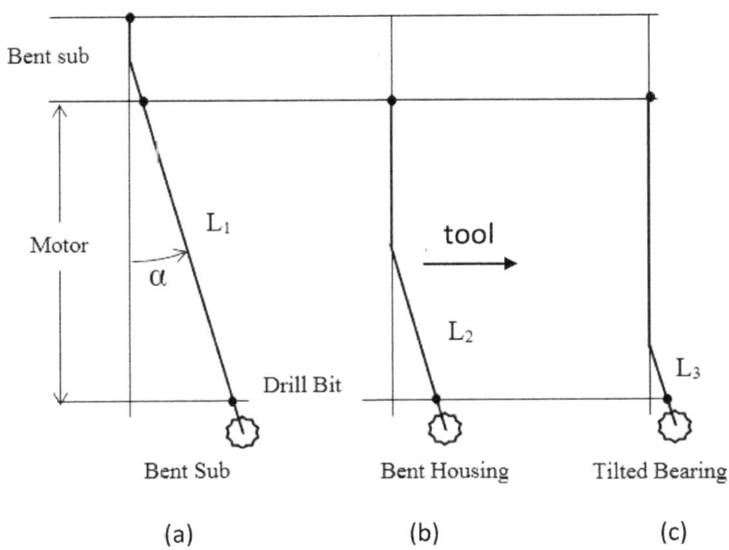

Figure 12-12. Effect of bend location on rate of build.

12.4.4 Rotary Steerable Tools

A vision for many years has been a technology that would allow directional drilling to proceed along a specified path without stopping to make directional changes. Essentially the drill bit would advance through layers of various formations much like airplanes flying and navigated along a specified flight path, with the ability to alter the course in the event of strong cross winds or bad weather. In each case three elements are essential:

- Power to move forward
- Navigation capabilities to define position and well path
- Control of the well path

Technology is now in place to direct drill bits in three-dimensional space. Tool face monitoring and control is central. The ability to monitor bit location and path through MWD and then bring about directional change is equally important. Each of these tasks is now achieved by a rotary steerable tool (RST) system. The only limitation appears to equipment reliability.

The schematic in Figure 12-13 illustrates an internal shaft, which is deflected by an internal force, F. The direction and magnitude of the force established tool face orientation and rate of change in hole angle. Note that the parameters, α and L_4 have the same role as the ones in Figure 12-13, however, the magnitude and orientation of α is adjusted while drilling. Many mechanical designs for this purpose have been patented. The author proposes another method to orient tool face and magnitude of angle α, which affects dog-leg severity.

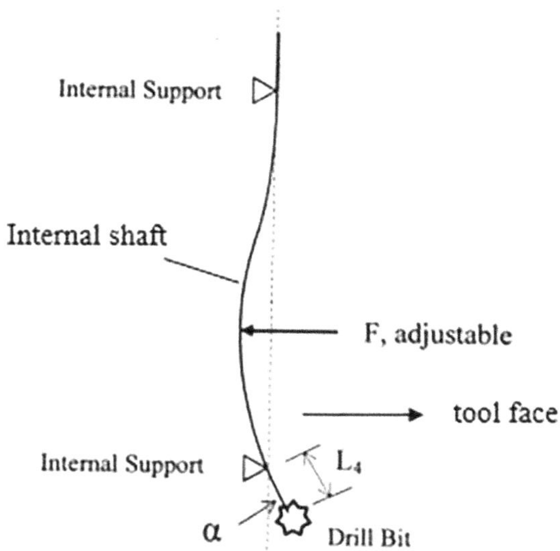

Figure 12-13. Schematic of rotary steerable tool.

12.4.5 Stabilized Bottom-Hole Assemblies

Downhole motors are commonly used to deflect a well at a kick off point toward a target. The deflection takes place in the plane of the target line. A bent sub/motor assembly is used until hole angle reaches about 6 to 8 deg.; a stabilized building assembly is not effective below these angles. In hard rock, a bent sub/motor assembly may have to establish even higher angles before a building assembly is place in the hole. Building assemblies are desirable as early as possible because tool face orientation is not a consideration. However, bent sub/motor assemblies have been used to build up to 45 deg. from the kick off point.

Bent sub/motor assemblies are used to

- Kick off
- Establish rate of build
- Make hole corrections

After hole angle is achieved, drilling proceeds with a stabilized assembly: building, holding or dropping. These assemblies differ primarily in the number and location of stabilizers.

Stabilizers are special tools that centralize drill collars at specific locations to establish a desired side load to drill bits. Stabilizer blades are roughly 2 ft long and allow fluid to flow upward in the annulus with minor flow restriction. Stabilized assemblies can build, hold, or drop hole angle, depending on stabilizer number and spacing.

12.4.6 Analytical Method for Reaction Forces

The effect of stabilizer location on magnitude and direction of drill bit side force can be calculated by treating the stabilized section as a multi-spanned beam column. One method is outlined in ref [8]. The goal is to produce the needed side force on drill bits to maintain or affect a desired well bore direction. This particular example is based on the following input data.

Hole Size = 8½ in.
Drill Collar Size = 6¾ × 2 13/16 in.
Drilling Mud Wt. = 10 ppg
Inclination of the Well Bore = 5 deg.
WOB = 50,000 lb

The output gives and accurate prediction of drill collar deflection from the drill bit to point of contact with the wall. Collar deflections between drill bit and the first stabilizer are shown in Figures 12-14 and 12-15.

Predicted side forces developed at drill bit, the stabilizers and wall contact are listed in Table 12-1.

This stabilized arrangement creates a holding to mild dropping assembly.

Prediction of bit side forces produced by various stabilizer assemblies is a guide for predicting directional performance. Ultimately stabilizer assemblies have to

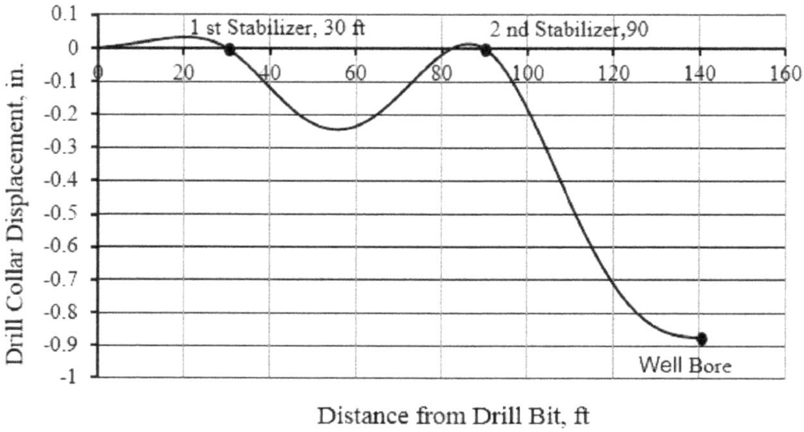

Figure 12-14. Deflections over a drill collar span.

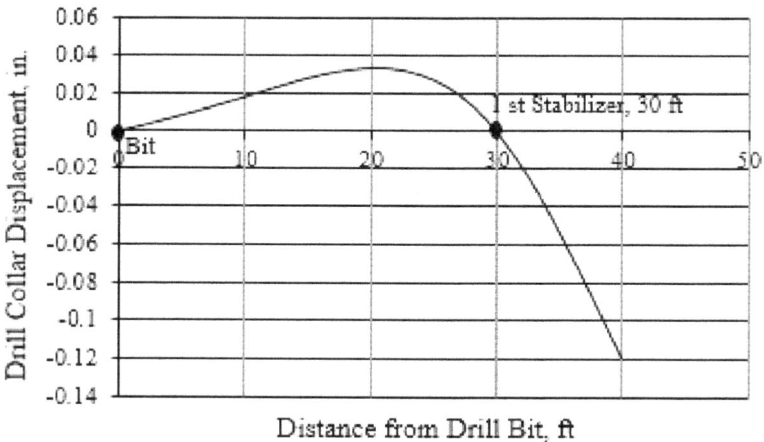

Figure 12-15. Drill collar deflections between drill bit and first stabilizer.

Table 12-1. Transverse side forces.

Location	Transverse Force, lb
Drill Bit	38.8
1st Stabilizer	289.9
2nd Stabilizer	352.2
Well Bore	157.4*

*Actually, minus shear.

be tried, tested, and usually modified in a given area until they perform as needed. A good data base of BHA performance in a given area is useful in this regard.

Building Assemblies — Building assemblies develop an upward side force on drill bits. A near bit stabilizer acts as a fulcrum point for the collar weight to pry against (Figure 12-16a). The mechanical advantage of this assembly is greatest when the first stabilizer is close to the drill bit. A second stabilizer is positioned some distance to establish a large free span distance to increase the strength of building assemblies. Positioning of the second stabilizer depends on drill collar size. Stiffer or larger drill collars will allow a longer free span distance.

Holding Assemblies — Stabilizers are positioned within drill collars so that relatively small side forces (usually less than 200 lbs) are developed at the bit. When drill bits having long shanks or large gauge protection areas are used, it is desirable to minimize the tilt angle with stabilizer placement to alleviate bit binding. A typical holding assembly is illustrated in Figure 12-16b. Calculations show that hole angle directly affects the magnitude of the side force for each of these assemblies. Side forces increase with hole angle; examples are given in the Appendix.

Dropping Assemblies — Unlike building and holding assemblies, dropping assemblies have no near bit stabilizers. The drill bit stands alone and is pushed downward by the weight of drill collars suspended between the bit and the first stabilizer, which is roughly 60 ft from the bit. The first stabilizer in this case acts much like a hinge allowing the suspended collars to swing downward like a pendulum.

To make maximum use of the pendulum assembly, the freely suspended length should be as long as possible. If the first stabilizer is located too far from the bit, the suspended collars can sag too much, touch the inside of the well bore

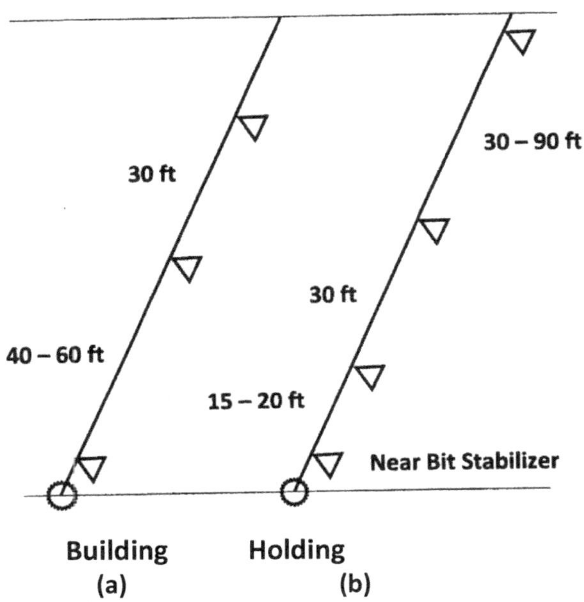

Figure 12-16. Bottom hole assemblies.

and lose the strength of the pendulum. Stiffer or larger drill collars weigh more and allow a greater suspended distance for a stronger dropping assembly.

12.5 MONITORING WELL BORE PROGRESS

As a directional well progress, decisions are made on whether to make a correction run or stay with a given bottom hole assembly, if the present assembly is not performing as expected. The real dilemma arises when a bit is still green and cannot be used as a reason for tripping out of the hole. The more hole drilled with an ineffective assembly, the shorter the remaining distance for a hole correction. It is good practice to project ahead after each survey to make sure the hole can be turned to the target, still keeping the well path within specified limits.

12.5.1 Projecting Ahead

The change in well bore inclination or direction required to turn a well toward a target cannot be achieved instantaneously. These changes have to be achieved over an interval of drilling while staying within a specified maximum dog-leg severity. The actual angle change will be greater than the estimated angle because of the footage drilled continually shortens the distance to the target. It is, therefore, important to make corrections as early as possible.

Each survey station, in a way, is a new kick-off point. If the last survey station has a unit target vector, $\bar{a}$, then it is essential to be able to turn the hole in the plane of drilling toward the target and stay within maximum dogleg severity limits. The situation depicted in Figure 12-17 shows that the target can still

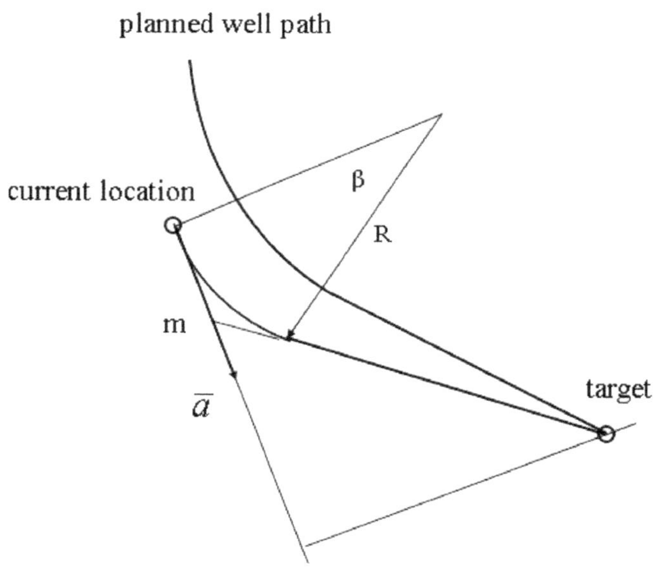

Figure 12-17. Needed correction to reach target.

be reached with a correction run having a radius of curvature, R. However, the maximum DLS must not the exceeded.

If drilling continues along direction, $\bar{a}$, a greater dogleg will be required to turn the hole toward the target. However, if drilling continues along direction $\bar{a}$ and the hole reaches point m, the hole cannot be turned toward the target. Point "m" is a critical point and prior to reaching this critical point, the well would have to be plugged back and side tracked.

12.5.2 Tool Face Orientation for Hole Correction

The tool face of any deflection devices, such as whipstocks, jet bits or motors, is the side of the device that points toward the desire hole change. Hole corrections are made by properly orienting the tool face in a required plane of drilling. There are various techniques for calculating this tool face orientation. The method given below is the vector method (Figure 12-18). Note that the x,y,z reference frame show the z axis upward to be consistent with right hand unit vector (i,j,k) system.

The objective for the formulation is to determine angle λ, which is measured from the high side of the hole to the desired plane of drilling. The plane of drilling contains the target line, the unit vector, $\bar{a}_T$, and tangent vector, $\bar{a}_2$, at current location, station 2. Unit vector, $\bar{b}$, is perpendicular to the unit tangent vector, $\bar{a}_2$ and points upward toward the high side of the hole. Unit vector, $\bar{c}$,

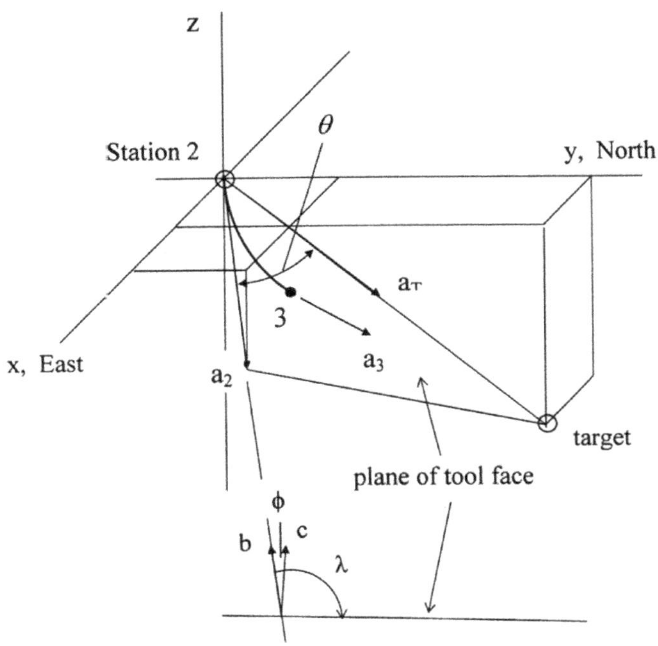

Figure 12-18. Vector method for determining tool face orientation.

is perpendicular to the desired plane of drilling or the plane containing the oriented tool face. These unit vectors are defined by

$$\bar{a}_2 = (\sin\alpha_2 \sin\varepsilon_2)\bar{i} + (\sin\alpha_2 \cos\varepsilon_2)\bar{j} - \cos\alpha_2\bar{k} \quad (12\text{-}34)$$

$$\bar{b} = (\cos\alpha_2 \sin\varepsilon_2)\bar{i} + (\cos\alpha_2 \cos\varepsilon_2)\bar{j} + \sin\alpha_2\bar{k} \quad (12\text{-}35)$$

$$\bar{c}\sin\theta = \bar{a}_2 x \bar{a}_T \quad (12\text{-}36)$$

where $\bar{i}, \bar{j}, \bar{k}$ are unit vectors in x,y,z directions. Also note that the unit vector, $\bar{c}$, is

$$\bar{c} = \frac{\bar{a}_2 \times \bar{a}_T}{\sin\theta} \quad (12\text{-}37)$$

The angle, θ, is the projected angle toward the target and is determined from

$$\cos\theta = \bar{a}_2 \bullet \bar{a}_T \quad (12\text{-}38)$$

The actual dog leg angle (β) required to turn the hole toward the target will be greater than θ (see Figure 12-19) because the hole cannot be turned instantaneously. Space between station 2 and the target will also be required.

The desired tool face orientation referenced from the top side of the hole is

$$\lambda = 90 \pm \phi \quad (12\text{-}39)$$

where ϕ is determined from

$$\cos\varphi = \bar{b} \bullet \bar{c} \quad (12\text{-}40)$$

Consider the following directional drilling situation, which requires a correction to the target.

Inclination, α = 60 deg.
Azimuth, ε = 18 deg.
Coordinates of target relative to Station 2
x = 1000 ft
y = 5000 ft
z = −2000 ft

The tangent vector, target vector and target coordinates are displayed graphically in Figure 12-19. These vectors are defined as follows:

$$\bar{a}_2 = \sin 60 \ \sin 18 \ \ \mathrm{i} \ + \sin \ 60 \ \ \cos 18 \ \mathrm{j} - \cos \ 60 \ \mathrm{k}$$

$$\bar{a}_2 = 0.2676\,\mathrm{i} + 0.8236\,\mathrm{j} - 0.5000\,\mathrm{k}$$

$$\bar{a}_T = \frac{1}{\sqrt{30}}(1i + 5j - 2k)$$

$$\bar{a}_T = 0.1826\ \mathrm{i} \ + 0.9129\,\mathrm{j} - 0.3651\ \mathrm{k}$$

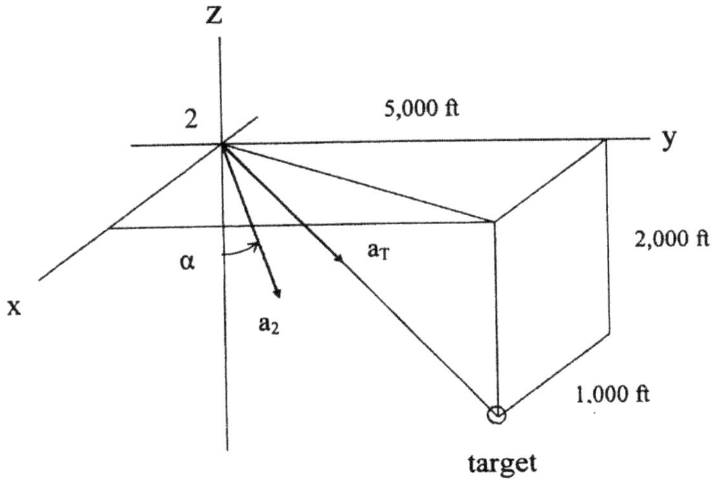

Figure 12-19. Numerical example.

Calculated parameters are

$$\cos\theta = 0.9833$$
$$\theta = 10.49^0$$
$$\bar{a}_2 x \bar{a}_T = 0.1558\ i + 0.0064\ j + 0.0939\ k$$
$$\sin\theta = 0.1821$$
$$\bar{c} = 0.8556\ i + 0.0351\ j + 0.5157\ k$$
$$\bar{b} = \cos\ 60\ \sin\ 18\ i + \cos\ 60\ \cos\ 18\ j + \sin\ 60\ k$$
$$\bar{b} = 0.1545\ i + 0.4755\ j + 0.866\ k$$
$$\cos\phi = 0.1322 + 0.0167 + 0.4466 = 0.5955$$
$$\phi = 53.45\,\text{deg.}$$

The orientation of the tool face from the top side of the hole is

$$\lambda = 90 + 53.45 = 143.45\ \text{deg.}$$

counterclockwise rotation from the surface.

Historically these calculations were made over a relative large increment of the well path. With the use of rotary steerable tools, these measurements and hole corrections are made over short drilling increments, allowing shorter response time for navigation toward the target.

Once tool face orientation is determined, the tool face in the bottom hole assembly is set from the rig floor by angular displacements of the rotary table. Two important factors affect the direct drive transmission of rotary table displacements to the tool face: 1) friction along the drillstring and 2) reactive torque across the downhole motor.

The output shaft in downhole motors applies right hand torque to drill bits. An equal but reverse torque is created in the motor housing to react against bit torque. This reverse torque or reactive torque tends to twist the entire drillstring, which causes the tool face to rotate counterclockwise. The amount of counterclockwise rotation depends on

- bit weight and bit torque
- size and length of drill pipe
- wall friction

In the absence of wall friction, the amount of angular movement of the tool face can be calculated by classical torsion equation,

$$\theta = \frac{TL}{GJ} \tag{12-41}$$

For example, each 1000 ft of 4½ (16.6) drill pipe will twist 180 deg. or ½ turn when 1500 ft-lb of reactive torque is applied. Since the reactive torque is applied to the motor housing and the tool face is coupled to motor housing, the tool face moves counterclockwise with the motor housing. Field experience indicates that as a rule of thumb, the tool face twists about 10 deg. per 1000 ft of drill pipe, showing that friction affects tool face setting.

Once tool face orientation has been established a reference mark is made on the rotary table to capture the position of the rotary table relative to the tool face orientation. When adding joints of drill pipe, the driller is careful not to lose "trapped torque" in the drillstring so that adjustments to tool face orientation can be made with slight angular movement of the rotary table. Working the drillstring up and down tends to release trapped torque causing loss of tool face setting.

12.5.3 Advancing the Well Bore in the Plane of Drilling

Once tool face orientation has been set within the new plane of drilling (Figure 12-20), the question becomes how to best reach the target from the

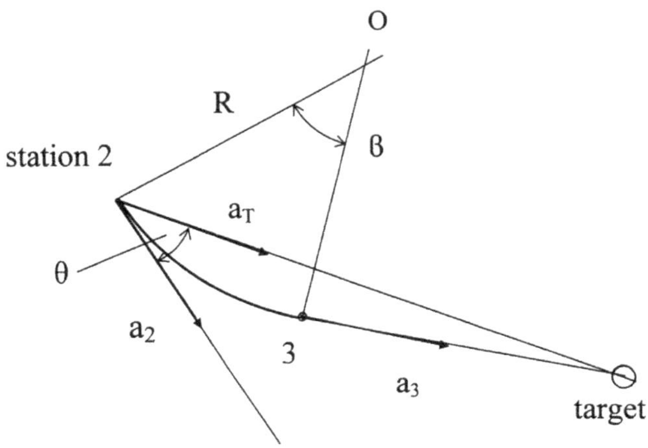

Figure 12-20. 3-D Plane of hole correction.

current location (station 1). Obviously, the well bore cannot be turned (θ) toward the target immediately. Distance will be required to make the turn in order to stay within a specified DLS. If the current location is too close to the target, there may not be enough space to turn the well bore toward the target. It is desirable to make a straight-line approach into the target, much like an airplane approaching a runway for landing.

The radius, R, of the turn is established from the specified DLS. The drilling path from the current location to the target can be determined from the given information. Recasting the drilling plane as shown in Figure 12-21, arc distances and linear distance ℓ (as well as the turning angle, β) are found as follows:

The coordinates of the target referenced from the η,ξ reference frame are

$$\eta_T = L \sin\theta \tag{12-42}$$

$$\xi_T = L \cos\theta \tag{12-43}$$

The equation defining the straight-line portion of the new well path is

$$\eta = \xi \tan\beta - R\tan\frac{\beta}{2}\tan\beta \tag{12-44}$$

All variables are known except the angle, β, once the target coordinates are substituted.

In the previous example

$$\eta_T = 5477.2 \sin 10.49 = 997.2 \text{ ft}$$

$$\xi_T = 5477.2 \cos 10.49 = 5385.6 \text{ ft}$$

The angle, β, was determined by solving the following equation numerically.

$$\tan\beta\left(5{,}383.6 - 1432\tan\frac{\beta}{2}\right) - 997.2 = 0$$

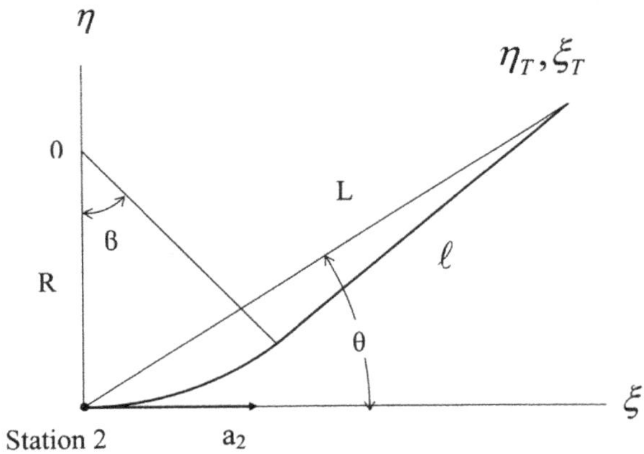

Figure 12-21. Required dog leg to hit target.

giving $\beta = 10.75^0$. The corresponding arc length, s, is

$$s = (1432)10.75\frac{\pi}{180} = 268.67 \text{ ft}$$

The DLS is calculated to be

$$\text{DLS} = \frac{10.75}{268.67}100 = 4$$

or 4 deg./100 ft, which agrees with the specified DLS. The distance (ℓ) of the straight portion of the hole is determined as follows:

$$\ell^2 = [997.2 - 1432(1 - \cos 10.75)]^2 + [5385.6 - 1432 \sin 10.75]^2$$

$$\ell^2 = 944{,}918 + 26{,}199{,}017$$

$$\ell^2 = 5210 \text{ ft}$$

Total footage to be drilled from station 2 to the target is

$$\ell + s = 5210 + 269 = 5479 \text{ ft}$$

Note the hole correction in this case could be made with a DLS = 4. Had the hole been drilled further without make this correction, a point would be reached where it would impossible to reorient the well bore toward the target and stay within DLS limits.

Another method sometimes used to determine DLS is based on Eq. (12-45), which has the form of the Law of Cosines.

$$\beta^2 = \alpha_1^2 + \alpha_2^2 - 2\alpha_1\alpha_2 \cos \Delta\varepsilon \qquad (12\text{-}45)$$

where

β = angular sweep of well bore between stations 1 and 2
α_1 = hole drift at station 1
α_2 = hole drift at station 2
$\Delta\varepsilon$ = change in azimuth

Equation (12-45) is developed from Eq. (12-18) with simplifications to Eqs. (12-16) and (12-17) using

$$\cos \beta = \bar{a}_1 \bullet \bar{a}_2 \qquad (12\text{-}46)$$

$$\cos \beta = 1 - \frac{\beta^2}{2!} + \frac{\beta^4}{4!} - \ldots\ldots$$

$$\sin \alpha = \alpha - \frac{\alpha^3}{3!} + \frac{\alpha^5}{5!} - \ldots\ldots$$

$$\cos(\varepsilon_1 - \varepsilon_2) = \cos \varepsilon_1 \cos \varepsilon_2 + \sin \varepsilon_1 \sin \varepsilon_2$$

where

$$\bar{a}_1 = \sin\alpha_1 \sin\varepsilon_1 i + \sin\alpha_1 \cos\varepsilon_1 j + \cos\alpha_1 k \tag{12-47}$$

$$\bar{a}_2 = \sin\alpha_2 \sin\varepsilon_2 i + \sin\alpha_2 \cos\varepsilon_2 j + \cos\alpha_2 k \tag{12-48}$$

The application of Eq. (12-45) is limited to small values of drift angles and direction because of these simplifications.

12.5.4 Effect of a Given Building Assembly on Hole Movement

Another question of interest is the movement of the well path caused by a given building assembly and its tool face oriented from the top side of the hole. If the building assembly is quantified by $\frac{\beta^0}{s(ft)}$, then its build radius is $R(ft) = \frac{s(ft)}{\beta(rad)}$.

With reference to Figure 12-22, Station 1 represents the current hole location having a direction defined by vector, $\bar{a}_1$.

First consider the hole change taking place in the vertical plane (x_1 z_1 plane). After drilling "s" ft with a build angle of β deg., the bit moves upward by (see Figure 12-23)

$$r = R(1 - \cos\beta) \tag{12-49}$$

Now assume the drillstring is rotated counterclockwise by ψ degrees. The position of the bit relative to the $x_1y_1z_1$ reference frame becomes

$$\Delta x_1 = R\sin\beta$$
$$\Delta y_1 = r\sin\psi$$
$$\Delta z_1 = r\cos\psi$$

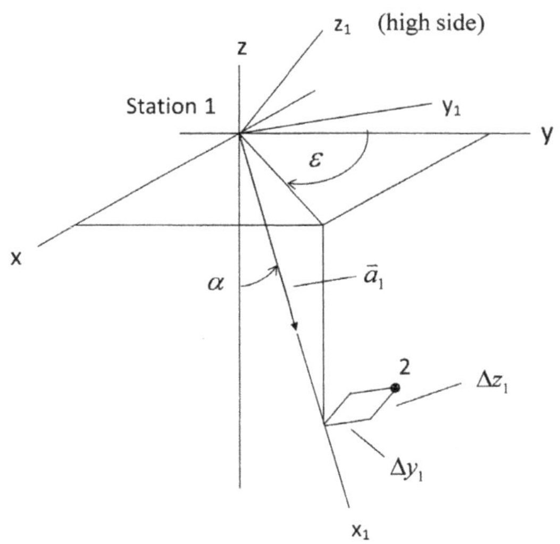

Figure 12-22. Turning hole with a given building assembly.

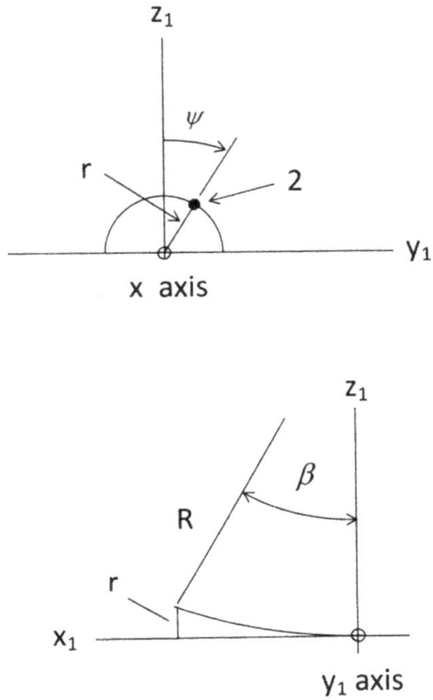

Figure 12-23. Movement of drill bit.

The location of point 2 relative to the fixed reference frame is

$$\vec{d} = \vec{d}_1 + \vec{d}_2 \tag{12-50}$$

where

$\vec{d}_1 = xi + yj + zk$
$\vec{d}_2 = \Delta x_1 i_1 + \Delta y_1 j_1 + \Delta z_1 k_1$

Expressions for the unit vectors $i_1 j_1 k_1$ in terms of unit vectors ijk are

$$i_1 = \bar{a}_1 = \sin\alpha \sin\varepsilon(i) + \sin\alpha \cos\varepsilon(j) + \cos\alpha(k) \tag{12-51}$$

$$j_1 = -\cos\varepsilon(i) + \sin\varepsilon(j) \tag{12-52}$$

$$k_1 = \cos\alpha \sin\varepsilon(i) + \cos\alpha \sin\varepsilon(j) + \sin\alpha(k) \tag{12-53}$$

12.6 FRICTION IN DIRECTIONAL WELLS

Directional wells may extend laterally several thousand feet. The well plan may call for defined paths to be drilled to reach a specific target in a reservoir. Complex well paths can be drilled and navigated by rotary steerable tools and MWD monitoring. Friction forces created by complex well paths can create high drillstring torque and pull out forces. These loads can sometimes exceed the capability of the surface equipment and strength of the drillstring. During directional drilling operations, drill pipe is usually limited by rotating torque and not by direct pull. One practice is to allow rotary torque to reach 80% of makeup torque.

Through experience, it has been determined that friction forces can be greatly reduced by rotating the pipe and maintaining mud circulation during tripping. Top drives have been designed for this purpose.

12.6.1 Coefficient of Friction

Johancsik, Friesen, Dawson [9] measured axial pipe force and rotary torque directly below the Kelly. This information, along with a computer model of drillstring friction, was used to determine the coefficient of friction in three different wells (9790 ft, 15,573 ft, and 12,200 ft). The coefficient of frictions for each well is given in Table 12-2. In each well, rotary torque at the surface was: 15,900, 18,300, and 24,500 ft-lb.

The friction model for computational purposes is typically based on Coulomb type frictions where the normal force is the resultant contact force over a 100 ft drilled interval. Torque predictions were tangential friction multiplied by pipe radius.

An interesting aspect of the measured data is that the total friction torque as measured at the surface is about 10 times the magnitude of typical drilling torque (~1500 ft-lbs). In the third well, the pick-up force was about 75% greater than the calculated string weight. While these numbers will vary from well to well, they do show the significance of friction in directional wells.

12.6.2 Effect of Friction on WOB

Weight on bit (WOB) is typically developed by gravity. In a vertical well the required length of drill collars is determined by

$$L_{dc} = \frac{BW}{0.85wBF} \tag{12-54}$$

as discussed in Ch 5. When drilling on a inclined plane, drill collar forces change. Considering the freebody diagram in Figure 12-24, the longitudinal and normal forces yield

$$\sum F_x = 0 \tag{12-55}$$

Table 12-2. Coefficients of frictions.

	Well #1	Well #2	Well #3*
Pick-up	0.28	0.31	0.40
Slack-off	0.27	0.31	0.41
Rotating	0.27	0.29	0.41

* Well #3 had several severe dog legs in the lower portion of the build zone.

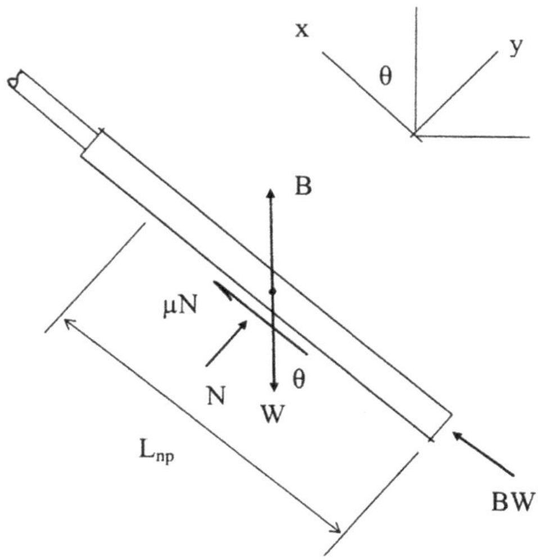

Figure 12-24. Drill collars on an inclined plane.

$$BW + \mu N - wL_{np}BF\cos\theta = 0 \tag{12-56}$$

$$\sum F_y = 0 \tag{12-57}$$

$$N - wL_{np}BF\sin\theta = 0 \tag{12-58}$$

Combining gives

$$BW = wL_{np}BF(\cos\theta - \mu\sin\theta) \tag{12-59}$$

$$\frac{BW}{L_{np}wBF} = \cos\theta - \mu\sin\theta \tag{12-60}$$

$$Ratio = \frac{L_{np}wBF}{BW} = \frac{1}{\cos\theta - \mu\sin\theta} \tag{12-61}$$

This ratio (Figure 12-25) illustrates the effect of inclination and coefficient of friction on length of drill collars required to develop a given WOB. If hole inclination is 50 deg. and $\mu = 0.2$, drill collar length must be doubled in order to develop in order to develop the same bit force as in a vertical well.

The coefficient of friction directly influences the amount of bit weight that can be derived from a given length of drill collars. The coefficient of friction can be reduced by drilling mud properties. Tests indicate that the coefficient of friction has values between 0.1 and 0.3 for oil base muds. Experience shows that friction is reduced further by drillstring rotation. It is common practice to rotate the drillstring, even though motors and turbines are supplying power the drill bits.

Figure 12-25 shows that as hole inclination approaches 90 deg., drill collar gravity becomes less and less effective in applying force to drill bits. In

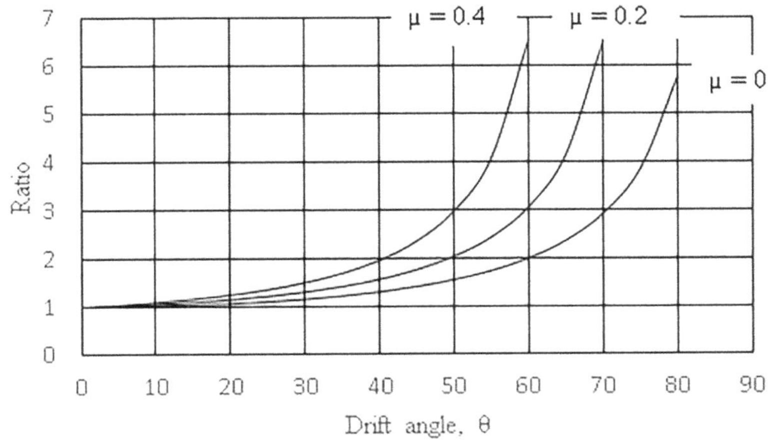

Figure 12-25. Effect of hole inclination on weight on bit.

horizontal drain hole drilling, gravity of pipe in the vertical section is often utilized to created force in the horizontal section.

12.6.3 Friction on Drillstrings While Tripping Out

The forces acting on a straight hole with inclination, θ, is shown in Figure 12-26. The direction of the friction force is downward indicating the pipe is being pulled out of the hole. The difference between F_1 and F_2 is the friction force applied over pipe length L.

The equations of statics lead to friction force per length of pipe.

$$\sum F_x = 0 \tag{12-62}$$

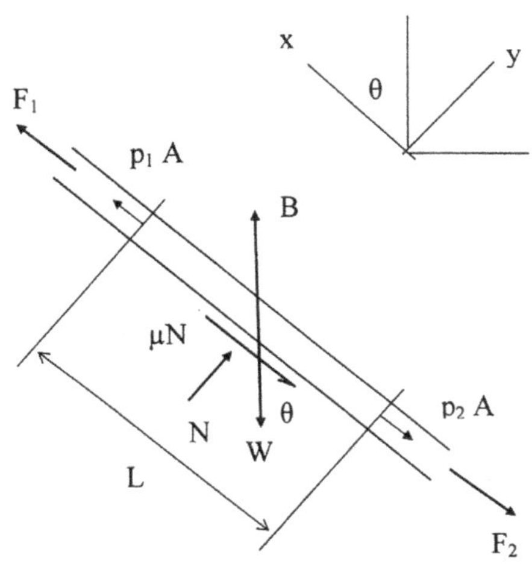

Figure 12-26. Friction force per unit length.

$$F_1 - F_2 + (p_1 - p_2)A - \mu N - (W - B)\cos\theta = 0 \tag{12-63}$$

The pressure terms are needed for equilibrium since the buoyancy B term assumes all surface are wetted.

$$\sum F_y = 0 \tag{12-64}$$

$$N - (W - B)\sin\theta = 0 \tag{12-65}$$

Combining gives

$$\frac{\Delta F}{L} = w_m \cos\theta + wBF(\cos\theta + \mu\sin\theta) \tag{12-66}$$

Friction during downward movement of pipe is determined by changing the sign on the coefficient of friction term, μ. Reference [9] shows a substantial difference between pick-up and slack-off weight.

12.7 FRICTION IN CURVED SECTION

12.7.1 Soft Pipe Model

The soft string model does not include the bending stiffness of the pipe. The free body diagram for this model is shown in Figure 12-27.

Definitions of terms in this figure are:

T = local tension
n = normal force per unit length
f = friction force per unit length

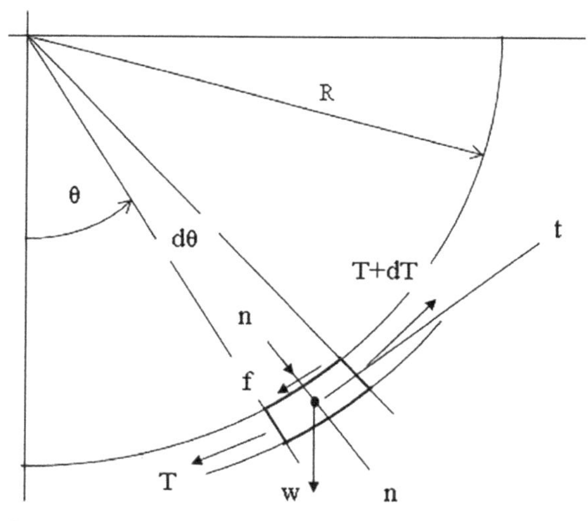

Figure 12-27. Freebody diagram of soft pipe.

w = weight per unit length
n,t = normal and tangent axes

Summation of the forces in the tangent direction gives

$$d(T) = fRd\theta + wRd\theta\sin\theta \qquad (12\text{-}67)$$

Summing forces in the normal direction gives

$$nR = T - wR\cos\theta \qquad (12\text{-}68)$$

Note that the internal shear force at any cross section can be shown to be equal to zero by summing moments about the center of curvature and combining this equation with the tangent force component.

Noting that $f = \mu n$ and combining Eqs. (12-67) and (12-68) gives

$$\frac{dT}{d\theta} - \mu T = Rw[\sin\theta - \mu\cos\theta]$$

The complimentary solution is

$$T_c = Ce^{\mu\theta} \qquad (12\text{-}69)$$

where C is an arbitrary constant.

The particular solution can be determined by using complex variables.

$$\overline{T}_p = \overline{A}e^{i\theta} \qquad (12\text{-}70)$$

and replacing

$$\sin\theta = -ie^{i\theta}$$

$$\cos\theta = e^{i\theta}$$

gives

$$\overline{T}_p = Rw\frac{-(i+\mu)}{i-\mu}e^{i\theta} \qquad (12\text{-}71)$$

The numerator is equal to $re^{i\phi}$ while the denominator is equal to $re^{i(\pi-\phi)} = -re^{-i\phi}$, where $\tan\phi = \frac{1}{\mu}$. Bringing these terms together gives

$$\overline{T}_p = Rwe^{i2\phi}e^{i\theta} \qquad (12\text{-}72)$$

Giving a particular solution of

$$T_p = Rw\cos(\theta + 2\phi) \quad (\text{real part of } \overline{T}_p) \qquad (12\text{-}73)$$

Combining Eqs. (12-69) and (12-73) gives the total solution

$$T = Ce^{u\theta} + Rw\cos(\theta + 2\phi) \qquad (12\text{-}74)$$

Imposing the boundary condition that $T(\theta_0) = T_0$ at $\theta = \theta_0$ gives

$$C = e^{-\mu\theta_0}\left[T_0 - Rw\cos(\theta_0 + 2\phi)\right] \tag{12-75}$$

Equation (12-74) represents the solution and defines tension in the pipe at each θ location.

When the weight term is dropped, Eq. (12-74) gives the belt friction solution [10].

$$T = T_0 e^{\mu(\theta - \theta_0)} \tag{12-76}$$

Consider the following example.

R = 1500 ft
w = 20 lb/ft (arbitrary)
$T_0 = 40{,}000\,\text{lbs}$
$\theta_0 = 0\,\text{deg.}$
$\mu = 0.3$

Internal tension along the pipe between θ = 0 deg. and θ = 90 deg. is shown in Figure 12-28. If pipe weight is ignored the solution resorts to a belt problem, whose solution is shown by the dashed line.

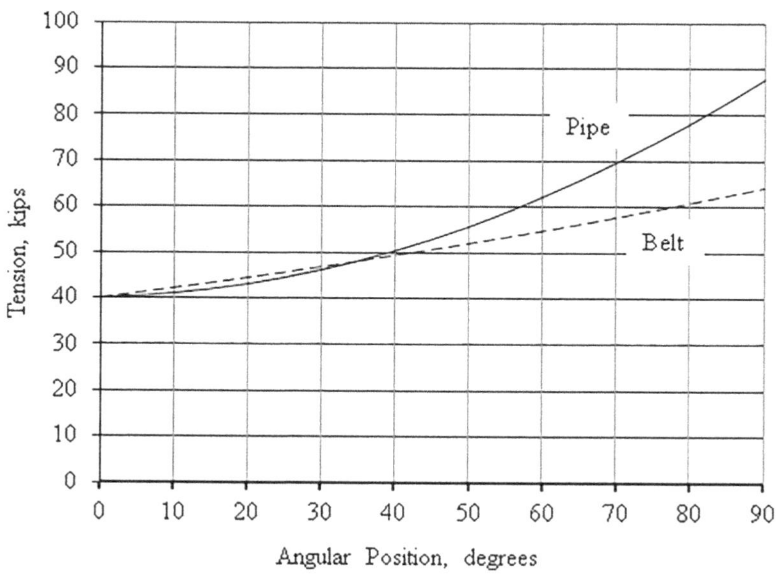

Figure 12-28. Tension around a curved well path (based on soft model).

12.7.2 Bending Equations in Polar Coordinates

The soft model does not include the effects of stiffness and while it may be a guide as to the level of friction in dog-legs, its accuracy is suspect. In reality, bending stiffness does affect pipe bending and friction within curved portions of directional wells. Polar coordinates are useful in analyzing affect.

The differential element of length ds = R dθ is used to establish the necessary bending equations in terms of polar coordinates (Figure 12-29).

Summing forces in the "n" direction gives

$$\frac{dV}{ds} = q + \frac{T}{R} \tag{12-77}$$

Summing forces in the "t" direction gives

$$\frac{dT}{ds} = -\frac{V}{R} \tag{12-78}$$

Summing moments about the center of curvature of the well bore gives

$$\frac{dM}{ds} - \frac{d(Tw)}{ds} + R\frac{dT}{ds} = 0 \tag{12-79}$$

By substitution

$$\frac{dM}{ds} - V - T\frac{dw}{ds} = 0 \tag{12-80}$$

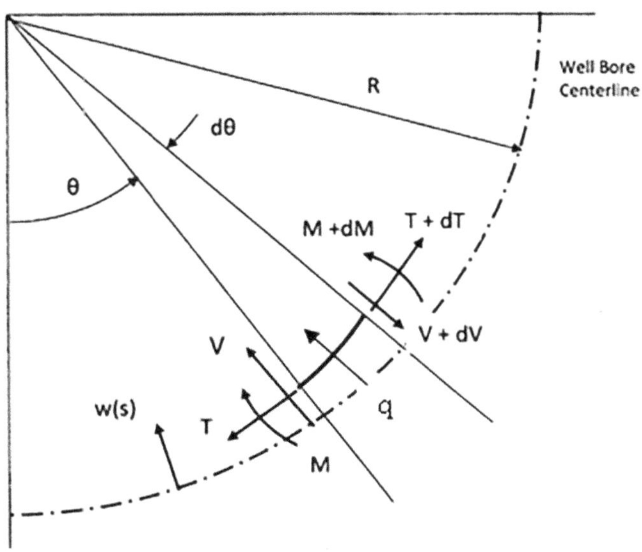

Figure 12-29. Equations of bending in terms of polar coordinates.

$$\frac{d^2M}{ds^2} = \frac{dV}{ds} + \frac{d}{ds}\left[T\frac{dw}{ds}\right]$$

$$\frac{d^2M}{ds^2} = q + \frac{T}{R} + \frac{d}{ds}\left[T\frac{dw}{ds}\right]$$

The curvature of the displaced beam is expressed in terms of beam deflection (see Timoshenko and Woinowsky-Krieger [11, pp. 503–504])

$$\frac{1}{\rho} = \frac{1}{R} + \frac{w}{R^2} + \frac{d^2w}{ds^2} \tag{12-81}$$

Assuming Euler bending

$$M = EI\frac{1}{\rho} \tag{12-82}$$

and combining the above equations gives

$$q + \frac{T}{R} + \frac{d}{ds}\left[T\frac{dw}{ds}\right] = \frac{EI}{R^2}\frac{d^2w}{ds^2} + EI\frac{d^4w}{ds^4} \tag{12-83}$$

$$EI\frac{d^4w}{ds^4} + \frac{EI}{R^2}\frac{d^2w}{ds^2} - \frac{d}{ds}\left[T\frac{dw}{ds}\right] = q + \frac{T}{R} \tag{12-84}$$

In general, T can vary with s, however, in many cases T can be assumed constant over stepwise pipe intervals. In which case Eq. (12-84) simplifies to

$$\frac{d^4w}{ds^4} - \lambda^2\frac{d^2w}{ds^2} = \zeta \tag{12-85}$$

where

$$\lambda^2 = \frac{T}{EI} - \frac{1}{R^2}$$

$$\zeta = \frac{Rq + T}{REI}$$

In this discussion q, the distributed load is assumed. The solution to the differential equation is

$$w(s) = -\frac{1}{2}\chi s^2 + As + B + C\cosh\lambda s + D\sinh\lambda s \tag{12-86}$$

Letting

$$\chi = \frac{T}{RT - \frac{EI}{R}}$$

Once w(s) has been determined, shear and bending moment can be established.

Shear force along the pipe

$$V = \frac{dM}{ds} - T\frac{dw}{ds} \tag{12-87}$$

Bending moment along the pipe is

$$M = EI\left(\frac{1}{R} + \frac{w}{R^2} + \frac{d^2w}{ds^2}\right) \tag{12-88}$$

In this case, λ^2 is assumed to be plus. In some cases, this coefficient could be minus making the bending equation

$$\frac{d^4w}{ds^4} + \lambda^2\frac{d^2w}{ds^2} = \frac{q}{EI} \tag{12-89}$$

For example, T could be a compressive force. The second term now has a (+) sign and this leads to a complimentary solution with sines and cosines.

12.7.3 Application to Pipe Bending in Dog Legs

A major challenge is analysis of bending and friction on drillstrings in dog legs. Several parameters factor into the complexity of this problem: location of points of contact, magnitude contact forces and resulting friction, shape of the dog leg, tension, flexural stiffness of pipe to mention a few.

Consider a pipe making contact at three points as shown in Figure 12-30. This problem is like a straight beam supported at three points. It is statically indeterminate and requires the solution to beam deflection as well as statics. The same is true of the model in the figure. In general, the center contact point will not be in the center between points A and B requiring solutions to sections on both sides of point C with matching boundary conditions at C.

For the sake of simplicity, we assume point C is midway between A and B, to take advantage of symmetry. The problem in this case becomes a cantilever beam problem, using polar coordinates.

The bending equations for the CB section are given as follows.

$$w(s) = -\frac{1}{2}\chi s^2 + As + B + C\cosh\lambda s + D\sinh\lambda s$$

The boundary conditions that apply are

$$w(0) = \delta \qquad w(\ell) = -\delta$$

$$\left.\frac{dw}{ds}\right|_{s=0} = 0 \qquad \left.\frac{1}{\rho}\right|_{s=\ell} = 0 \quad \text{which leads to} \quad \left.\frac{d^2w}{ds^2}\right|_{s=\ell} + \left.\frac{w}{R}\right|_{s=\ell} = -\frac{1}{R}$$

where ℓ is arc length between points C and B. Distance, δ, is inward (+) radial displacement of point C and outward (–) radial displacement of point B.

The boundary conditions quantify the constants, A, B, C, D. The moment at point C is determined by eq 12-88. Forces V and R_B are determined by statics. Reaction R_C is equal to 2V.

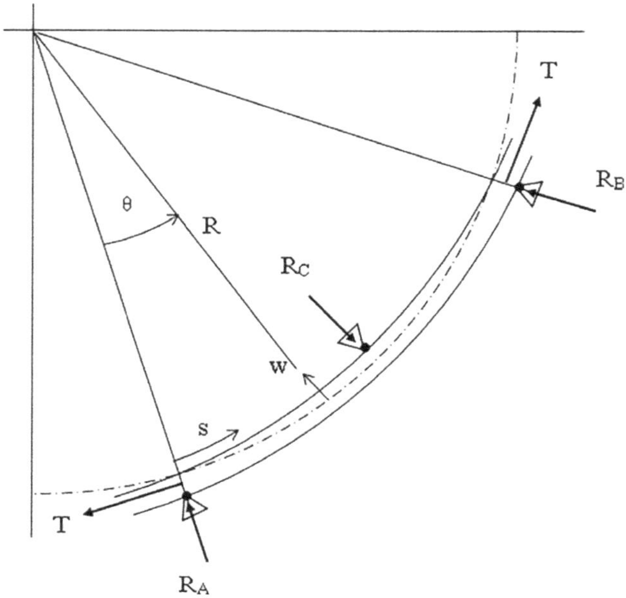

Figure 12-30. Model of pipe with three points of contact.

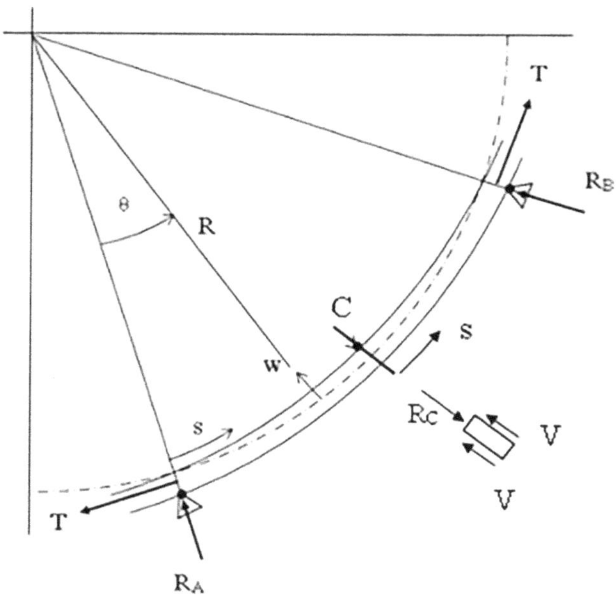

Figure 12-31. Isolation of pipe section between points C and B.

12.7.4 Multi-Spanned Beam Approach

Another approach for solving the multi-spanned beam problem is explained by Huang, Dareing, Beran [11]. Results are given for straight beams, but the method applies equally well to polar coordinates.

Another approach for solving the multi-spanned beam problem is explained by Huang, Dareing, Beam [11]. Results are given for straight beams, but the method applies equally well to polar coordinates.

A stepwise approach was used by Dareing and Ahlers [13] and Rocheleau and Dareing [14] predict drill pipe deflection between tool joints as well as contact forces and friction. The model in both papers assumes contact is made by each tool joint within a curved portion of the well bore. The solutions give insight as to how bending stiffness enters into the friction calculations. The basic equations for 1) pulling out of the well bore and 2) going in to the well bore are explained below.

In both papers, pipe is analysis by considering deflections of segments between tool joints. It is also assumed that tool joints make contact with the well bore (Figure 12-32).

12.7.4.1 Pulling Out of the Well Bore

The differential equation of bending for each pipe section is

$$\frac{d^4w}{ds^4} - \lambda^2 \frac{d^2w}{ds^2} = \zeta \qquad (12\text{-}90)$$

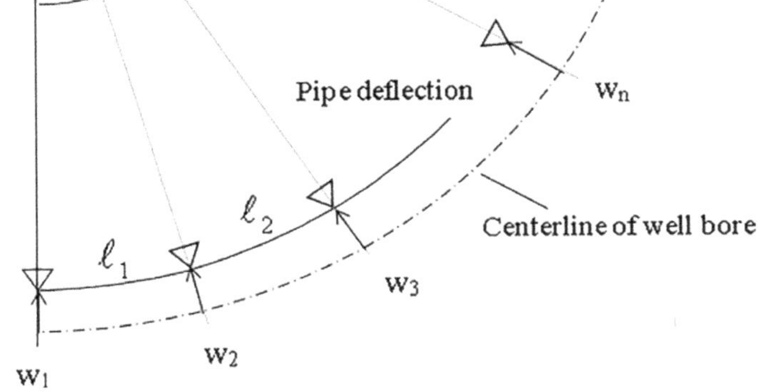

Figure 12-32. Multiple spanned pipe in a dogleg.

Table 12-3. Pull out force over a 90 deg. turn; R = 200 ft (4½ in. 16.5 lb/ft drill pipe).

Back pull (lbs)	Top Tension free-free (lbs)	Top Tension fixed-fixed (lbs)	Top Tension no stiffness (lbs)
2,000	6,162	5,544	4,234
4,000	9,421	8,773	7,438
6,000	12,794	12,361	10,642
8,000	16,164	15,849	13,846
10,000	19,614	19,337	17,050
12,000	23,117	22,826	20,254

where

$$\lambda^2 = \frac{T}{EI} - \frac{1}{R^2}$$

$$\zeta = \frac{Rq + T}{REI}$$

The general solution to Eq. (12-90) is

$$w(s) = -\frac{1}{2}\chi s^2 + As + B + C\cosh\lambda s + D\sinh\lambda s$$

Following the above procedure given in ref [12], pull-out forces corresponding to six (6) different pull back forces were calculated, and the results given in Table 12-3.

12.7.4.2 Going Into the Well Bore

The equation to be solved in this case is slightly different because of the change in direction friction.

$$\frac{d^4w}{ds^4} + \lambda^2\frac{d^2w}{ds^2} = \zeta \qquad (12\text{-}91)$$

where

$$\lambda^2 = \frac{T}{EI} + \frac{1}{R^2}$$

$$\zeta = \frac{Rq - P}{REI}$$

Note the sign change in the second term resulting from compression. The general solution in this case is

$$w(s) = -\frac{1}{2}\chi s^2 + As + B + C\cos\lambda s + D\sin\lambda s$$

The results of this analysis are given in ref. [13].

12.7.5 Summary of Equations

Average Angle Method

$$\alpha_a = 1/2(\alpha_1 + \alpha_2)$$

$$\varepsilon_a = 1/2(\varepsilon_1 + \varepsilon_2)$$

$$\Delta x = CL\sin\alpha_a \sin\varepsilon_a$$

$$\Delta y = CL\sin\alpha_a \sin\varepsilon_a$$

$$\Delta z = -CL\cos\alpha_a$$

Radius of Curvature Method

$$\Delta x = R_H(\sin\phi_2 - \sin\phi_1) \quad \text{(departure)}$$

$$\Delta y = R_H(\cos\phi_2 - \cos\phi_1) \quad \text{(latitude)}$$

$$\Delta z = R_v(\sin\alpha_1 - \sin\alpha_2)$$

Minimum Curvature Method

$$\bar{a}_1 = \sin\alpha_1 \sin\varepsilon_1 i + \sin\alpha_1 \cos\varepsilon_1 j + \cos\alpha_1 k$$

$$\bar{a}_2 = \sin\alpha_2 \sin\varepsilon_2 i + \sin\alpha_2 \cos\varepsilon_2 j + \cos\alpha_2 k$$

$$\bar{B} = \bar{A}_1 + \bar{A}_2, \qquad \bar{A}_1 = R\tan\frac{\beta}{2}\bar{a}_1, \qquad \bar{A}_2 = R\tan\frac{\beta}{2}\bar{a}_2$$

Incremental steps are components of vector, B.

$$DLS = \frac{18{,}000}{\pi R}\ \text{deg per 100 ft}$$

Bending of beams in terms of polar coordinates

$$\frac{d^4w}{ds^4} - \lambda^2 \frac{d^2w}{ds^2} = \zeta \quad \text{(sinh, cosh)}$$

where $\lambda^2 = \dfrac{T}{EI} - \dfrac{1}{R^2}$ (can be + or −)

$$\zeta = \frac{Rq + T}{REI}$$

If minus then $\dfrac{d^4w}{ds^4} + \lambda^2 \dfrac{d^2w}{ds^2} = \zeta$ (sin, cos)

12.8 PROBLEMS

1. Survey calculations indicate the location of station 1 is defined by

   Latitude = 1060 ft
   Departure = 680 ft
   Vertical depth = 5000 ft

   Determine the location of station 2 using the tangent method, a simple straight line projection, if the survey data at station 1 are

   Drift angle = 22 deg.
   Hole Direction = 35 deg. (Azimuth)

   The distance drilled between stations 1 and 2 is 95 ft.

2. Determine the radius of curvature over the angle building portion of a directional well if the rate of build is 2° per 100 ft. Determine the nominal bending stress in 9 5/8 in. (47 lb/ft) casing set in the building portion of the well.

3. Determine the dog-leg severity (DLS) between Station 1 and Station 2 if the survey data at the two stations are

   Station 1: Measured depth = 4800 ft
   Inclination = 8 deg.
   Azimuth = 81 deg.

   Station 2: Measured depth = 4890 ft
   Inclination = 10 deg.
   Azimuth = 77 deg.

   Use the minimum curvature method.

4. A Type 1 directional well is to be drilled in a vertical plane according to the following conditions:

    - Vertical depth of target from rig = 10,000 ft
    - Kick off point = 3000 ft
    - Maximum dogleg severity = 3 deg./100 ft
    - Target deviation = 5000 ft

    Determine:

    a. Footage (s) to be drilled in the angle building portion
    b. Inclination (ϕ) of the build section
    c. Total footage to be drilled

    Draw the well plan on graph paper.

5. Determine the latitude, departure, and vertical distances between survey Stations 1 and 2 if

$$\alpha_1 = 40^0 \quad \alpha_2 = 38^0$$

$$\varepsilon_1 = 25^0 \quad \varepsilon_2 = 29^0$$

    Assume a Course Length of 93 ft. Determine these distances using the following methods:

    1. Average Angle
    2. Radius of Curvature
    3. Minimum Curvature

6. A given well bore is straight down to a depth of 1000 ft. Below this depth, surveys show that inclination and direction change according to the values given in the table below.

Measured depth (ft)	Inclination (deg.)	Direction (deg.)
1500	1.0	S12E
1750	2.0	S10E
2000	2.0	S
2250	3.0	S10W
2500	4.0	S12W
2750	3.75	S13W
3000	3.5	S15W

Determine the incremental steps (Δx, Δy, Δz) from station 1. Use the average angle method.

7. The previous drift angle is 13 deg. and the current drift angle is 10 deg. Over 100 ft of drilling, the hole direction changed 2.0 deg. Determine the dog leg severity using the law of cosine method.

8. Use the law of cosine method to determine the **DLS** for the following case;

$$\alpha_1 = 15^0 \quad \alpha_2 = 20^0$$

$$\varepsilon_1 = 45^0 \quad \varepsilon_2 = 50^0$$

Distance drilled is 100 ft. Compare the answer with results using the minimum curvature method.

9. What angle must a rig be tilted if the maximum rate of build is 4 deg. per 100 ft and the directional well is to horizontally penetrate a reservoir located 1000 ft below the surface?

10. The previous drift angle is 45 deg. A 4 deg. per 100 ft correction assembly is oriented 55 deg. right of high side. After 100 ft of hole is drilled, how much does the hole turn and how much does it build?

11. What is the nominal bending stress in a 7-in. (23 lb/ft) casing within a 2.5 deg./100ft dog leg?

12. Determine the drag friction force on a 1000 ft of drill pipe (5½ in, 19.20 lb/ft) inclined in a straight well bore, assuming

- Inclination, α = 60 deg.
- Coefficient of friction, μ = 0.15
- Mud weight, γ_m = 11 ppg

13. Determine the distance to the neutral point in (7 in. × 2 in.; 120 lb/ft) drill collars if

- Inclination, α = 30 deg.
- Coefficient of friction, μ = 0.2
- Mud weight, γ_m = 12 ppg
- WOB = 40,000 lbs

14. Determine the tool face orientation required to direct the well bore at Station 2 toward the target for the following information.

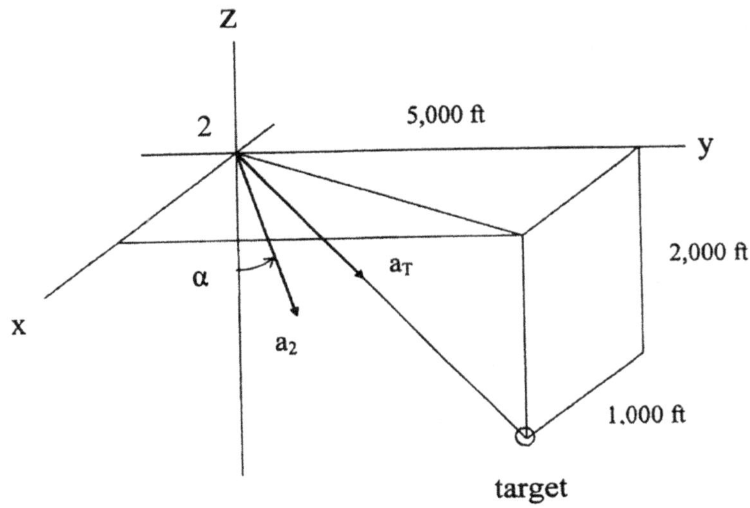

Assume: $\alpha_2 = 55$ deg. and $\varepsilon_2 = 20$ deg.

12.9 LIST OF REFERENCES

1. A. Lubinski, "A study of the Buckling of Rotary Drilling Strings", *Drilling and Production Practice* (API), **17**, 1950, pp. 178–214.
2. MacDonald, G. C. and Lubinski, A.: "Straight-Hole Drilling in Crooked-Hole Country", *Drill. And Prod. Prac.*, API (1951) 80.
3. A. Lubinski and H. B. Woods, "Factors Affecting the Angle of Inclination and dog-legging in rotary bore holes", *Drilling and Production Practice* (API), **20**, 1953, pp. 222–250.
4. Woods, H. and Lubinski, A.: "Use of Stabilizers in Controlling Hole Deviation", *Drill. and Prod. Prac.*, API (1955) 165.
5. Bruce H. Walker; "Some Technical and Economic Aspects of Stabilizer Placement", *Journal of Petroleum Technology*, June 1973.
6. Lear, W. E. and D. W. Dareing; "Effect of Drillstring Vibrations on MWD Pressure Pulse Signals", ASME, *Journal of Energy Resources Technology*, Vol. 112, June 1990, 84–89.
7. Garrett, W. R. and Rollins, H. M.; "Steering Wheel for Rock Bits", Presented at the 19th Annual Petroleum Mechanical Engineering Conference, Los Angeles, Calif. (Sept. 20, 1964).

8. D. W. Dareing; *Mechanics of Drillstrings and Marine Risers*, ASME Press, 2012.
9. C. A. Johancsik, D. B. Friesen, Rapier Dawson, "Torque and Drag in Directional Wells — Prediction and Measurement". *Journal of Petroleum Technology*, June 1984, pp. 987–992.
10. Shigley, J. E. and Mitchell, L. D.; *Mechanical Engineering Design*, McGraw-Hill, 1983.
11. Timoshenko, S. and Woinowsky-Krieger, S; *Theory of Plates and Shells*, McGraw-Hill, 1959.
12. Huang, T., Dareing, D. W., and Beran, W. T.: "Bending of Tubular Bundles Attached to Marine Risers", Trans. ASME, *Journal of Energy Resources Technology*, Vol. 102, March 1980, pp. 24–29.
13. Dareing, D. W., and Ahlers, C. A., "Tubular Bending and Pull Out Forces in High Curvature Well Bores," ASME, *Journal of Energy Resources Technology*, Vol. 113, pp. 133–139, June 1991.
14. Rocheleau, David N. and Dareing, Don W., "Effect of Drag Forces on Bit Weight in High Curvature Well Bores," Trans ASME, *Journal of Energy Resources Technology*, Vol. 114, No. 3, pp. 175–180, Sept. 1992.

12.10 APPENDIX

12.10.1 Stabilized Bottom Hole Assemblies

This appendix includes several stabilized bottom hole assemblies each showing side forces at drill bits. Drill collar sizes and operating conditions are given in each case. A plus side force indicates a building assembly. A minus side force indicates a dropping assembly. Near zero side force indicates a holding assembly. These data were developed sometime ago and are given here as a guide for stabilized assemblies.

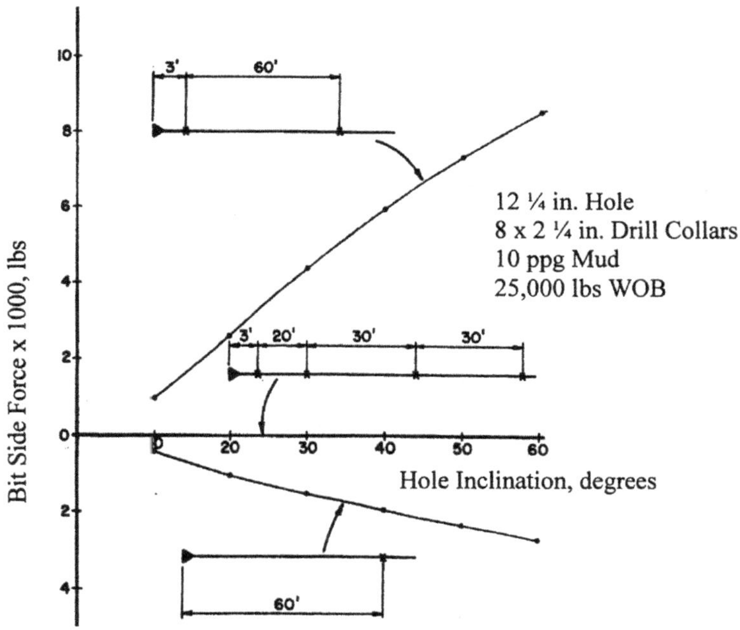

Figure 12-33. Effect of hole inclination on side force.

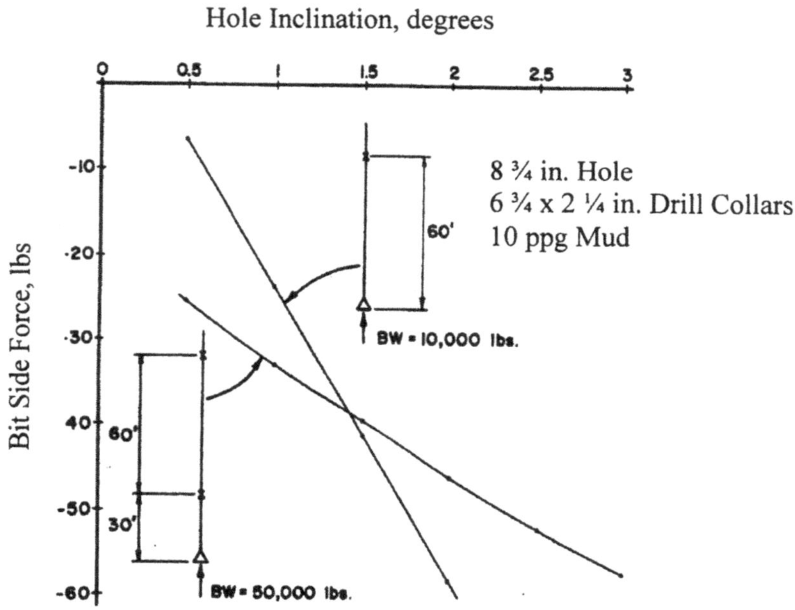

Figure 12-34. Comparison of 30–60 pendulum assembly with conventional pendulum assembly.

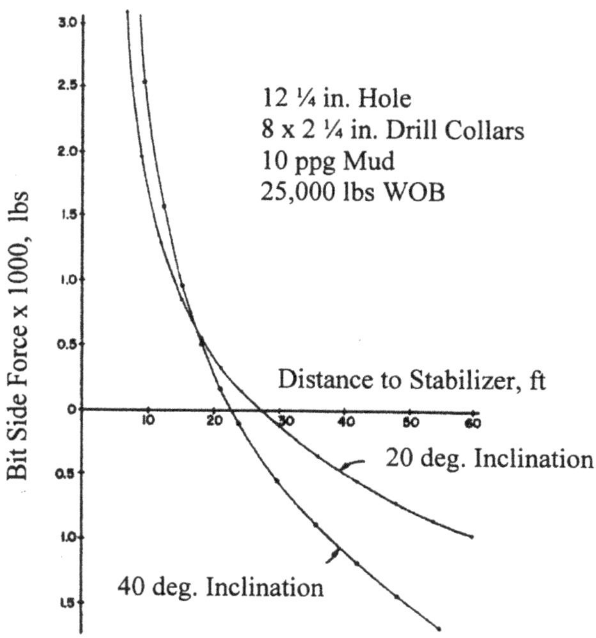

Figure 12-35. Effect of single stabilizer positions on bit side force.

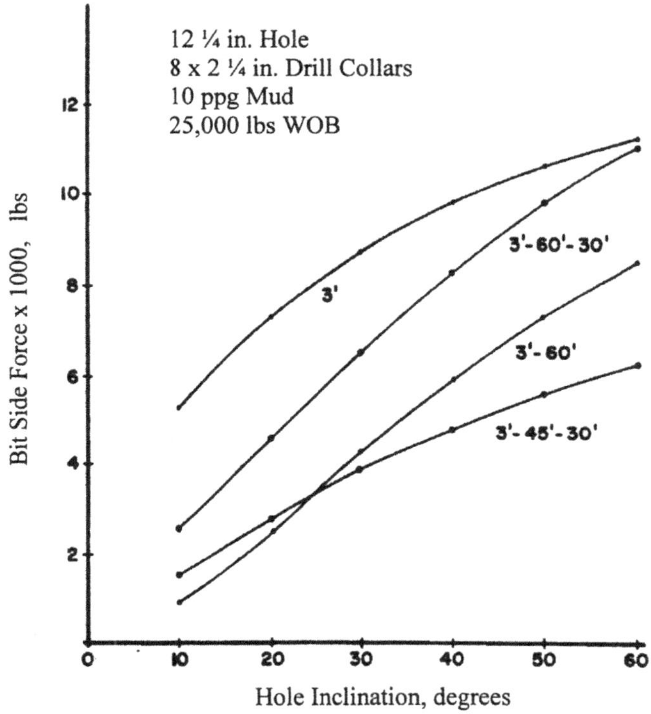

Figure 12-36. Response of various building assemblies.

INDEX

Page numbers followed by f and t indicate figures and tables, respectively.

R

www.ingramcontent.com/pod-product-compliance
Lightning Source LLC
LaVergne TN
LVHW011532200926
848484LV00001B/244